More information about this series at http://www.springer.com/series/4318

H.G. Dales • F.K. Dashiell, Jr.
A.T.-M. Lau • D. Strauss

Banach Spaces of Continuous Functions as Dual Spaces

Springer

H.G. Dales
Department of Mathematics
and Statistics
University of Lancaster
Lancaster, UK

F.K. Dashiell, Jr.
Center of Excellence in Computation
Algebra, and Topology (CECAT)
Chapman University
Orange, CA, USA

A.T.-M. Lau
Department of Mathematical Sciences
University of Alberta
Edmonton, AB, Canada

D. Strauss
Department of Pure Mathematics
University of Leeds
Leeds, UK

ISSN 1613-5237 ISSN 2197-4152 (electronic)
CMS Books in Mathematics
ISBN 978-3-319-81263-2 ISBN 978-3-319-32349-7 (eBook)
DOI 10.1007/978-3-319-32349-7

Mathematics Subject Classification (2010): 46-02, 46B03, 46B04, 46B10, 46B22, 46B25, 46B42, 28A33, 28A60, 46L05, 46L10

Printed on acid-free paper

This Springer imprint is published by Springer Nature
The registered company is Springer International Publishing AG
The registered company address is: Gewerbestrasse 11, 6330 Cham, Switzerland

This volume is dedicated to the memory of

William G. Bade
29 May, 1924–10 August, 2012

with our affection and respect.

Preface

Let K be a locally compact space, and denote by $C_0(K)$ the Banach space of all continuous functions on K that vanish at infinity, taken with the uniform norm. This fundamentally important and very familiar Banach space has been studied for many decades, and it arises in a vast number of applications in mathematical analysis. This book is devoted to the study of certain aspects of this space.

Indeed we shall address the rather specific questions:

(I) *When is $(C_0(K), |\cdot|_K)$ isometrically isomorphic or isomorphic (i.e. linearly homeomorphic) to the dual of a Banach space? If so, how unique is the predual?*

(II) *When is $(C_0(K), |\cdot|_K)$ isometrically isomorphic or isomorphic to the bidual of a Banach space?*

A more general theme that informs our work is the following question:

(III) *What are the relations between topological properties of the locally compact space K and Banach-space properties of $C_0(K)$?*

These questions have a long history, developed over around 80 years, some of which we shall record. Nevertheless, it seems that answers to even some basic questions are not complete, and at best are rather scattered in the literature. Sometimes existing proofs seem to be more complicated than is necessary.

We aim to give a coherent survey account of these matters; we shall recall, and sometimes clarify, necessary background from topology, measure theory, functional analysis, and other relevant areas of mathematics. Our approach will be close in spirit to the theory of Boolean algebras and ultrafilters, and there will be little mention of approaches through the theory of representations of C^*-algebras as algebras of operators on Hilbert spaces.

As we shall recount, some of the seminal work on these topics was brought together by Professor William Bade of Berkeley in mimeographed, unpublished lecture notes as early as around 1957.

We shall include some new results and examples. (However, some unattributed remarks are 'well known' and not original to us.) We shall offer more straightforward proofs of some known theorems and shall raise a number of open questions, some of which have withstood the test of time. We shall recall quickly some quite elementary results in topology, measure theory, and functional analysis, and a reader may wish to skim these pages, referring back to them when necessary. In general, we do not repeat proofs of theorems that are available in easily accessible and standard texts, but we do give proofs of some results that are basic to our work or which seem to be somewhat obscure or in less accessible sources.

We now give some more information on particular results that appear in the text; each chapter has a more detailed introduction to its contents.

Chapter 1 gives background in topology and Boolean algebras. Themes that will emerge include those of Stonean spaces, ultrafilters and the Stone space of a lattice, covers of locally compact spaces, the Stone–Čech compactification of a completely regular topological space, Gleason's characterization of projective compact spaces, and the Boolean algebras of regular-open and Borel subsets of a topological space.

Chapter 2 recalls background in Banach spaces and Banach lattices; we are particularly concerned to determine when two Banach spaces are mutually isomorphic and when they are mutually isometrically isomorphic, noting that there is a large difference between these two notions. We shall define what it means for a Banach space to be isomorphically/isometrically a (bi)dual space, and we shall discuss representations of the bidual of a given Banach space.

Let K be a compact space. A key fact for us is the classic result that the Banach lattice $C_{\mathbb{R}}(K)$ is Dedekind complete if and only if K is Stonean. In §2.5, we shall introduce λ-injective Banach spaces; later, in §6.8, we shall prove the famous result that $C(K)$ is 1-injective if and only if K is Stonean; indeed, every 1-injective Banach space is isometrically isomorphic to $C(K)$ for some Stonean space K. It is one of the long-standing open questions to determine whether every injective Banach space is isomorphic to a 1-injective space.

The Krein–Milman property that is introduced in §2.6 will mainly be used to show that $C(K)$ is not isometrically a dual space for certain compact spaces K.

The topic of Chapter 3 is that of Banach algebras and C^*-algebras; of course the spaces $C_0(K)$, for locally compact spaces K, are the generic examples of commutative C^*-algebras. Let A be a Banach algebra. Then the bidual space A'' is a Banach algebra containing A as a closed subalgebra for two, sometimes distinct, products, $\Box$ and $\Diamond$, called the Arens products; the algebra A is said to be Arens regular if these two products coincide on A''. It is another famous classical result that each C^*-algebra A is Arens regular and that $(A'', \Box)$ is itself a C^*-algebra; it is the enveloping von Neumann algebra of A. By using the Gel'fand–Naimark theorem, it follows that the bidual space $C_0(K)''$ has the form $C(\widetilde{K})$ for a uniquely determined compact space $\widetilde{K}$, which we shall call the hyper-Stonean envelope of K in §5.4. For example, the hyper-Stonean envelope of $\mathbb{N}$ is the Stone–Čech compactification $\beta\mathbb{N}$. However, we do not follow this abstract approach: in §5.4, we shall give a more explicit 'construction' of $\widetilde{K}$. Indeed we give three somewhat different constructions of $\widetilde{K}$. This will enable us, in §6.5, to give a topological characterization of $\widetilde{K}$ for each uncountable, compact, metrizable space K, and, eventually, in §6.6, to determine the cardinalities of various subsets of this space $\widetilde{K}$. An earlier, rather simple, proof of the fact that each space $C_0(K)$ is Arens regular will be given in §4.5.

In §3.3, we shall discuss some commutative C^*-algebras that are the Baire classes on certain topological spaces; this topic will be developed further in §6.7. In the final section of Chapter 3, we shall make some remarks on the extensions of our theory from the commutative C^*-algebras $C_0(K)$ to more general, non-commutative C^*-algebras; however, we shall say very little on the vast topic of the representation theory of C^*-algebras on Hilbert spaces.

In Chapter 4, we shall first recall some theory of measures on a locally compact space, with a brief mention of more general measure spaces. Let K be a locally compact space. Then the Banach space $M(K)$ of complex-valued, regular Borel measures on K is identified with the dual space of $C_0(K)$. In §4.4, we shall recall properties of the Banach spaces $L^p(K,\mu)$, where K is a locally compact space, $\mu \in M(K)^+$, and $1 \leq p \leq \infty$, and in §4.5 we shall consider rather briefly when spaces of the form $C(K)$, for K a compact space, are Grothendieck spaces; we shall also note that all injective Banach spaces are Grothendieck spaces. Maximal singular families of measures on K, defined in §4.6, are a key ingredient in one construction of $\widetilde{K}$. In §4.7, we shall define the space $N(K)$ of normal measures on a locally compact space K, and we shall give various examples of compact spaces K for which $N(K)$ is and is not equal to $\{0\}$; in particular, we shall prove a new result of Plebanek that shows that there is connected, compact space K such that $N(K) \neq \{0\}$.

The hyper-Stonean spaces of Chapter 5 are Stonean spaces with 'many' normal measures. Let K be a locally compact space, and take $\mu \in M(K)^+$. Then the compact character space of the unital C^*-algebra $L^\infty(K,\mu)$ is denoted by Φ_μ; the Gel'fand transform $\mathscr{G}_\mu : L^\infty(K,\mu) \to C(\Phi_\mu)$ is a C^*-isomorphism and a Banach-lattice isometry, and Φ_μ is a hyper-Stonean space identified with the Stone space of the Boolean algebra $\mathfrak{B}_\mu$ that is a natural quotient of the Boolean algebra of Borel subsets of K. We shall use the spaces Φ_μ to 'build' the space $\widetilde{K}$. We shall then give a new 'construction' of $\widetilde{K}$ as βS_K, where S_K is the Stone space of a Boolean ring $M(K)^+/\sim$; we shall also give two further representations of the space $\widetilde{K}$, one involving L-decompositions, whose theory is introduced in §5.5. The analogous theory for general C^*-algebras is sketched in §5.6.

Our main study of the Banach spaces $C_0(K)$ is given in Chapter 6. As a preliminary, we shall give in §6.1 some isomorphic invariants of $C(K)$-spaces; these will include the cardinality of K. Then we shall give in §6.2 some easy examples of locally compact spaces K such that $C_0(K)$ is not (either isometrically or isomorphically) a dual space. In §6.3, many Banach spaces that are isomorphic preduals of the Banach space ℓ^1, but are not isomorphic to each other, will be described. The question when a space $C(K)$, for compact K, is isometrically a dual space is then fully determined in §6.4. For example, this is the case if and only if the space K is hyper-Stonean, and then the unique isometric predual $C(K)_*$ of $C(K)$ is identified with the space $N(K)$ of normal measures on K. This result combines classic theorems of Dixmier and of Grothendieck; aspects of the proof were first expounded by William Bade.

It is apparently much harder to characterize the locally compact spaces K such that $C_0(K)$ is isomorphically a dual space, and we do not have a full answer to this question. Of course this holds whenever K is hyper-Stonean. We shall give several examples of compact spaces, even of Stonean spaces, K such that $C(K)$ is not isomorphically a dual space, and other examples, even of a totally disconnected space which is not Stonean, such that $C(K)$ is isomorphically a dual space. Each injective space of the form $C_0(K)$, and hence each space $C_0(K)$ that is isomorphically a dual space, is such that K contains a dense, open, extremely disconnected subset and so has infinitely many components, but we do not know whether K must be totally disconnected.

In §6.10, we shall discuss when a space $C(X)$, for an infinite compact space X, is isometrically a bidual space. Our aim in this section was to establish the conjecture that there is then a compact space K such that X is homeomorphic to $\widetilde{K}$; we can at least show that there is a compact space K such that X is homeomorphic to a clopen subspace of $\widetilde{K}$. In the case where $C(X)$ is isometrically the bidual of a separable Banach space, we can resolve the conjecture. Indeed, there are only two possibilities for X: either X is homeomorphic to $\beta\mathbb{N}$, and $C(X)$ is isometrically isomorphic to $C(\beta\mathbb{N}) = c_0''$ or X is homeomorphic to $\widetilde{\mathbb{I}}$, and $C(X)$ is isometrically isomorphic to $C(\mathbb{I})''$.

In §6.11, we shall summarize some results that we have obtained concerning the question when a space $C_0(K)$ is injective, when it is (isomorphically or isometrically) the dual of a Banach space, and when it is the bidual of a Banach space. There is a list of open questions in §6.12.

We have striven to eliminate errors in our text, but some are likely to remain. Readers are invited to send comments or errors to CK-Banachduals-book@cecat.chapman.edu. Corrections and some new results will be posted on the CECAT home page, http://mathcs.chapman.edu/CECAT.

We trust that this volume will stimulate new research in this attractive area, by graduate students and many researchers.

H.G. Dales, Lancaster, UK
F.K. Dashiell, Jr., Los Angeles, CA, USA
A.T.-M. Lau, Edmonton, AB, Canada
D.Strauss, Leeds, UK

Acknowledgements

We have dedicated this work to the memory of William G. (Bill) Bade of Berkeley, who died in August, 2012; we acknowledge the inspiration of his work on Banach spaces of continuous functions from a long time ago; we express our deep thanks to Bill and to the whole Bade family for their generous and stimulating hospitality in Berkeley over many years. An obituary of Professor Bade has appeared in [69].

First thoughts on the possibility of a book related to the present one came to us whilst we were part of 'Research in Teams' at the *Banff International Research Station* in July 2010. We are very grateful to BIRS for the chance to work together in such a lovely environment on this and several other occasions, including May 2012 and February 2016.

In the subsequent years, we have met at the University of Sussex (thanks to Charles Goldie), in Berkeley (thanks to Marc Rieffel), at the Fields Institute in Toronto, in Lancaster (thanks to Andrey Lazarev for arranging funding), and at the University of Alberta; we are grateful for the generous support of the NSERC Grant MS100 of Anthony To-Ming Lau during our visits to the University of Alberta. We have benefited from the generous hospitality of the Department of Mathematics at the University of California, Los Angeles, and from the collegial environment at the Center of Excellence in Computation, Algebra, and Topology (CECAT) at Chapman University, California.

We are very grateful to several people for valuable comments on early versions of the manuscript. These include James Hagler (Denver), Tomasz Kania (Lancaster), and Grzegorz Plebanek (Wrocław). We are particularly grateful to Ajit Iqbal Singh (Delhi), who read a draft very carefully and provided a long list of suggested changes. An anonymous referee read a draft with great care and made a number of very valuable comments and suggestions; we thank him/her for the substantial amount of work involved in this task.

We thank Professor Keith Taylor of Dalhousie University, Halifax, Nova Scotia, an editor of the *Canadian Mathematical Society Books in Mathematics*, for accepting our work in this series and for his thoughtful care in moving the project forward.

Contents

Chapter 1
Introduction

In this chapter, we shall begin by introducing some basic notations. This will be followed by a discussion in §1.4 of some topological concepts, including those of locally compact spaces and Stonean spaces. We shall later frequently refer to the Stone–Čech compactification βX of a completely regular space X, and this is introduced in §1.5. In §1.6, we shall prove Gleason's theorem characterizing projective topological spaces as the Stonean spaces, and, in §1.7, we shall also recall some basic theory of Boolean algebras, generalizing this slightly to cover Boolean rings; we shall discuss the Stone space of a Boolean ring and give various important examples of Boolean rings.

We are aware that many (but not all) of the definitions and remarks made within the present introductory chapter are quite elementary and will be well known to many readers; they are included rather briefly to standardize notation and can be skimmed quickly by those knowledgeable about the relevant topic; we include a few justifications of results that we shall quote later and which are perhaps not quite standard. The index and cross-references will enable the reader to check basic definitions in this introductory material when they are used.

1.1 The spaces $C^b(K)$ and $C_0(K)$

Our objects of study will be the Banach spaces $C^b(K)$ and $C_0(K)$ of bounded, continuous functions on a locally compact space K and, in particular, of those that vanish at infinity on K, and we shall first describe these spaces.

Let K be a non-empty, locally compact space (always assumed to be Hausdorff). Then $C^b(K)$ denotes the collection of all the bounded, continuous, complex-valued functions on K, and $C_0(K)$ denotes the subset of all the functions f in $C^b(K)$ that vanish at infinity, in the sense that $\{x \in K : |f(x)| \geq \varepsilon\}$ is a compact subspace of K for each $\varepsilon > 0$. Clearly $C^b(K)$ is a linear space with respect to the pointwise

H.G. Dales et al., *Banach Spaces of Continuous Functions as Dual Spaces*,
CMS Books in Mathematics, DOI 10.1007/978-3-319-32349-7_1

operations, and $C_0(K)$ is a linear subspace of $C^b(K)$. The *uniform norm* on K is denoted by $|\cdot|_K$, so that

$$|f|_K = \sup\{|f(x)| : x \in K\} \quad (f \in C^b(K)).$$

Then $(C^b(K), |\cdot|_K)$ is a Banach space, and $C_0(K)$ is a closed subspace of $C^b(K)$. Further, with respect to the product given by pointwise multiplication of functions, $C^b(K)$ is a Banach algebra (see §3.1); indeed, it is a commutative, unital C^*-algebra (see §3.2). The space $C_0(K)$ is a closed ideal in $C^b(K)$, and so also a commutative C^*-algebra. We take $C_{00}(K)$ to be the subalgebra of functions in $C_0(K)$ of compact support, so that $C_{00}(K)$ is a dense ideal in $(C_0(K), |\cdot|_K)$.

In the case where the space K is compact, we write $C(K)$ for the unital algebra $C_{00}(K) = C_0(K) = C^b(K)$.

The spaces of real-valued functions in $C^b(K)$ and $C_0(K)$ are denoted by $C^b_{\mathbb{R}}(K)$ and $C_{0,\mathbb{R}}(K)$, respectively, and we write $C_{\mathbb{R}}(K)$ in the case where K is compact. Thus $C^b_{\mathbb{R}}(K)$ and $C_{0,\mathbb{R}}(K)$ are real Banach lattices (see §2.3) with respect to the pointwise ordering of functions, and $C^b(K)$ and $C_0(K)$ are (complex) Banach lattices. We set

$$C^b(K)^+ = \{f \in C^b_{\mathbb{R}}(K) : f \geq 0\} \quad \text{and} \quad C_0(K)^+ = \{f \in C_{0,\mathbb{R}}(K) : f \geq 0\};$$

these are the cones of positive elements in $C^b_{\mathbb{R}}(K)$ and $C_{0,\mathbb{R}}(K)$, respectively.

For a discussion of $C^b(K)$ and $C_0(K)$ as Banach spaces, see the comprehensive and seminal early monograph of Semadeni [225] from 1971 and the survey of Rosenthal [215] from 2003; for $C^b(K)$ and $C_0(K)$ as Banach algebras, see the monograph [68, §4.2] of Dales from 2000; for $C^b(K)$ and $C_0(K)$ as Banach lattices, see [1, 184, 223], for example.

1.2 Notation

We shall use the following standard notation: $\mathbb{N} = \{1, 2, 3, \ldots\}$ is the set of natural numbers; $\mathbb{Z}$ is the set of integers; $\mathbb{Z}^+ = \{n \in \mathbb{Z} : n \geq 0\}$; $\mathbb{R}$ is the real line; $\mathbb{Q}$ is the set of rational numbers; $\mathbb{R}^+ = \{t \in \mathbb{R} : t \geq 0\}$; $\mathbb{C}$ is the complex plane; $\mathbb{I} = [0, 1]$ is the closed unit interval in $\mathbb{R}$; $\mathbb{D} = \{z \in \mathbb{C} : |z| < 1\}$ is the open unit disc in the plane; $\mathbb{T} = \{z \in \mathbb{C} : |z| = 1\}$ is the unit circle. The real and imaginary parts of $z \in \mathbb{C}$ are $\Re z$ and $\Im z$, respectively; the conjugate and argument of z are $\bar{z}$ and $\arg z$, respectively. Let $n \in \mathbb{N}$. Then $\mathbb{N}_n = \{1, \ldots, n\}$. However we set

$$\mathbb{Z}_n = \{0, 1, \ldots, n-1\};$$

this set is a group, denoted by $(\mathbb{Z}_n, +)$, with respect to addition modulo n.

Take $p > 1$. Then the *conjugate index* to p is q, where $1/p + 1/q = 1$; we also regard ∞ and 1 as the conjugates of 1 and ∞, respectively.

Let S be a non-empty set. The cardinality of S is denoted by $|S|$, and the characteristic function of a subset T of S is denoted by χ_T; however we shall sometimes write 1 or 1_S for χ_S. The complement of a subset T of S is denoted by $S \setminus T$ or by T^c, and the *symmetric difference* of two subsets T_1 and T_2 of S is

$$T_1 \Delta T_2 = (T_1 \setminus T_2) \cup (T_2 \setminus T_1).$$

The *power set*, $\mathcal{P}(S)$, of S is the set of all subsets of S, so that $|\mathcal{P}(S)| = 2^{|S|}$. A non-empty subset $\mathcal{F}$ of $\mathcal{P}(S)$ has the *finite intersection property* if the intersection of any non-empty, finite subfamily of $\mathcal{F}$ is non-empty. The set of functions from S to a set T is denoted by T^S, and we write $\mathbb{Z}_2^n$ for $\{0,1\}^{\mathbb{N}_n}$ for $n \in \mathbb{N}$; the restriction of $f \in \mathbb{C}^S$ to a subset T of S is $f \mid T \in \mathbb{C}^T$.

Let S be a non-empty set. A *partial order* on S is a binary relation $\leq$ such that:

(i) for each $r,s,t \in S$, we have $r \leq t$ whenever $r \leq s$ and $s \leq t$;

(ii) $s \leq s$ for each $s \in S$;

(iii) for each $r,s \in S$, we have $r = s$ whenever $r \leq s$ and $s \leq r$.

A partial order $\leq$ on S is a *total order* if $r \leq s$ or $s \leq r$ whenever $r,s \in S$. A pair $(S, \leq)$ is a *partially ordered set*, respectively, a *totally ordered set*, when $\leq$ is a partial order, respectively, a total order, on S. For example, $(\mathcal{P}(S), \subset)$ is a partially ordered set for each non-empty set S. A totally ordered set $(S, \leq)$ is *well-ordered* if each non-empty subset of S has a minimum element.

A partially ordered set $(D, \leq)$ is a *directed set* if, for each $\alpha, \beta \in D$, there exists $\gamma \in S$ with $\alpha \leq \gamma$ and $\beta \leq \gamma$. A *net* in a set S is a function f from a directed set D into S; a *subnet* of f is a map of the form $f \circ \psi : E \to S$, where E is a directed set and, for each $d \in D$, there exists $e \in E$ with $\psi(e) \geq d$. A net is often denoted by $(x_\gamma : \gamma \in D)$ or (x_γ).

Let $(S, \leq)$ be a partially ordered set. For $s,t \in S$, we write $s < t$ whenever $s \leq t$ and $s \neq t$. Now take $s,t \in S$. Then the *order interval* $[s,t]$ in S is defined to be the subset $\{r \in S : s \leq r \leq t\}$. Let (s_α) be a net in S. Then (s_α) is *decreasing* if $s_\alpha \leq s_\beta$ when $\alpha \geq \beta$ and *increasing* if $s_\alpha \geq s_\beta$ when $\alpha \geq \beta$. Take $s \in S$. Then we write $s_\alpha \searrow s$ when (s_α) is decreasing and $\inf s_\alpha = s$ and $s_\alpha \nearrow s$ when (s_α) is increasing and $\sup s_\alpha = s$.

Let $(S, \leq)$ and $(T, \leq)$ be two partially ordered sets. A map $\theta : S \to T$ is an *order homomorphism* if $\theta(s_1) \leq \theta(s_2)$ whenever $s_1 \leq s_2$ in S; θ is an *order isomorphism* if it is a bijection such that both θ and θ^{-1} are order homomorphisms; the partially ordered sets are *order isomorphic* if there is an order isomorphism from S onto T.

For each non-zero ordinal α, the interval $[0, \alpha]$ is a well-ordered set. The first infinite and first uncountable ordinals are ω and ω_1, respectively, and we denote these as $\aleph_0$ and $\aleph_1$, respectively, when we regard them as cardinals.

The axiom scheme ZFC consists of the Zermelo–Fraenkel axioms of set theory, together with the Axiom of Choice; the letters 'CH' denote the Continuum Hypothesis. The cardinality of the continuum is $\mathfrak{c} = 2^{\aleph_0}$, so that CH is the statement that $\mathfrak{c} = \aleph_1$. Results that are only proved to hold in the theory ZFC + CH are marked with the symbol '(CH)'.

A family $\mathcal{F}$ of subsets of a non-empty set S is a *σ-algebra* on S if $S \in \mathcal{F}$, if $F^c \in \mathcal{F}$ whenever $F \in \mathcal{F}$, and if the union of each countable family of members of $\mathcal{F}$ belongs to $\mathcal{F}$. It follows that the intersection of each countable family of members of $\mathcal{F}$ also belongs to $\mathcal{F}$. For each subfamily $\mathcal{F}$ of $\mathcal{P}(S)$, the intersection of all the σ-algebras in $\mathscr{P}(S)$ containing $\mathscr{F}$ is a σ-algebra; it is the σ-algebra *generated by* $\mathscr{F}$ and denoted by $\sigma(\mathscr{F})$.

Let $\mathscr{F}_0$ be the family of subsets of S which are in $\mathscr{F}$, together with their complements. Suppose that $0 < \beta < \omega_1$, and assume that $\mathscr{F}_\alpha$ has been defined for every α with $0 < \alpha < \beta$. Then $\mathscr{F}_\beta$ is defined to be the family of all subsets of S which are the unions or intersections of countable subfamilies of $\bigcup\{\mathscr{F}_\alpha : 0 < \alpha < \beta\}$. Then $\sigma(\mathscr{F}) = \bigcup\{\mathscr{F}_\beta : 0 < \beta < \omega_1\}$. Suppose that $|\mathscr{F}| \leq \mathfrak{c}$. Then, by induction, $|\mathscr{F}_\beta| \leq \mathfrak{c}$ for each β with $0 < \beta < \omega_1$, and so $|\sigma(\mathscr{F})| \leq \mathfrak{c}$.

We write $\mathbb{Z}_2^\omega$ for $\{0,1\}^\omega$, and regard elements of $\mathbb{Z}_2^\omega$ as sequences $\varepsilon = (\varepsilon_j)$, where $\varepsilon_j \in \{0,1\}$ $(j \in \mathbb{N})$, so that $\mathbb{Z}_2^\omega$ is a compact, abelian group with respect to the coordinatewise operations and product topology; we write $\mathbb{Z}_2^{<\omega}$ for the subgroup consisting of the sequences in $\mathbb{Z}_2^\omega$ that are eventually 0, so that we can identify $\mathbb{Z}_2^{<\omega}$ with the collection of finite sequences $(\varepsilon_1, \dots, \varepsilon_n)$, where $\varepsilon_j \in \{0,1\}$ $(j \in \mathbb{N}_n)$. For $\varepsilon = (\varepsilon_j) \in \mathbb{Z}_2^\omega$ and $n \in \mathbb{N}$, we set $\varepsilon \upharpoonright n = (\varepsilon_1, \dots, \varepsilon_n) \in \mathbb{Z}_2^{<\omega}$; for $\varepsilon = (\varepsilon_1, \dots, \varepsilon_n) \in \mathbb{Z}_2^n$, we set

$$\varepsilon^- = (\varepsilon_1, \dots, \varepsilon_n, 0) \quad \text{and} \quad \varepsilon^+ = (\varepsilon_1, \dots, \varepsilon_n, 1), \tag{1.1}$$

so that $\varepsilon^-, \varepsilon^+ \in \mathbb{Z}_2^{n+1}$.

Let S be a non-empty set. The linear spaces of all functions from S to $\mathbb{C}$ and to $\mathbb{R}$ are denoted by $\mathbb{C}^S$ and $\mathbb{R}^S$, respectively; $\mathbb{C}^S$ and $\mathbb{R}^S$ are complex and real algebras, respectively, for the pointwise operations. A function in $\mathbb{C}^S$ or $\mathbb{R}^S$ is a *simple function* if its range is finite; the set A of simple functions is the linear span of the set of characteristic functions χ_D for $D \subset S$, so that A is a subalgebra of $\mathbb{C}^S$ or $\mathbb{R}^S$, respectively. There is an obvious ordering on the space $\mathbb{R}^S$: for $f, g \in \mathbb{R}^S$, we set

$$f \leq g \quad \text{if} \quad f(s) \leq g(s) \quad (s \in S),$$

so that $(\mathbb{R}^S, \leq)$ is a partially ordered set. For a subset F of $\mathbb{R}^S$, we set

$$F^+ = \{f \in F : f \geq 0\}.$$

Functions such as $|f|$ and $\exp f$ for $f \in \mathbb{C}^S$, and $f \vee g$ and $f \wedge g$ for $f, g \in \mathbb{R}^S$ are defined pointwise, so that

$$(f \vee g)(s) = \max\{f(s), g(s)\}, \quad (f \wedge g)(s) = \min\{f(s), g(s)\} \quad (s \in S);$$

we then define the functions $f^+ = f \vee 0$, $f^- = (-f) \vee 0$, and

$$|f| = f^+ + f^- = f \vee (-f)$$

for $f \in \mathbb{R}^S$, so that $f = f^+ - f^-$ and $f^+ f^- = 0$.

Let $f \in \mathbb{C}^S$. Then we shall sometimes write $\Re f$ and $\Im f$ for the real and imaginary parts of f, respectively, so that $\Re f, \Im f \in \mathbb{R}^S$ and $f = \Re f + \mathrm{i}\Im f$; the *conjugate function* of f is

$$\overline{f} = \Re f - \mathrm{i}\Im f.$$

The space $\ell^\infty(S)$ of bounded, complex-valued functions on S is a subalgebra of $\mathbb{C}^S$; it is a Banach space (see Chapter 2) with respect to the uniform norm $|\cdot|_S$, now defined by

$$|f|_S = \sup\{|f(s)| : s \in S\} \quad (f \in \ell^\infty(S)).$$

We take $c_0(S)$ to be the closed ideal in $\ell^\infty(S)$ consisting of the functions f such that $\{s \in S : |f(s)| \geq \varepsilon\}$ is finite for each $\varepsilon > 0$. The spaces of real-valued functions in $\ell^\infty(S)$ and $c_0(S)$ are $\ell^\infty_{\mathbb{R}}(S)$ and $c_{0,\mathbb{R}}(S)$, respectively.

The spaces $c_0 = c_0(\mathbb{N})$ and c are the Banach spaces of all null sequences and of all convergent sequences, respectively; the spaces $(\ell^p, \|\cdot\|_p)$, for $1 \leq p < \infty$, are the standard Banach spaces of all p-summable sequences, so that

$$\|(\alpha_n)\|_p = \left(\sum_{n=1}^{\infty} |\alpha_n|^p\right)^{1/p} \qquad ((\alpha_n) \in \ell^p).$$

For $n \in \mathbb{N}$, we denote by δ_n the sequence $(\delta_{m,n} : m \in \mathbb{N})$ in $\mathbb{R}^{\mathbb{N}}$, where $\delta_{m,n} = 1$ when $m = n$ and $\delta_{m,n} = 0$ when $m \neq n$. Thus $\delta_n \in c_0$ and $\delta_n \in \ell^p$, with norm equal to 1, for each $n \in \mathbb{N}$. More generally, we define $(\ell^p(S), \|\cdot\|_p)$ for a non-empty set S and $1 \leq p < \infty$ to be the set of functions $f \in \mathbb{C}^S$ such that

$$\|f\|_p = \left(\sum\{|f(s)|^p : s \in S\}\right)^{1/p} < \infty,$$

so obtaining a Banach space. Indeed, the spaces $\ell^\infty(S)$, $\ell^p(S)$, c_0, and c are commutative Banach algebras with respect to the pointwise product, and $\ell^\infty(S)$, c_0, and c are C^*-algebras; see §§ 3.1, 3.2. The spaces of real-valued functions in $\ell^p(S)$ are $\ell^p_{\mathbb{R}}(S)$, etc.

Similarly, $L^\infty(\mathbb{I})$ consists of the equivalence classes of the essentially bounded, (Lebesgue) measurable functions f on $\mathbb{I}$ with

$$\|f\|_\infty = \operatorname{ess\,sup}\{|f(t)| : t \in \mathbb{I}\} \quad (f \in L^\infty(\mathbb{I})),$$

and, for $1 \leq p < \infty$, the Banach space $L^p(\mathbb{I})$ consists of the equivalence classes of the (Lebesgue) measurable functions f on $\mathbb{I}$ such that

$$\|f\|_p = \left(\int_0^1 |f(s)|^p \,\mathrm{d}s\right)^{1/p} < \infty.$$

We shall briefly allude to a *measure space*; this is a triple (Ω, Σ, μ), where Ω is a set, Σ is a σ-algebra of subsets of Ω, and μ is a positive measure, so that μ takes values in $[0, \infty]$ and is countably additive on Σ. The measure space is a *probability measure space* if $\mu(\Omega) = 1$, is a *finite measure space* if $\mu(\Omega) < \infty$, and is a σ-*finite measure space* if Ω is the union of a sequence (Ω_n) of sets in Σ such that $\mu(\Omega_n) < \infty$ $(n \in \mathbb{N})$. The obvious example of a σ-finite measure space arises from Lebesgue measure on $\mathbb{R}$. See [59, §1.2], [132, §7], and [217, Definition 1.18], for example.

A measure space (Ω, Σ, μ) is *decomposable* if there is a subfamily $\mathscr{F}$ of Σ that partitions Ω such that:

(i) $0 \leq \mu(F) < \infty$ $(F \in \mathscr{F})$;

(ii) $\mu(S) = \Sigma\{\mu(S \cap F) : F \in \mathscr{F}\}$ for each $S \in \Sigma$ with $\mu(S) < \infty$;

(iii) $S \in \Sigma$ for each $S \subset \Omega$ such that $S \cap F \in \Sigma$ $(F \in \mathscr{F})$.

We also say that the measure μ is *decomposable* in this case. See [138, Definition (19.25)], for example. A σ-finite measure is decomposable.

Let (Ω, Σ, μ) be a measure space. For p such that $1 \leq p < \infty$, we define

$$L^p(\Omega, \mu) = \left\{ f : \Omega \to \mathbb{C} : \ f \text{ is } \Sigma\text{−measurable}, \int_\Omega |f|^p \, \mathrm{d}\mu < \infty \right\}$$

and

$$\|f\|_p = \left(\int_\Omega |f|^p \, \mathrm{d}\mu \right)^{1/p} \qquad (f \in L^p(\Omega, \mu)).$$

Also, $L^\infty(\Omega, \mu)$ is the Banach space of the essentially bounded Σ-measurable functions. Suppose that $1 \leq p \leq \infty$. Then, as usual, we identify functions f and g such that $\|f - g\|_p = 0$ to obtain $(L^p(\Omega, \mu), \|\cdot\|_p)$ as a Banach lattice. See [132, §42], [217, Definition 3.6], and §4.4.

Suppose that (Ω, Σ, μ) is a σ-finite measure space. Then there is a Σ-measurable function φ on Ω such that $\varphi(x) > 0$ $(x \in \Omega)$ and $\int_\Omega \varphi \, \mathrm{d}\mu = 1$; define a positive measure ν on Σ by setting $\nu = \varphi \cdot \mu$, so that (Ω, Σ, ν) is a probability measure space. Then $L^1(\Omega, \nu)$ is Banach-lattice isometric (see page 70) to $L^1(\Omega, \mu)$. We shall discuss the uniqueness of such spaces in §4.4.

Further background on Banach spaces will be given in §2.1; for an introduction to measures and a more formal definition of the Banach space $L^p(K, \mu)$ for a locally compact space K and a positive measure μ, see §§4.1,4.4.

1.3 Lattices and linear spaces

A partially ordered set $(L, \leq)$ is a *lattice* if, for each pair $\{x, y\}$ of elements of L, there is a supremum, denoted by $x \vee y$, and an infimum, denoted by $x \wedge y$; the two maps $(x, y) \mapsto x \vee y$ and $(x, y) \mapsto x \wedge y$ from $L \times L$ to L are the *lattice operations*. The supremum and infimum of a non-empty subset F of a lattice (if they exist) are denoted by

$$\bigvee F \quad \text{and} \quad \bigwedge F,$$

respectively. A lattice L is *distributive* if

$$x \wedge (y \vee z) = (x \wedge y) \vee (x \wedge z) \quad (x, y, z \in L).$$

Definition 1.3.1. A lattice is *complete* (respectively, *σ-complete*) if every non-empty subset (respectively, every countable, non-empty) subset has a supremum and an infimum; it is *Dedekind complete* (respectively, *Dedekind σ-complete*) if every non-empty (respectively, every countable, non-empty) subset which is bounded above has a supremum and every non-empty (respectively, every countable, non-empty) subset which is bounded below has an infimum.

For example, for each non-empty set S, $(\mathscr{P}(S), \subset)$ is a complete, distributive lattice with a minimum element, namely $\emptyset$, and a maximum element, namely S.

Let L_1 and L_2 be lattices. A map $T : L_1 \to L_2$ is a *lattice homomorphism* if

$$T(x \vee y) = Tx \vee Ty, \quad T(x \wedge y) = Tx \wedge Ty \quad (x, y \in L_1);$$

the map T is a *lattice isomorphism* if, further, it is a bijection; in this case, the inverse map $T^{-1} : L_2 \to L_1$ is also a lattice homomorphism. Clearly a lattice homomorphism is an order homomorphism.

We shall be concerned with ultrafilters in various contexts. We give a general definition here; we shall discuss more specific contexts and the topology of Stone spaces later.

Definition 1.3.2. Let L be a lattice with a minimum element 0. An *ultrafilter* on L is a non-empty subset p of L that is maximal with respect to the property that $\bigwedge F \neq 0$ whenever F is a non-empty, finite subset of p. The *Stone space* of L is the set of ultrafilters on L; it is denoted by $St(L)$.

The Stone space of a lattice L has been defined by other authors as the space of *prime* filters on L [28, Chapter 4]. A filter on L is prime if and only if it is an ultrafilter whenever L is a Boolean ring (see §1.7) [28, Theorem III.6.3].

Let E be a linear space, always taken to be over the complex field, $\mathbb{C}$, unless otherwise indicated; a linear space over $\mathbb{R}$ is a *real-linear* space. Each linear space can be regarded as a real-linear space by restricting scalar multiplication to elements of $\mathbb{R}$; this is the *underlying real-linear space*. The zero element of E is 0_E or just 0.

Let F and G be linear subspaces of a linear space E. Then

$$F + G = \{y + z : y \in F, z \in G\}.$$

We write $E = F \oplus G$ when $F + G = E$ and $F \cap G = \{0\}$, and we say that F is *complemented* (by G) in E. In this case, each $x \in E$ can be expressed uniquely as $x = y + z$, where $y \in F$ and $z \in G$. Each subspace of a linear space E is complemented by some linear subspace of E. We also write $\bigoplus E_\alpha$ for the linear space that is the direct sum of a family $\{E_\alpha\}$ of linear spaces.

A real-linear space V has a standard *complexification* $E = V \oplus \mathrm{i}V$, where

$$(\alpha + \mathrm{i}\beta)(x + \mathrm{i}y) = \alpha x - \beta y + \mathrm{i}(\beta x + \alpha y) \quad (\alpha, \beta \in \mathbb{R}, x, y \in V),$$

so that E is a (complex) linear space; we set $E_{\mathbb{R}} = V$.

A subset C of a real-linear or linear space is *convex* if the segment

$$[x, y] = \{tx + (1 - t)y : t \in \mathbb{I}\}$$

is contained in C whenever $x,y \in C$; a subset C of a linear space is *absolutely convex* if $\alpha x + \beta y \in C$ whenever $\alpha, \beta \in \mathbb{C}$ with $|\alpha| + |\beta| \leq 1$ and $x,y \in C$. Let S be a subset of a linear space E. The subspace *spanned* by S, the *convex hull* of S, and the *absolutely convex hull* of S are denoted by

$$\operatorname{lin} S, \quad \operatorname{co} S, \quad \text{and} \quad \operatorname{aco} S,$$

respectively; for $\zeta \in \mathbb{C}$, we set $\zeta S = \{\zeta x : x \in S\}$; the *circled hull* of S is

$$\operatorname{ci} S = \{\zeta x : \zeta \in \mathbb{T}, x \in S\},$$

and S is *circled* if $\operatorname{ci} S = S$; the set S is *absorbent* if $\bigcup\{\alpha S : \alpha \geq 0\} = E$.

Let L be a convex set in a real-linear space E. A *face* of L is a convex subset F of L such that $x,y \in F$ whenever $x,y \in L$ and $(x+y)/2 \in F$; a point $x \in L$ is an *extreme point* of L if $\{x\}$ is a face. The sets of faces and of extreme points in L are denoted by

$$\mathfrak{F}(L) \quad \text{and} \quad \operatorname{ex} L,$$

respectively. Let F be a face of L. It is an elementary exercise to verify that, if $x,y \in L$ with $tx + (1-t)y \in F$ for some $t \in (0,1)$, then $x,y \in F$. A face F in L is *complemented* in L if there is a face G in L such that $F \cap G = \emptyset$ and $\operatorname{co}(F \cup G) = L$; this face G is uniquely specified by these conditions, and it is often denoted by $F^{\perp}$. The set of complemented faces of L is denoted by

$$\operatorname{Comp}_L .$$

A face F is a *split face* if F is complemented by the face G and if, further, every point $x \in L \setminus (F \cup G)$ is *uniquely* represented by a convex combination $x = ty + (1-t)z$, where $0 < t < 1$, $y \in F$, and $z \in G$.

The monograph of Goodearl [118] of 1986 is a superb source for infinite-dimensional faces and contains appropriate references to earlier sources. See also [4] and [104].

Let K be a convex subset of a linear space. A function h from K into another linear space is *affine* if

$$h(sx+ty) = sh(x) + th(y) \quad \text{whenever} \quad x,y \in K \text{ and } s,t \in \mathbb{I} \text{ with } s+t = 1,$$

and a real-valued function h is *convex* if

$$h(sx+ty) \leq sh(x) + th(y) \quad \text{whenever} \quad x,y \in K \text{ and } s,t \in \mathbb{I} \text{ with } s+t = 1.$$

Let E and F be linear spaces over $\mathbb{K}$, where $\mathbb{K} = \mathbb{C}$ or $\mathbb{K} = \mathbb{R}$. A map $T : E \to F$ is *linear* if

$$T(\alpha x + \beta y) = \alpha Tx + \beta Ty \quad (\alpha, \beta \in \mathbb{K}, x,y \in E).$$

The collection of all the linear maps from E to F is $\mathscr{L}(E,F)$; it is also a linear space over $\mathbb{K}$ in the obvious way. The space $\mathscr{L}(E,E)$ is denoted by $\mathscr{L}(E)$; it is an algebra with respect to the composition of operators, and its identity is I_E, the identity operator on E. An *idempotent* in an algebra is an element a such that $a^2 = a$; an idempotent P in $\mathscr{L}(E)$ is a *projection* onto $P(E)$, and $P(E)$ is complemented by the subspace $(I_E - P)(E)$.

A bijection in $\mathscr{L}(E,F)$ is a *linear isomorphism.* Take $T \in \mathscr{L}(E,F)$. Then $\ker T = \{x \in E : Tx = 0\}$, and T induces a linear map

$$\overline{T} : x + \ker T \mapsto Tx, \quad E/\ker T \to F. \tag{1.2}$$

Of course, $\overline{T}$ is a linear isomorphism from $E/\ker T$ onto $T(E)$; this statement is the *fundamental isomorphism theorem.*

Let E be a linear space over $\mathbb{K}$. The set of linear functionals on E is denoted by E^*; it is also a linear space over $\mathbb{K}$, called the *algebraic dual space* of E.

Let E be a real-linear space such that $(E, \leq)$ is also a partially ordered set for an order $\leq$. Then E is an *ordered linear space* if the linear space and order structures are compatible, in the sense that:

(i) $x + z \leq y + z$ whenever $x, y, z \in E$ and $x \leq y$;

(ii) $\alpha x \leq \alpha y$ whenever $\alpha \in \mathbb{R}^+$ and $x, y \in E$ with $x \leq y$.

An element e in an ordered linear space E is an *order unit* for E if, for each $x \in E$, there exists $\alpha \in \mathbb{R}^+$ such that $-\alpha e \leq x \leq \alpha e$.

For example, $\mathbb{R}^S$ and $\ell^\infty_{\mathbb{R}}(S)$ are ordered linear spaces for each non-empty set S, and 1_S is an order unit for $\ell^\infty_{\mathbb{R}}(S)$. Further, $fg \geq 0$ whenever $f, g \geq 0$ in $\mathbb{R}^S$, and so $(\mathbb{R}^S, \leq)$ may be termed a partially ordered algebra.

Let E be an ordered linear space. The *positive cone* of E is

$$E^+ = \{x \in E : x \geq 0\}.$$

The ordering on E is determined by E^+: $x \leq y$ in E if and only if $y - x \in E^+$. A linear functional λ on E is *positive* if $\lambda(x) \geq 0$ $(x \in E^+)$. Let $\lambda, \mu \in E^*$. Then $\lambda \geq \mu$ if $\lambda - \mu$ is a positive linear functional; with respect to this ordering, E^* is also an ordered linear space. In the case where E has an order unit e, a positive linear functional λ is a *state* (with respect to e) if $\lambda(e) = 1$; the set K_E of states on E is then a convex subset of E^*. A state is a *pure state* if it is an extreme point of K_E.

We shall use the following lemma in §3.1.

Lemma 1.3.3. *Let E be an ordered linear space with an order unit e, and suppose that $\lambda \in K_E$. Then λ is a pure state (with respect to e) if and only if each positive linear functional μ on E with $\mu \leq \lambda$ is a scalar multiple of λ.*

Proof. Suppose that each positive linear functional μ on E with $\mu \leq \lambda$ is a scalar multiple of λ. Take $\lambda_1, \lambda_2 \in K_E$ and $t \in (0,1)$ with $\lambda = t\lambda_1 + (1-t)\lambda_2$. Then $t\lambda_1 \leq \lambda$ and so $t\lambda_1 = s\lambda$ for some $s \in \mathbb{R}$. Since $\lambda_1(e) = \lambda(e) = 1$, it follows that $s = t$ and so $\lambda_1 = \lambda$. Hence λ is a pure state.

Conversely, suppose that λ is a pure state, and take a positive linear functional μ with $\mu \leq \lambda$, so that $0 \leq \mu(e) \leq \lambda(e) = 1$. For each $x \in E$, take $\alpha_x \in \mathbb{R}^+$ such that $-\alpha_x e \leq x \leq \alpha_x e$. If $\mu(e) = 0$, then $0 = \mu(-\alpha_x e) \leq \mu(x) \leq \mu(\alpha_x e) = 0$, and so

$\mu(x) = 0$, whence $\mu = 0 \cdot \lambda$. If $\mu(e) = 1$, then similarly $\mu = 1 \cdot \lambda$. If $0 < \mu(e) < 1$, set $t = \mu(e)$, $\lambda_1 = (\lambda - \mu)/(1-t)$, and $\lambda_2 = \mu/t$, so that λ_1 and λ_2 are states and $\lambda = (1-t)\lambda_1 + t\lambda_2$, so that $\lambda_2 = \lambda$ and $\mu = t\lambda$. □

Definition 1.3.4. An ordered linear space E is a *Riesz space* if $(E, \leq)$ is a lattice. A *complex Riesz space* is the complexification of a Riesz space. A Riesz space is *Dedekind complete* if it has this property as a lattice; a complex Riesz space is *Dedekind complete* if it is the complexification of a Dedekind complete Riesz space.

A Riesz space is a distributive lattice; indeed, Riesz spaces satisfy an infinite distributive law, in the sense that, for each subset F of E such that $\bigvee F$ exists and each $x \in E$, it is also true that $\bigvee(x \wedge F)$ exists and equals $x \wedge \bigvee F$ [5, Theorem 1.8].

Let $\{E_\alpha : \alpha \in A\}$ be a family of Riesz spaces. Then it is clear that the linear space $\bigoplus\{E_\alpha : \alpha \in A\}$ is a Riesz space with respect to the order $\leq$, where $(x_\alpha) \leq (y_\alpha)$ in $\bigoplus\{E_\alpha : \alpha \in A\}$ whenever $x_\alpha \leq y_\alpha$ in E_α for each $\alpha \in A$. A similar remark applies to complex Riesz spaces.

For a comprehensive account of Riesz spaces, see the texts [180] of Luxemburg and Zaanen from 1971 and Zannen [245] from 1997.

Let $(E, \leq)$ be a Riesz space. For $x \in E$, set

$$x^+ = x \vee 0, \quad x^- = (-x) \vee 0, \quad |x| = x \vee (-x);$$

thus, x^+, x^-, and $|x|$ are the *positive part*, the *negative part*, and the *modulus* of x, respectively. Two elements x and y of E are *disjoint*, written $x \perp y$, if $|x| \wedge |y| = 0$, and two subsets S and T of E are *disjoint*, written $S \perp T$, if $x \perp y$ whenever $x \in S$ and $y \in T$. A linear subspace F of a Riesz space E is a *sublattice* if $x \vee y, x \wedge y \in F$ whenever $x, y \in F$ and a *lattice ideal* if $x \in F$ whenever $x \in E$ and $|x| \leq |y|$ for some $y \in F$, so that each lattice ideal is a sublattice.

For example, for each non-empty set S, the space $\mathbb{R}^S$ is a Riesz space with the pointwise lattice operations, and the definitions of $|f|$, etc., coincide with the ones given on page 4; the space $\mathbb{C}^S$ is a complex Riesz space.

Let $(E, \leq)$ and $(F, \leq)$ be two Riesz spaces. A *Riesz homomorphism* or *Riesz isomorphism* is a linear map $T : E \to F$ that is a lattice homomorphism or a lattice isomorphism, respectively; the two Riesz spaces are *Riesz isomorphic* if there is such a Riesz isomorphism from E onto F. Now suppose that E and F are complex Riesz spaces that are the complexifications of the Riesz spaces $E_{\mathbb{R}}$ and $F_{\mathbb{R}}$, respectively. Then $T \in \mathscr{L}(E, F)$ is a *Riesz homomorphism* or *Riesz isomorphism* if $T(E_{\mathbb{R}}) \subset F_{\mathbb{R}}$ and $T \mid E_{\mathbb{R}} : E_{\mathbb{R}} \to F_{\mathbb{R}}$ is a Riesz homomorphism or Riesz isomorphism, respectively.

Let E be a linear space. A map $* : x \mapsto x^*$, $E \to E$, is a *linear involution* on E if

$$(\alpha x + \beta y)^* = \overline{\alpha} x^* + \overline{\beta} y^* \quad (\alpha, \beta \in \mathbb{C}, x, y \in E) \quad \text{and} \quad (x^*)^* = x \quad (x \in E).$$

An element $x \in E$ is *self-adjoint* (or *hermitian*) if $x^* = x$; the real-linear subspace of self-adjoint elements in E is denoted by E_{sa}. Let E and F be linear spaces with linear involutions, both denoted by $*$. A linear map $T : E \to F$ is $*$*-linear* if

$$T(x^*) = (Tx)^* \quad (x \in E).$$

1.4 Topological notions

We recall some standard notions from topology. For background in topology, see the two great classic texts [99] and [155] of Engelking and Kelley, respectively, and many modern sources.

Let X be a topological space. Suppose that $S \subset X$. Then the *interior* of S in X is denoted by $\mathrm{int}_X S$ or $\mathrm{int}\, S$, the *closure* of S by $\mathrm{cl}_X S$ or $\overline{S}$, and the *frontier* of S by $\partial_X S$; S is *dense* in X if $\overline{S} = X$. The subset S is a *G_δ-set* if it is a countable intersection of open sets and an *F_σ-set* if it is a countable union of closed sets. A subset of X is *clopen* if it is both open and closed; the family of clopen subsets of X is denoted by $\mathfrak{U}_X$. A *neighbourhood* of $x \in X$ is a set that contains x in its interior; the family of open neighbourhoods of a point $x \in X$ is denoted by $\mathscr{N}_x$. A point $x \in X$ is *isolated* if $\{x\}$ is an open set; we write D_X for the set of isolated points of X, so that D_X is an open subset of X and the relative topology from X on D_X is the discrete topology. A point $x \in X$ is a *P-point* if each G_δ-set in X which contains x is a neighbourhood of x, and a *limit point* of a subset S of X if $U \cap (S \setminus \{x\}) \neq \emptyset$ for each $U \in \mathscr{N}_x$; the subspace S is *dense-in-itself* if each point of S is a limit point of S, equivalently, if $D_S = \emptyset$; a subset of X is *perfect* if it is closed and dense-in-itself. Clearly the closure of a subspace of X that is dense-in-itself is a perfect subspace.

A Hausdorff topological space X is *regular* if, for each closed subset F of X and each $x \in X \setminus F$, there are disjoint open sets U and V with $x \in U$ and $F \subset V$, and X is *normal* if, for each pair $\{F, G\}$ of disjoint closed subsets, there are disjoint open sets U and V with $F \subset U$ and $G \subset V$; every metrizable space is normal.

A set S taken with the discrete topology (so that every subset of S is open) is sometimes denoted by S_d. We shall quite often use implicitly the fact that every infinite, Hausdorff topological space contains a countable, infinite discrete subspace.

Of course, each metric space is a topological space. A metric space (X, d) is *complete* if every Cauchy sequence with respect to the metric d converges to a point of X. A topological space is *(completely) metrizable* if its topology is given by a (complete) metric.

Let (X, τ) be a Hausdorff topological space. Then τ is a complete lattice. Indeed, let $\mathscr{F}$ be a subset of τ. Then the supremum and infimum of $\mathscr{F}$ in τ are

$$\bigcup\{U : U \in \mathscr{F}\} \quad \text{and} \quad \mathrm{int}\left(\bigcap\{U : U \in \mathscr{F}\}\right),$$

respectively.

A *base* for a topological space (X, τ) is a subset $\mathscr{B}$ of τ such that each non-empty set in τ is a union of a subfamily of sets in $\mathscr{B}$. Equivalently, $\mathscr{B}$ is a base for τ if and only if, for each $x \in X$ and $V \in \mathscr{N}_x$, there exists $B \in \mathscr{B}$ with $x \in B \subset V$. A *subbase* for (X, τ) is a subset $\mathscr{S}$ of τ such that the family of sets which are finite intersections of members of $\mathscr{S}$ form a base.

A topological space is *compact* if it is Hausdorff and every open cover has a finite subcover; equivalently, every non-empty family of closed subsets with the finite intersection property has a non-empty intersection; a subset is *relatively compact* if its closure is compact. The space is *locally compact* if each point has a compact neighbourhood (and so each locally compact space is Hausdorff); a locally compact space is *σ-compact* if it is a countable union of compact subspaces, and it satisfies CCC, the *countable chain condition*, if each pairwise-disjoint family of non-empty, open sets in the space is countable. The family of all compact and open subsets of X is denoted by $\mathfrak{C}_X$, so that $\mathfrak{C}_X \subset \mathfrak{U}_X$. A σ-compact space is normal and a locally compact space is regular (but not necessarily normal).

A *compactification* of a topological space X is a compact space that contains X as a dense subspace. In the case where a locally compact space K is not compact, the *one-point* (or *Alexandroff*) *compactification* of K is denoted by K_∞, so that $K_\infty = K \cup \{\infty\}$ for a point ∞.

A net $(x_\gamma : \gamma \in D)$ in a topological space X *converges* to $x \in X$, written

$$x_\gamma \to x \quad \text{or} \quad \lim_\gamma x_\gamma = x,$$

if, for each $U \in \mathscr{N}_x$, there exists $\delta \in D$ such that $x_\gamma \in U \ (\gamma \geq \delta)$, and $x \in X$ is an *accumulation point* of the net (x_γ) if, for each $U \in \mathscr{N}_x$ and each $\delta \in D$, there exists $\gamma \geq \delta$ such that $x_\gamma \in U$; the latter holds if and only if there is a subnet of (x_γ) that converges to x. A Hausdorff topological space X is compact if and only if each net in X has an accumulation point, and so each net in a compact space has a convergent subnet.

We shall use the following collection of subsets of a locally compact space rather often.

Definition 1.4.1. Let K be a locally compact space. Then $\mathscr{K}_K$ is the family of compact subsets L of K such that $\operatorname{int}_K L = \emptyset$.

A topological space X is *separable* if it contains a countable, dense subset. The *weight* of X is the minimum cardinal of a base for the topology; it is denoted by $w(X)$. It is easy to see that $|X| \leq 2^{w(X)}$ whenever X is Hausdorff. The *density character* of X is the minimum cardinal of a dense subset of X; it is denoted by $d(X)$. It is clear that $d(X) \leq w(X)$ and that $d(X) \leq \aleph_0$ if and only if X is separable. Further, $d(K) = w(K) = \aleph_0$ and $|K| \leq 2^{\aleph_0} = \mathfrak{c}$ whenever K is infinite, compact, and metrizable. Each subspace of a separable, metrizable space is separable.

Definition 1.4.2. Let (X, τ) be a topological space. Then the *Borel sets* in X are the members of the σ-algebra $\sigma(\tau)$ generated by the family τ of open subsets of X; we set $\mathfrak{B}_X = \sigma(\tau)$.

Equivalently, $\mathfrak{B}_X$ is the σ-algebra generated by the closed subsets of X. We note that $|\mathfrak{B}_X| \geq \mathfrak{c}$ whenever X is infinite and Hausdorff.

A subset S of a topological space X is *nowhere dense* if $\operatorname{int}_X \overline{S} = \emptyset$ (these spaces are said to be 'rare' by Dixmier [91] and others); a *meagre* (or *first category*) set in X is a countable union of nowhere dense sets. For example, $\mathbb{Q}$ is meagre as a subset of $\mathbb{R}$. Two subset Y_1 and Y_2 of X are *congruent* if $Y_1 \Delta Y_2$ is meagre, and then we write $Y_1 \equiv Y_2$. Clearly $\equiv$ is an equivalence relation on $\mathscr{P}(X)$. Further, suppose that $Y_1 \equiv Z_1$ and $Y_2 \equiv Z_2$. Then $Y_1 \cup Y_2 \equiv Z_1 \cup Z_2$ and $Y_1 \cap Y_2 \equiv Z_1 \cap Z_2$.

A subset Y of X has the *Baire property* if there is an open subset U of X such that $Y \equiv U$; the family of sets with the Baire property is denoted by $\mathfrak{BP}_X$. It is easy to check that $\mathfrak{BP}_X$ is a σ-algebra of subsets of X containing all the open sets, and so $\mathfrak{B}_X \subset \mathfrak{BP}_X$; in fact, $|\mathfrak{B}_{\mathbb{R}}| = \mathfrak{c}$, but $|\mathfrak{BP}_{\mathbb{R}}| = 2^{\mathfrak{c}}$, and so there are many subsets of $\mathbb{R}$ which have the Baire property, but which are not Borel sets. There are also subsets of $\mathbb{R}$ which do not have the Baire property. See [68, Appendix A.4] for details of these remarks.

A subset U in a topological space X is *regular–open* if $U = \text{int}_X\overline{U}$ and *regular–closed* if its complement is regular–open, so that a subset F is regular–closed if and only if $F = \overline{\text{int}F}$; the collection of regular–open subsets of X is denoted by $\mathfrak{R}_X$. Since $U \equiv \overline{U} \equiv \text{int}_X\overline{U}$, every set in $\mathfrak{BP}_X$, and, in particular, every Borel set in X, is congruent to a regular–open subset of X; in the special case where X is completely metrizable or locally compact, this regular–open set is uniquely defined.

A topological space X is *connected* if $\mathfrak{U}_X = \{\emptyset, X\}$, so that the only clopen subspaces of X are $\emptyset$ and X; X is *locally connected* if the family of connected and open subsets is a base for the topology; a *component* of X is a maximal connected subspace; X is *totally disconnected* if the only components are singletons; X is *zero-dimensional* if the family $\mathfrak{U}_X$ of clopen sets is a base for the topology; X is *extremely disconnected* if the closure of every open set is itself open, or, equivalently, if pairs of disjoint open subsets of X have disjoint closures. A regular, extremely disconnected space is zero-dimensional; a zero-dimensional Hausdorff space is totally disconnected, and the two notions are equivalent for locally compact spaces.

We shall use the fact that open subspaces and dense subspaces of an extremely disconnected space are also extremely disconnected. For example, suppose that Y is an extremely disconnected space and that X is dense in Y. Take U to be an open subset of X, so that $\text{cl}_X U = \overline{U} \cap X$, say $U = V \cap X$, where V is open in Y. Then $\text{cl}_X U = \overline{V \cap X} \cap X$, and the latter set is $\overline{V} \cap X$ because $\overline{X} = Y$. Since $\overline{V}$ is clopen in Y, the space $\text{cl}_X U$ is clopen in X, and so X is extremely disconnected. However, closed subspaces of an extremely disconnected space are not necessarily extremely disconnected; see Example 1.7.14.

Let $\{X_\alpha : \alpha \in A\}$ be a family of non-empty topological spaces, and set

$$X = \prod\{X_\alpha : \alpha \in A\},$$

with the product topology. Then X is compact, respectively, connected if and only if each space X_α is compact, respectively, connected.

Let X and Y be topological spaces, and take $x \in X$. A map $f : X \to Y$ is: *continuous at* x if $\lim_\gamma f(x_\gamma) = f(x)$ whenever $\lim_\gamma x_\gamma = x$; *continuous* if f is continuous at each $x \in X$, equivalently, if $f^{-1}(U)$ is open in X for each open set U in Y; *open* if $f(U)$ is open in Y for each open set U in X; *closed* if $f(F)$ is closed in Y for each closed set F in X. The family of continuous maps from X into Y is denoted by $C(X,Y)$. The two spaces X and Y are *homeomorphic* if there is a bijection $f : X \to Y$ such that f and f^{-1} are both continuous, and then f is a *homeomorphism*; in the case where X is compact and Y is Hausdorff, it is sufficient to require that the bijection f be continuous, and then it is necessarily a homeomorphism. A subspace Y (with the relative topology) of a topological space X is a *retract* of X if there is a continuous map of X onto Y that is the identity map on Y; the map is a *retraction* (from X onto Y). In this case, Y is closed in X. A topological space X is *homogeneous* if, given $x,y \in X$, there is a homeomorphism $\eta : X \to X$ such that $\eta(x) = y$.

A function $f : X \to \mathbb{R}$ is *lower semi-continuous* if $f^{-1}((-\infty, r))$ is open in X for each $r \in \mathbb{R}$.

Let X be a non-empty topological space. The algebras (with respect to the pointwise operations) of all complex-valued, continuous functions on X and of all bounded, continuous functions on X are denoted by $C(X)$ and $C^b(X)$, respectively, so that $(C^b(X), |\cdot|_X)$ is a Banach space; the corresponding spaces of real-valued functions are $C_{\mathbb{R}}(X)$ and $C^b_{\mathbb{R}}(X)$, respectively. Clearly $C(X)$ and $C^b(X)$ are the complexifications of $C_{\mathbb{R}}(X)$ and $C^b_{\mathbb{R}}(X)$, respectively, and both are complex Riesz spaces; the spaces $C^b_{\mathbb{R}}(X)$ and, in the case where X is locally compact, $C_{0,\mathbb{R}}(X)$, are sublattices of $\mathbb{R}^X$. Further, $C_{0,\mathbb{R}}(X)$ is a lattice ideal in $C^b_{\mathbb{R}}(X)$.

The *support* of a function $f \in C(X)$, denoted by $\operatorname{supp} f$, is the closure of the set $\{x \in X : f(x) \neq 0\}$; clearly, $\operatorname{int}\operatorname{supp} f \neq \emptyset$ whenever $f \neq 0$.

For a non-empty, compact space K, we have $C^b(K) = C(K)$; for $f \in C(K)$, there exists $x \in K$ with $|f(x)| = |f|_K$. A non-empty, locally compact space K is *pseudo-compact* if $C^b(K) = C(K)$, so that every compact space is pseudo-compact. Manifestly each infinite discrete space is not pseudo-compact: if $\{x_n : n \in \mathbb{N}\}$ is a set of distinct points in a discrete space, the function $\sum_{n=1}^{\infty} n\chi_{\{x_n\}}$ is an unbounded, continuous function on the space.

Let X be a totally ordered set. Then the *order topology* on X is specified by taking as a subbase the family of sets of the form $\{y \in X : y > x\}$ and $\{y \in X : y < x\}$ for $x \in X$. In particular, let σ be an ordinal regarded as the totally ordered interval $[0, \sigma)$. Then the order topology on σ is specified by taking as a subbase the family of sets of the form $[0, \tau)$ and (τ, σ) for $\tau \in (0, \sigma)$. A non-zero point in σ is isolated if and only if it is not a limit ordinal; the spaces $\sigma + 1 = [0, \sigma]$ are each compact. We recall that, for each countable subset $\{\sigma_n : n \in \mathbb{N}\}$ of $[0, \omega_1)$, there exists $\sigma \in [0, \omega_1)$ with $\sigma > \sigma_n$ $(n \in \mathbb{N})$, and we may suppose that $\sigma = \lim_{n\to\infty} \sigma_n$.

Set $K = [0, \omega_1)$, a locally compact space that is not compact, and take $f \in C(K)$. Assume that, for each $n \in \mathbb{N}$, there exists $\sigma_n \in K$ with $|f(\sigma_n)| > n$, and take $\sigma \in K$ with $\sigma > \sigma_n$ $(n \in \mathbb{N})$. Then $f \mid [0, \sigma]$ is unbounded, a contradiction. Thus K is pseudo-compact.

Now let $f \in C(K)$, where $K = [0, \omega_1)$. We *claim* that there exist $z_0 \in \mathbb{C}$ and $\sigma \in K$ with $f(\tau) = z_0$ $(\tau \in [\sigma, \omega_1))$. For this, we may suppose that $f \in C_{\mathbb{R}}(K)$. First note that, given $a < b$ in $\mathbb{R}$, either $\{\sigma \in K : f(\sigma) > b\}$ or $\{\sigma \in K : f(\sigma) < a\}$ is bounded in K, for otherwise there is a strictly increasing sequence (σ_n) in K with $f(\sigma_{2n}) < a$ and $f(\sigma_{2n-1}) > b$ for all $n \in \mathbb{N}$; set $\sigma = \lim_{n\to\infty} \sigma_n$, so that $f(\sigma) \geq b$ and $f(\sigma) \leq a$, a contradiction. Using this, we inductively construct strictly increasing sequences (σ_k) in K and $([a_k, b_k])$ of closed intervals in $\mathbb{R}$ with $a_k \leq a_{k+1} < b_{k+1} \leq a_k$, with $b_{k+1} - a_{k+1} \leq 2(b_k - a_k)/3$, and with $f(\tau) \in [a_k, b_k]$ $(\tau \in [\sigma_k, \omega_1))$ for each $k \in \mathbb{N}$. Set $z_0 = \sup a_k = \inf b_k$ and $\sigma = \lim_{n\to\infty} \sigma_k$. Then $f(\tau) = z_0$ $(\tau \in [\sigma, \omega_1))$, giving the claim. By setting $f(\omega_1) = z_0$ we extend f to a continuous function on $[0, \omega_1]$.

We now give a formal definition of a class of spaces of particular interest to us; properties of this class will emerge in a few pages, and first examples will be given in Example 1.5.10.

Definition 1.4.3. A compact, extremely disconnected space is a *Stonean space.*

Clearly a clopen subspace of a Stonean space is Stonean and a homeomorphic image of a Stonean space is Stonean.

We shall use the following easy remark.

Proposition 1.4.4. *Let K be a Stonean space. Then every regular–open set in K is clopen, and, for every $B \in \mathfrak{B}_K$, there is a unique set $C \in \mathfrak{C}_K$ with $B \equiv C$.* □

We define Δ to be the *Cantor set*, so that Δ is a subset of $\mathbb{I}$. Indeed, for $n \in \mathbb{N}$ and $r = 1, \dots, 3^{n-1}$, set

$$U_{n,r} = \left(\frac{3r-2}{3^n}, \frac{3r-1}{3^n}\right),$$

an open subinterval of $\mathbb{I}$, and set $U = \bigcup\{\bigcup\{U_{n,r} : r = 1, \dots, 3^{n-1}\} : n \in \mathbb{N}\}$. Then $\Delta = \mathbb{I} \setminus U$. The space $\mathbb{Z}_2^\omega$ is homeomorphic to Δ by the map

$$(\varepsilon_n) \mapsto 2\sum_{n=1}^{\infty} \frac{\varepsilon_n}{3^n}, \quad \mathbb{Z}_2^\omega \to \Delta\,. \tag{1.3}$$

The space Δ is compact, perfect, a complete metric space, and totally disconnected, and $|\Delta| = \mathfrak{c}$. However Δ is not extremely disconnected.

We make a further remark about the Cantor set, identified with $\mathbb{Z}_2^\omega$.

Proposition 1.4.5. *The space $\mathbb{Z}_2^\omega$ contains $\mathfrak{c}$ pairwise disjoint, closed subspaces, each homeomorphic to $\mathbb{Z}_2^\omega$.*

Proof. Clearly there is a continuous bijection $\theta : \mathbb{Z}_2^\omega \to \mathbb{Z}_2^\omega \times \mathbb{Z}_2^\omega$. For each sequence $\varepsilon \in \mathbb{Z}_2^\omega$, we set $F_\varepsilon = \theta^{-1}(\{\varepsilon\} \times \mathbb{Z}_2^\omega)$, so that F_ε is a compact subset of $\mathbb{Z}_2^\omega$ and F_ε is homeomorphic to $\mathbb{Z}_2^\omega$. The family $\{F_\varepsilon : \varepsilon \in \mathbb{Z}_2^\omega\}$ is pairwise disjoint and so has the required properties. □

We shall use the following facts; for a neat proof of (i), see [210]; for a full proof of (ii), see [141, Corollary 2-98] or [237, Section 27].

Proposition 1.4.6. (i) *Each non-empty, compact, metrizable space is the continuous image of the Cantor set.*

(ii) *Each compact, perfect, metrizable, and totally disconnected space is homeomorphic to the Cantor set.* □

The surjection in (i) can sometimes taken to be ‘irreducible’; see Proposition 1.4.23, below.

More generally, we define the *Cantor cube of weight* κ, where κ is an infinite cardinal, to be the space $\{0,1\}^\kappa = \mathbb{Z}_2^\kappa$. The space $\mathbb{Z}_2^\kappa$ is compact, totally disconnected, and perfect. Take $k \in \mathbb{N}$ and $\alpha = (\alpha_1, \dots, \alpha_k) \in \mathbb{Z}_2^k$, and then define

$$U_{F,\alpha} = \{(\varepsilon_\tau) \in \mathbb{Z}_2^\kappa : \varepsilon_{t_i} = \alpha_i \ (i \in \mathbb{N}_k)\}, \tag{1.4}$$

where $t_1 < t_2 < \cdots < t_k < \kappa$ and $F = \{t_1, \dots, t_k\}$, so that $U_{F,\alpha}$ is a clopen subset of $\mathbb{Z}_2^\kappa$. The sets $U_{F,\alpha}$ form a base of cardinality κ for the topology of $\mathbb{Z}_2^\kappa$; each clopen set is a finite, pairwise disjoint union of these basic clopen sets, and so $w(\mathbb{Z}_2^\kappa) \leq \kappa$. For each base for the topology and each $\sigma < \kappa$, the base contains a subset of the open set $\{(\varepsilon_\tau) \in 2^\kappa : \varepsilon_\sigma = 0\}$, and so $w(\mathbb{Z}_2^\kappa) \geq \kappa$. Thus

$$|\mathbb{Z}_2^\kappa| = 2^\kappa, \quad \left|\mathfrak{C}_{\mathbb{Z}_2^\kappa}\right| = w(\mathbb{Z}_2^\kappa) = \kappa. \tag{1.5}$$

We mention the following rather surprising result; it is a special case of the *Hewitt–Marczewski–Pondiczery theorem*, proved in [99, 2.3.15]. Our argument uses polynomials with rational coefficients to establish separability; this technique became well known from [112, Problem 9.O].

Proposition 1.4.7. *The space $\mathbb{Z}_2^\mathfrak{c}$ is separable in the product topology, and so*

$$d(\mathbb{Z}_2^\mathfrak{c}) = \aleph_0 < \mathfrak{c} = w(\mathbb{Z}_2^\mathfrak{c}).$$

Proof. It suffices to show that $\{0,1\}^\mathbb{R}$ (with the product topology, τ) is separable.

For each $f \in \mathbb{R}^\mathbb{R}$, define $Tf \in \{0,1\}^\mathbb{R}$ by setting $(Tf)(t) = 0$ if $f(t) \leq 1/2$ and $(Tf)(t) = 1$ if $f(t) > 1/2$, where $t \in \mathbb{R}$.

Let $\mathscr{P}$ denote the set of real-valued polynomials with coefficients in $\mathbb{Q}$, so that $\mathscr{P}$ is a countable set. We *claim* that $\{Tp : p \in \mathscr{P}\}$ is dense in $(\{0,1\}^\mathbb{R}, \tau)$. But this is almost immediate. For each $f \in \{0,1\}^\mathbb{R}$ and $t_1, \dots, t_n \in \mathbb{R}$, choose a real-valued polynomial q on $\mathbb{R}$ with $q(t_i) = f(t_i)$ $(i \in \mathbb{N}_n)$; clearly, there exists $p \in \mathscr{P}$ such that $|p(t_i) - q(t_i)| < 1/3$ $(i \in \mathbb{N}_n)$. Then $(Tp)(t_i) = f(t_i)$ $(i \in \mathbb{N}_n)$, giving the claim.

Hence $\mathbb{Z}_2^\mathfrak{c}$ is separable, and so $d(\mathbb{Z}_2^\mathfrak{c}) = \aleph_0$.

We have remarked that $w(\mathbb{Z}_2^\mathfrak{c}) = \mathfrak{c}$. □

We now return to the Stone space of a lattice; this was defined in Definition 1.3.2.

Definition 1.4.8. Let L be a lattice with a minimum element 0. Then the *Stone topology* on $St(L)$ is formed by taking the sets of the form

$$S_x = \{p \in St(L) : x \in p\},$$

where $x \in L$, as a base for the topology.

The family $\{S_x : x \in L\}$ is closed under finite intersections, and so is indeed a base for a topology.

Proposition 1.4.9. *Let L be a lattice with a minimum element 0. Then S_x is a clopen subset of $St(L)$ for each $x \in L$, and so $St(L)$ is zero-dimensional and Hausdorff.*

In the case where L is distributive, the map

$$x \mapsto S_x\,, \quad L \to \mathfrak{U}_{St(L)}\,,$$

is a lattice homomorphism.

Proof. Take $x \in L$. By the definition of the topology on $St(L)$, the set S_x is open. For each $p \in St(L) \setminus S_x$, there exists $y \in p$ with $x \wedge y = 0$, and so $p \in S_y \subset St(L) \setminus S_x$. Thus $St(L) \setminus S_x$ is open, and hence S_x is closed. We have shown that S_x is clopen.

Now suppose that L is distributive. Take $x_1, x_2 \in L$. To see that $S_{x_1 \vee x_2} = S_{x_1} \cup S_{x_2}$, take $p \in St(L) \setminus (S_{x_1} \cup S_{x_2})$, so that there exist $y_1, y_2 \in p$ with $x_1 \wedge y_1 = x_2 \wedge y_2 = 0$. Then $y_1 \wedge y_2 \in p$ and $(x_1 \vee x_2) \wedge (y_1 \wedge y_2) = 0$, and so $p \notin S_{x_1 \vee x_2}$. Thus we have $S_{x_1 \vee x_2} \subset S_{x_1} \cup S_{x_2}$; the reverse inclusion is obvious. We have shown that the map $x \mapsto S_x$, $L \to \mathfrak{U}_{St(L)}$, is a lattice homomorphism. □

In the case where L is a distributive lattice, $St(L)$ is compact if and only if, for each $x \in L$, there exists an element $y \in L$ that is maximal with respect to the property that $x \wedge y = 0$.

Definition 1.4.10. A *Polish space* is a topological space that is separable and completely metrizable.

For example, each compact metric space is a Polish space.

The *diameter* of a subset S of a metric space (X, d) is denoted by diam S, so that diam $S = \sup\{d(x,y) : x, y \in S\}$. For $x \in X$ and a non-empty subset F of X, set

$$d(x, F) = \inf\{d(x,y) : y \in F\}\,;$$

this is the *distance* from x to F; the function $y \mapsto d(y, F)$ is continuous on X. In a complete metric space, each nested sequence of closed subsets whose diameters form a null sequence has a non-empty intersection that is a singleton.

Let (E, d_E) and (F, d_F) be metric spaces. A map $T : E \to F$ is an *isometry* if

$$d_F(Tx, Ty) = d_E(x, y) \quad (x, y, \in E)\,.$$

The following is a form of the *Baire category theorem*.

Theorem 1.4.11. *Let X be a non-empty space that is completely metrizable or locally compact. Suppose that (F_n) is a sequence of closed sets in X with union X. Then there exists $n \in \mathbb{N}$ such that $\mathrm{int}_X F_n \neq \emptyset$. The complement of each meagre set in X is dense; each G_δ-set in X that has no isolated points is uncountable.* □

We shall later use the following classical fact.

Proposition 1.4.12. *Every G_δ-subset of a complete, metrizable space is homeomorphic to a complete metric space. Further, each G_δ-subset of a Polish space is a Polish space.*

Proof. Let (X,d) be a complete metric space, and let U be a G_δ-subset of X, say $U = \bigcap_{n\in\mathbb{N}} G_n$, where the G_n are open in X. It suffices to describe a complete metric on U which generates its subspace topology. If $U = X$ the result is trivial, so we shall suppose that $G_n \neq X$ $(n \in \mathbb{N})$. For $n \in \mathbb{N}$, set $F_n = X \setminus G_n$, and set

$$d_n(x,y) = |d(x,F_n)^{-1} - d(y,F_n)^{-1}| \quad (x,y \in G_n).$$

Finally, set

$$\delta(x,y) = d(x,y) + \sum_{n=1}^{\infty} \frac{1}{2^n} \frac{d_n(x,y)}{(1+d_n(x,y))} \quad (x,y \in U).$$

It is an elementary exercise to verify that δ is a metric on U.

Clearly $d(x,y) \leq \delta(x,y)$ $(x,y \in U)$. Now suppose that $x_k \to x$ in (U,d). Then $d_n(x_k,x) \to 0$ as $k \to \infty$ for each $n \in \mathbb{N}$, and so $x_k \to x$ in (U,δ). It follows that (U,d) is homeomorphic to (U,δ).

As to the completeness of (U,δ), let (x_k) be a Cauchy sequence in (U,δ). Then (x_k) is Cauchy in (X,d), and hence converges, say $x_k \to x \in X$. Assume that $x \notin U$, so that $x \in F_n$ for some $n \in \mathbb{N}$. Then $d(x_k,F_n) \to 0$ and $d_n(x_j,x_k) \to \infty$ as $k \to \infty$ for each $j \in \mathbb{N}$. Since $\delta(x_j,x_k) \geq 2^{-n} d_n(x_j,x_k)/(1+d_n(x_j,x_k))$, we have $\limsup_{k\to\infty} \delta(x_j,x_k) \geq 2^{-n}$ $(j \in \mathbb{N})$, contradicting the assumption that (x_k) is δ-Cauchy. Thus $x \in U$ and $\delta(x_k,x) \to 0$, showing that (U,δ) is complete.

In the case where (X,d) is a Polish space, the subspace (U,d) is separable, and so (U,δ) is a Polish space. □

We shall also use the following result concerning the topology of Polish spaces; the proof is adapted from that of [228, Theorem 2.6.7].

Proposition 1.4.13. *Let E be an equivalence relation on a Polish space X such that there are uncountably many equivalence classes, and suppose that E is a closed subspace of the product $X \times X$. Then there is a continuous map $f : \Delta \to X$ such that $f(s)$ and $f(t)$ belong to distinct equivalence classes whenever $s,t \in \Delta$ with $s \neq t$. Further, E has exactly $\mathfrak{c}$ equivalence classes.*

Proof. For $s \in \mathbb{Z}_2^{<\omega}$, we write $|s|$ for the length of s, and, for $s,t \in \mathbb{Z}_2^{<\omega}$, we write $s \perp t$ to mean that neither s nor t is an extension of the other.

Fix a complete metric d on X that defines the topology and, further, is such that $d(x,y) \leq 1$ $(x,y \in X)$. We propose to construct a system $\{U_s : s \in \mathbb{Z}_2^{<\omega}\}$ of non-empty, open subsets of X with the following properties:

(i) $\operatorname{diam} U_s \leq 2^{-|s|}$ for $s \in \mathbb{Z}_2^{<\omega}$;

(ii) $\overline{U_{s^+}} \cup \overline{U_{s^-}} \subset U_s$ for $s \in \mathbb{Z}_2^{<\omega}$;

(iii) $E \cap (\overline{U_s} \times \overline{U_t}) = \emptyset$ for $s,t \in \mathbb{Z}_2^{<\omega}$ with $s \perp t$.

Assume that such a system has been defined. For each $s \in \mathbb{Z}_2^\omega$, the set $\bigcap_{n\in\mathbb{N}} \overline{U_{s\restriction n}}$ is, by (i) and (ii), a singleton, say x_s in X. Suppose that $s,t \in \mathbb{Z}_2^\omega$ with $s \neq t$. Then there exists $n \in \mathbb{N}$ with $s \restriction n \neq t \restriction n$. Since $x_s \in \overline{U_{s\restriction n}}$ and $x_t \in \overline{U_{t\restriction n}}$, it follows from (iii) that x_s and x_t are not equivalent. Clearly the map $f : s \mapsto x_s$, $\Delta = Z_2^\omega \to X$, is a continuous map, and so has the required properties.

The space X has a countable base, say $\{V_n : n \in \mathbb{N}\}$; set

$$V := \bigcup\{V_n : V_n \text{ intersects only countably many equivalence classes}\},$$

and set $Y := X \setminus V$. Then $Y \neq \emptyset$ because, by assumption, E has uncountably many equivalence classes; Y is closed, and therefore complete; $E \cap (Y \times Y)$ is closed in $Y \times Y$, and every non-empty, relatively open set in Y has non-empty intersection with uncountably many equivalence classes, and, in particular, contains two non-equivalent elements.

To construct the system $\{U_s : s \in \mathbb{Z}_2^{<\omega}\}$, we shall further require that $U_s \cap Y \neq \emptyset$ for each $s \in \mathbb{Z}_2^{<\omega}$. Take $n \in \mathbb{N}$, and assume that U_s has been defined for each $s \in \mathbb{Z}_2^n$. For any such s, the set $U_s \cap Y$ contains two non-equivalent elements, say y_s and z_s. Then $(y_s, z_s) \notin E$. Since E is closed in $X \times X$, we can take open neighbourhoods U_{s^+} and U_{s^-} of y_s and z_s, respectively, with diameters at most $1/2^{n+1}$ and with $\overline{U_{s^+}} \cup \overline{U_{s^-}} \subset U_s$ and also with $E \cap \left(\overline{U_{s^+}} \times \overline{U_{s^-}}\right) = \emptyset$. This defines U_t for all $t \in \mathbb{Z}_2^{n+1}$. By recursion, the set U_s is defined for all $s \in \mathbb{Z}_2^{<\omega}$, and (i), (ii), (iii) hold.

Since the space X is separable, the number of equivalence classes is at most $|X| \leq |\mathscr{P}(\mathbb{N})| = \mathfrak{c}$, and so there are exactly $\mathfrak{c}$ equivalence classes. □

In fact, Proposition 1.4.13 is a particular case of a deeper theorem in which the same conclusion holds under the more general assumption that the equivalence relation E is a Borel set (or even a coanalytic set) in $X \times X$. A proof of this more general result is given in [145, Theorem 32.1].

Proposition 1.4.13 is a generalization of the following well-known classical fact; the result and its corollaries are given in greater generality in [59, §8.2] and [164, §33].

Proposition 1.4.14. *Every uncountable G_δ-set in a Polish space contains a homeomorphic copy of Δ, and so has cardinality $\mathfrak{c}$. In particular, every uncountable, compact, metrizable space K has $|K| = \mathfrak{c}$.*

Proof. Let X be an uncountable Polish space, and then set $E = \{(x,x) : x \in X\}$. By Proposition 1.4.13, E has exactly $\mathfrak{c}$ equivalence classes, which in this case are just the singleton sets $\{x\}$ for $x \in X$. Hence X contains a homeomorphic copy of Δ, and $|X| = \mathfrak{c}$. □

Corollary 1.4.15. *Let X be an uncountable Polish space. Then $|X| = |\mathfrak{B}_X| = \mathfrak{c}$, and so there is a subset of X that is not a Borel set.*

Proof. By Proposition 1.4.14, $|X| = \mathfrak{c}$, and $|\mathfrak{B}_X| \geq \mathfrak{c}$. Since X has a countable base for its topology and each open set is a countable union of basic open sets, so that

$\mathfrak{B}_X$ is the σ-algebra generated by the basic open sets, it follows as on page 4 that $|\mathfrak{B}_X| \le \mathfrak{c}$. Hence $|\mathfrak{B}_X| = \mathfrak{c}$.

Since $|\mathscr{P}(X)| = 2^{\mathfrak{c}} > \mathfrak{c}$, there are $2^{\mathfrak{c}}$ subsets of X which are not Borel sets. □

Corollary 1.4.16. *Let K be a compact, metrizable space, and suppose that S is an uncountable subset of K. Then the set S contains an uncountable subset T such that T is dense-in-itself. Further, $\overline{T}$ is a perfect set which contains a homeomorphic image of Δ and $|\overline{T}| = \mathfrak{c}$.*

Proof. Since K is compact and metrizable, the topology of K has a countable base.

Let T be the subset of S consisting of the points $x \in S$ such that $U \cap S$ is uncountable for each $U \in \mathscr{N}_x$. Clearly T is dense-in-itself, and hence $\overline{T}$ is perfect. Further, $S \setminus T$ is countable, and so T is uncountable.

By Proposition 1.4.14, $\overline{T}$ contains a homeomorphic image of Δ and $|\overline{T}| = \mathfrak{c}$. □

The following related result and remarks concerning Borel sets will be used later for a couple of (rather minor) remarks; the proofs can be found in [59, §§8.2, 8.3].

Proposition 1.4.17. *Each Borel subset B of a Polish space is the image of a zero-dimensional Polish space; in the case where B is uncountable, B contains a subset that is homeomorphic to Δ, and so $|B| = \mathfrak{c}$.* □

An *analytic* space is a continuous image of a Polish space. Every Borel subset of an analytic space is analytic, and it follows from *Souslin's separation theorem* that a subspace Y of an analytic space X is Borel if and only if both Y and $X \setminus Y$ are analytic. Let X be an uncountable Polish space. Then there are analytic subsets of X that are not Borel. Proposition 1.4.17 can be generalized to show that each uncountable analytic subset of a Polish space has cardinality $\mathfrak{c}$; see [209, §3.3] for these remarks.

Definition 1.4.18. Let X and Y be topological spaces, and let $\eta : X \to Y$ be a continuous map. Then η is *irreducible* if $\eta(X) = Y$ and the image of each proper, closed subspace of X is a proper subspace of Y, and η is *perfect* if it is closed and the inverse image of each point in X is compact.

Suppose that $\eta : X \to Y$ is a continuous surjection. For $y \in Y$, the set

$$F_y := \eta^{-1}(\{y\}) = \{x \in X : \eta(x) = y\}$$

is the *fibre* above y.

Let X and Y be Hausdorff topological spaces. Then every irreducible surjection from X onto Y maps D_X onto D_Y bijectively.

Let $\eta : X \to Y$ be a continuous surjection. Then η is irreducible if and only if, for each non-empty, open set U in X, there exists $x \in U$ with $F_{\eta(x)} \subset U$. For suppose that η is irreducible, and let U be a non-empty, open subspace in X. Assume that $F_{\eta(x)} \not\subset U$ $(x \in U)$. Then $X \setminus U$ is a proper, closed subspace of X with $\eta(X \setminus U) = Y$, a contradiction. Thus there exists $x \in U$ with $F_{\eta(x)} \subset V$. The converse is immediate. It follows easily that, for each non-empty, open set U in X, the set $\{x \in U : F_{\eta(x)} \subset U\}$ is dense in U whenever η is irreducible.

Proposition 1.4.19. *Let X and Y be Hausdorff topological spaces, and suppose that $\eta : X \to Y$ is an irreducible surjection.*

(i) *Suppose that Y is separable. Then X is also separable.*

(ii) *Suppose that η is an open map. Then η is a homeomorphism.*

Proof. (i) Let $\{y_n : n \in \mathbb{N}\}$ be a countable, dense subset of Y and, for each $n \in \mathbb{N}$, choose $x_n \in X$ with $\eta(x_n) = y_n$. Then $F := \overline{\{x_n : n \in \mathbb{N}\}}$ is a closed subspace of X with $\eta(F) = Y$, and so $F = X$, whence the countable set $\{x_n : n \in \mathbb{N}\}$ is dense in X.

(ii) It suffices to show that η is an injection.

Assume towards a contradiction that there exist two points $x_1, x_2 \in X$ with $x_1 \neq x_2$ and $\eta(x_1) = \eta(x_2)$, and take U_1 and U_2 to be disjoint neighbourhoods of x_1 and x_2, respectively. Since η is open, the set $V := \eta(U_1) \cap \eta(U_2)$ is an open neighbourhood of $\eta(x_1)$ in Y. Set $U = \eta^{-1}(V) \cap U_1$. Then $x_1 \in U$ and U is open in X, and so there exists $x \in U$ with $F_{\eta(x)} \subset U \subset U_1$. This is a contradiction because $\eta(x) \in \eta(U_2)$, and so $F_{\eta(x)} \cap U_2 \neq \emptyset$. Thus η is an injection and hence a homeomorphism. □

Definition 1.4.20. Let L be a compact space. A *cover* of L is a pair (K, π) such that K is compact and $\pi : K \to L$ is a continuous surjection.

In this case, π is a closed map. Examples of covers of a compact space K are the pairs $(\beta K_d, \pi)$ of page 26, below, and $(\widetilde{K}, \pi)$, to be described in detail in §6.5.

Proposition 1.4.21. *Let L be a non-empty, compact space, and suppose that (K, π) is a cover of L.*

(i) *For each open subset U of K, the set $\{x \in U : F_{\pi(x)} \subset U\}$ is also open in K.*

(ii) *The following conditions on π are equivalent:*

(a) *π is irreducible;*

(b) *for each non-empty, open set U in K, the set $\{y \in L : F_y \subset U\}$ is a non-empty, open set in L;*

(c) *for each non-empty, open set U in K, the set $\bigcup\{F_y : F_y \subset U\}$ is open in K and dense in U.*

(iii) *There is a closed subset F_0 of K such that $\pi \mid F_0 : F_0 \to L$ is an irreducible surjection.*

Proof. (i) The set U^c is closed in K, and so $\pi(U^c)$ is closed in L, $(\pi(U^c))^c$ is open in L, and $\pi^{-1}((\pi(U^c))^c)$ is open in K; the latter set is $\{x \in U : F_{\pi(x)} \subset U\}$.

(ii) (a) $\Rightarrow$ (b) Set $V = \{y \in L : F_y \subset U\}$. Then $V \neq \emptyset$ by a preliminary remark. Since $V = (\pi(U^c))^c$, the set V is open in L.

(a) $\Rightarrow$ (c) Set $W = \bigcup\{F_y : F_y \subset U\}$, so that $W = \{x \in U : F_{\pi(x)} \subset U\}$. Then W is open in K by (i) and dense in U by a preliminary remark.

(b), (c) $\Rightarrow$ (a) Let F be a proper, closed subspace of K. In both cases, there exists $y \in L$ with $F_y \subset F^c$. Then $y \notin \pi(F)$, and so $\pi(F)$ is a proper subspace of L. Hence π is irreducible.

(iii) Let $\mathscr{F}$ be the family of closed subsets F of K such that $\pi(F) = L$, so that $\mathscr{F} \neq \emptyset$. It is easy to see that each chain in $(\mathscr{F}, \subset)$ is bounded below, and so, by Zorn's lemma, $\mathscr{F}$ has a minimal element, and this is the required set F_0. □

Proposition 1.4.22. *Let K and L be compact spaces, and suppose that $\pi : K \to L$ is an irreducible surjection. Then $\pi(K_0) \in \mathscr{K}_L$ for each $K_0 \in \mathscr{K}_K$.*

Proof. Take $K_0 \in \mathscr{K}_K$, and set $L_0 = \pi(K_0)$. Since π is continuous, L_0 is a compact subspace of L.

Assume towards a contradiction that $\operatorname{int}_L L_0 \neq \emptyset$, say $U = \operatorname{int}_L L_0$, and then set $K_1 = \pi^{-1}(L \setminus U)$, so that $K_0 \cup K_1$ is a closed subspace of K with $\pi(K_0 \cup K_1) = L$. Since π is irreducible, $K_0 \cup K_1 = K$, and so $\pi^{-1}(U) \subset K_0$, whence $\operatorname{int}_K K_0 \neq \emptyset$, a contradiction. Thus $L_0 \in \mathscr{K}_L$. □

Proposition 1.4.23. *Let K be a non-empty, compact, metric space without any isolated points. Then there is an irreducible surjection from the Cantor set Δ onto K.*

Proof. By Proposition 1.4.6(i), there is a continuous surjection $\eta : \Delta \to K$. By Proposition 1.4.21(iii), there is a closed subspace F_0 of Δ such that $\eta : F_0 \to K$ is an irreducible surjection. The space F_0 is compact, metrizable, and totally disconnected, and it is perfect because it has no isolated points. By Proposition 1.4.6(ii), F_0 is homeomorphic to Δ. □

Proposition 1.4.24. *Let K and L be compact spaces such that L is Stonean, and suppose that $\pi : K \to L$ is an irreducible surjection. Then π is a homeomorphism, and so K is Stonean.*

Proof. It suffices to show that π is an injection. Assume towards a contradiction that there exist $x_1, x_2 \in K$ with $x_1 \neq x_2$ and $\pi(x_1) = \pi(x_2)$, and take U_1 and U_2 to be disjoint neighbourhoods of x_1 and x_2, respectively. For $i = 1, 2$, set

$$V_i = \{y \in L : F_y \subset U_i\},$$

so that $V_1 \cap V_2 = \emptyset$. By Proposition 1.4.21(ii), each V_i is open in L and $\pi^{-1}(V_i)$ is a dense subset of U_i. It follows that

$$\pi(x_i) \in \overline{\pi(\pi^{-1}(V_i))} = \overline{V_i} \quad (i = 1, 2),$$

and so $\overline{V_1} \cap \overline{V_2} \neq \emptyset$, a contradiction of the assumption that L is Stonean. Thus π is an injection. □

We shall use the following versions of *Urysohn's lemma* and the *Stone–Weierstrass theorem*, respectively.

Theorem 1.4.25. *Let K be a non-empty, locally compact space. Suppose that L is compact and U is open in K such that $L \subset U$. Then there exists $f \in C_{00}(K)^+$ with $\chi_L \leq f \leq \chi_U$.* □

Let K be a non-empty, locally compact space, and suppose that E is a subset of $C_0(K)$. Then E *separates the points* of K if, for each $x, y \in K$ with $x \neq y$, there exists $f \in E$ with $f(x) \neq f(y)$; E *separates strongly* the points of K if E separates the points of K and, for each $x \in K$, there exists $f \in E$ with $f(x) \neq 0$. The set E is *self-adjoint* if $\overline{f} \in E$ whenever $f \in E$.

Theorem 1.4.26. *Let K be a non-empty, locally compact space.*

(i) *Suppose that A is a real subalgebra of $C_{0,\mathbb{R}}(K)$ that separates strongly the points of K. Then A is dense in $(C_{0,\mathbb{R}}(K), |\cdot|_K)$.*

(ii) *Suppose that A is a self-adjoint subalgebra of $C_0(K)$ that separates strongly the points of K. Then A is dense in $(C_0(K), |\cdot|_K)$.* □

Corollary 1.4.27. *Let K be a non-empty, locally compact space, and suppose that E is a linear subspace of $C_0(K)$ (respectively, $C_{0,\mathbb{R}}(K)$) which separates strongly the points of K. Then $d(E) = d(C_0(K))$ (respectively, $d(E) = d(C_{0,\mathbb{R}}(K))$).*

Proof. We treat the complex case $C_0(K)$; the result for $C_{0,\mathbb{R}}(K)$ is similar.

Let $\mathscr{F}$ be a norm-dense subset of E with $|\mathscr{F}| = d(E)$. Necessarily $\mathscr{F}$ is infinite. The smallest self-adjoint subalgebra, $A(E)$, of $C(K)$ containing E has the same density character as E because $\mathscr{F}$ can be extended to a dense subset of $A(E)$ without increasing the cardinality by appending all finite sums of finite products of elements of $\mathscr{F} \cup \{\overline{f} : f \in \mathscr{F}\}$, and hence $d(E) = d(A(E))$. By the Stone-Weierstrass theorem, Theorem 1.4.26(ii), $A(E)$ is dense in $C_0(K)$, i.e., $d(A(E)) = d(C_0(K))$. Thus $d(E) = d(C_0(K))$, as desired. □

We shall also use the following version of *Dini's theorem*.

Theorem 1.4.28. *Let K be a non-empty, locally compact space, and suppose that $(f_\alpha : \alpha \in A)$ is a net in $C_{0,\mathbb{R}}(K)$ such that $f_\alpha(x) \searrow g(x)$ $(x \in K)$, where $g \in C_{0,\mathbb{R}}(K)$. Then, for each $\varepsilon > 0$, there exists $\alpha_0 \in A$ such that $|f_\alpha - g|_K < \varepsilon$ $(\alpha \geq \alpha_0)$.*

Proof. Fix $\varepsilon > 0$ and $\alpha_1 \in A$, and then take a compact subset L of K such that $f_{\alpha_1}(x) < \varepsilon$ $(x \in K \setminus L)$. Set $K_\alpha = \{x \in K : |f_\alpha(x) - g(x)| \geq \varepsilon\}$ and $L_\alpha = K_\alpha \cap L$ for $\alpha \in A$, so that each L_α is a compact subset of L.

Assume towards a contradiction that each set L_α is non-empty. Since the family $\{L_\alpha : \alpha \in A\}$ has the finite intersection property, it follows that $\bigcap\{L_\alpha : \alpha \in A\} \neq \emptyset$, a contradiction of the fact that $f_\alpha(x) \searrow g(x)$ $(x \in L)$. Thus there exists $\alpha_2 \in A$ with $L_{\alpha_2} = \emptyset$. Set $\alpha_0 = \max\{\alpha_1, \alpha_2\}$. Since $f_\alpha \searrow g$, it follows that $K_\alpha = \emptyset$ $(\alpha \geq \alpha_0)$, giving the result. □

Let X be a non-empty topological space. For $f \in C(X)$, set

$$\mathbf{Z}(f) = \mathbf{Z}_X(f) = f^{-1}(\{0\}) = \{x \in X : f(x) = 0\},$$

so that $\mathbf{Z}(f)$ is the *zero set* of f; the family of zero sets in X is denoted by $\mathbf{Z}(X)$. A *z-filter* on X is a non-empty subfamily $\mathscr{F}$ of $\mathbf{Z}(X)$ with the following properties:

$\emptyset \notin \mathscr{F}$; $F \cap G \in \mathscr{F}$ whenever $F, G \in \mathscr{F}$; $G \in \mathscr{F}$ whenever $G \in \mathbf{Z}(X)$ and $G \supset F$ for some $F \in \mathscr{F}$. A *z-ultrafilter* on X is a maximal z-filter when the family of z-filters is ordered by inclusion; equivalently, a z-filter $\mathscr{F}$ is a z-ultrafilter if $F \in \mathscr{F}$ or $G \in \mathscr{F}$ whenever $F, G \in \mathbf{Z}(X)$ and $F \cup G \in \mathscr{F}$. In the case where X is a discrete space, all subsets of X are zero sets and we refer to *filters* and *ultrafilters* on X. A subset of X of the form $X \setminus \mathbf{Z}(f)$ for $f \in C(X)$ is a *cozero set*. It follows from Theorem 1.4.25 that each non-empty, locally compact space has a base consisting of cozero sets. Clearly a countable intersection of zero sets is a zero set and a countable union of cozero sets is a cozero set.

Let X be a non-empty topological space, and take subsets S and T of X. Then we set

$$S \prec T \quad \text{if} \quad \overline{S} \subset \operatorname{int} T\,.$$

Suppose that K is a non-empty, compact space and that $S \prec T$ in K. Then it follows from Theorem 1.4.25 that there is a cozero set U with $S \prec U \prec T$. Indeed, take $f \in C(K)^+$ with $\chi_{\overline{S}} \leq f \leq \chi_{\operatorname{int} T}$ and set $U = \{x \in K : f(x) > 1/2\}$.

A Hausdorff topological space X is *completely regular* if, for each $x \in X$ and each open neighbourhood U of x, there exists $f \in C(X)$ with $f(x) = 1$ and $\operatorname{supp} f \subset U$. Thus X is completely regular if and only if the cozero sets are a base for the topology; this is the case if and only if X has a compactification. Each locally compact space is completely regular.

The space X is *basically disconnected* if every cozero set in X has an open closure. Thus an extremely disconnected space is basically disconnected and a basically disconnected, completely regular space is zero-dimensional and hence totally disconnected.

A zero set in a topological space X is clearly a closed G_δ-set; it follows from Theorem 1.4.25 that a closed G_δ-subset of a compact space is a zero set. Note that a point x of a completely regular space X is a P-point if and only if each $f \in C(X)$ is constant on a neighbourhood of x, and so, in the case where X is infinite, there exists a function in $C(X)$ that takes infinitely many distinct values. Thus each infinite, compact set contains points that are not P-points.

Definition 1.4.29. Let X be a non-empty topological space. Then the σ-algebra generated by the zero sets in X is the family of *Baire sets* in X, denoted by $\mathfrak{Ba}_X$.

In the case where K is a compact space, $\mathfrak{Ba}_K$ is the σ-algebra generated by the closed G_δ-sets in K; the latter is a common definition of $\mathfrak{Ba}_K$.

Thus $\mathfrak{Ba}_X \subset \mathfrak{B}_X$; clearly $\mathfrak{Ba}_X = \mathfrak{B}_X$ when each closed subset of X is a zero set, and so this holds when X is metrizable. Conversely, suppose that K is a non-empty, σ-compact, locally compact space such that $\mathfrak{B}_K = \mathfrak{Ba}_K$, so that each closed set in K is a Baire set. Then each closed subset of K is a zero set [62, Corollary 9.16].

We shall use the following version of *Tietze's extension theorem*, from which it follows immediately that every normal space is completely regular.

Theorem 1.4.30. *Let F be a closed subset of a normal topological space X. Suppose that $f \in C^b_{\mathbb{R}}(F)$ with $|f|_F = 1$. Then there exists $\widetilde{f} \in C^b_{\mathbb{R}}(X)$ such that $\left|\widetilde{f}\right|_X = 1$ and $\widetilde{f} \mid F = f$.* □

Definition 1.4.31. Let κ be an ordinal. A system $(K_\alpha, \pi^\beta_\alpha : 0 \leq \alpha \leq \beta < \kappa)$ is an *inverse system of compact spaces* if K_α is a non-empty, compact space whenever $0 \leq \alpha < \kappa$, if $\pi^\beta_\alpha : K_\beta \to K_\alpha$ is a continuous surjection for $0 \leq \alpha \leq \beta < \kappa$, if π^α_α is the identity map on K_α whenever $0 \leq \alpha < \kappa$, and if $\pi^\beta_\alpha \circ \pi^\gamma_\beta = \pi^\gamma_\alpha$ whenever $0 \leq \alpha \leq \beta \leq \gamma < \kappa$.

The following result, contained in [99, §§2.5, 3.2], shows that such a system has an appropriate *inverse limit* K.

Theorem 1.4.32. *Let κ be an infinite ordinal, and let $(K_\alpha, \pi^\beta_\alpha : 0 \leq \alpha \leq \beta < \kappa)$ be an inverse system of compact spaces. Then there are a non-empty, compact space K and continuous surjections $\pi_\alpha : K \to K_\alpha$ for $0 \leq \alpha < \kappa$ such that $\pi^\beta_\alpha \circ \pi_\beta = \pi_\alpha$ for $0 \leq \alpha \leq \beta < \kappa$. Further, the space K is connected whenever each space K_α is connected.* □

In fact, the required space K is given by

$$K = \left\{ (x_\alpha) \in \prod K_\alpha : \pi^\beta_\alpha(x_\beta) = x_\alpha \ (0 \leq \alpha \leq \beta < \kappa) \right\},$$

with the relative product topology from $\prod K_\alpha$.

We write $(K, \pi_\alpha) = (K, \pi_\alpha : 0 \leq \alpha < \kappa)$ for the inverse limit so constructed.

Corollary 1.4.33. *Let $(K_\alpha, \pi^\beta_\alpha : 0 \leq \alpha \leq \beta < \omega_1)$ be an inverse system of compact spaces, with inverse limit (K, π_α). Suppose that Z is a closed G_δ-set in K. Then there exist $\alpha < \omega_1$ and a closed G_δ-set W in K_α such that $Z = \pi^{-1}_\alpha(W)$.*

Proof. Consider the family of sets of the form $\pi^{-1}_\alpha(V)$ for $0 \leq \alpha < \omega_1$ and for V an open subset of K_α. This family forms a base for the topology of K, and so we can write

$$Z = \bigcap \{ \pi^{-1}_{\alpha_n}(V_n) : n \in \mathbb{N} \},$$

where (α_n) is a sequence in $[0, \omega_1)$ and V_n is an open set in K_{α_n} for each $n \in \mathbb{N}$. Take α with $\alpha_n < \alpha < \omega_1$ $(n \in \mathbb{N})$. Then

$$Z = \bigcap \{ (\pi^{-1}_\alpha \circ (\pi^\alpha_{\alpha_n})^{-1})(V_n) : n \in \mathbb{N} \} = \bigcap \{ \pi^{-1}_\alpha(W_n) : n \in \mathbb{N} \},$$

where W_n is an open set in K_α for $n \in \mathbb{N}$. Set $W = \bigcap \{ W_n : n \in \mathbb{N} \}$, so that W is a G_δ-set in K_α and $\pi^{-1}_\alpha(W) = Z$. Further, $\pi_\alpha(Z) = W$, and so W is closed in K_α. □

1.5 The Stone–Čech compactification

Let X be a completely regular topological space. The fundamental compactification theorem [112, Theorem 6.5] states that X has a Stone–Čech compactification βX with the following equivalent properties:

(a) every continuous mapping τ from X into a compact space K has a continuous extension $\overline{\tau} : \beta X \to K$;

(b) each $f \in C^b(X)$ has an extension to a function $f^\beta \in C(\beta X)$;

(c) disjoint zero sets in X have disjoint closures in βX.

Further, βX is unique in the sense that any compactification with the stated properties is homeomorphic to βX by a homeomorphism that leaves each point of X fixed. There are several different constructions of βX; see [112] and [239], for example. In particular, we can regard βX as the space of z-ultrafilters on X or as the character space or as the maximal ideal space of the C^*-algebra $C^b(X)$. We shall see that the map

$$f \mapsto f^\beta, \quad C^b(X) \to C(\beta X),$$

is an isometric algebra isomorphism. In fact, this map is the Gel'fand transformation of the commutative C^*-algebra $C^b(X)$ and it is a C^*-isomorphism; see §§2.2, 3.1, 3.2, below, and [68, §4.2]. In particular, let S be an infinite set. Then the map

$$f \mapsto f^\beta, \quad \ell^\infty(S) \to C(\beta S),$$

is an isometric algebra isomorphism. For our proof of the existence of βX, see Proposition 2.1.5.

Definition 1.5.1. The *Stone–Čech compactification* of a completely regular topological space X is denoted by βX; the space $X^* = \beta X \setminus X$ is the *growth* of X.

For example, $\beta[0, \omega_1) = [0, \omega_1]$ by a remark on page 14. In particular, we shall consider $\beta\mathbb{N}$ and $\mathbb{N}^*$; we shall often identify ℓ^∞ with $C(\beta\mathbb{N})$ and the quotient Banach space (see page 48) ℓ^∞ / c_0 with $C(\mathbb{N}^*)$.

Let K be a compact space. Then the identity map $\iota : K_d \to K$ has a continuous extension to a map $\pi : \beta K_d \to K$, and π is a surjection, and so $(\beta K_d, \pi)$ is a cover of K, in the sense of Definition 1.4.20.

The seminal and outstanding early expository text on Stone–Čech compactifications is that of Gillman and Jerison [112] from 1960; a later text of Hindman and Strauss [139] from 1998 considers, in particular, the Stone–Čech compactification βS of a semigroup S as itself a semigroup (where βS has two distinct semigroup operations – see page 33). For other studies of Stone–Čech compactifications, see [71, 239]; in particular, [239, Chapter 1] describes many different approaches to Stone–Čech compactifications.

Once we have a construction of βX, we obtain the z-ultrafilters on X as follows. Let X be a completely regular space. For $p \in \beta X$, define

$$\mathscr{U}_p = \{F \in \mathbf{Z}(X) : p \in \mathrm{cl}_{\beta X} F\}, \quad M^p = \{f \in C(X) : p \in \mathrm{cl}_{\beta X} \mathbf{Z}(f)\}.$$

Then the *Gel'fand–Kolmogorov theorem* is the following; for details, see [68, §4.2] and [112, Chapter 7].

Theorem 1.5.2. *Let X be a completely regular topological space.*

(i) *For each $p \in \beta X$, the set $\mathscr{U}_p$ is a z-ultrafilter on X, and the map $p \mapsto \mathscr{U}_p$ is a bijection from βX onto the set of z-ultrafilters on X.*

(ii) *For each $p \in \beta X$, the set M^p is a maximal ideal in $C(X)$, and the map $p \mapsto M^p$ is a bijection from βX onto the set of maximal ideals of $C(X)$.*

(iii) *For each $p \in \beta X$, the set $M^p \cap C^b(X)$ is a maximal ideal in $C^b(X)$, and the map $p \mapsto M^p \cap C^b(X)$ is a bijection from βX onto the set of maximal ideals of $C^b(X)$.* □

Let S be an infinite set. For an infinite subset A of S, the space A^* is identified with $\overline{A} \setminus A$, where $\overline{A}$ is the closure of A in βS, and then A^* is regarded as a clopen subset of S^*; each non-empty, clopen subset of S^* has this form, and each such set contains a copy of $\mathbb{N}^*$. The family $\{A^* : A \in \mathscr{P}(S)\}$ is a base for the topology of S^*, and so $w(S^*) \leq |\mathscr{P}(S)| = 2^{|S|}$. In fact, it is easily seen that

$$w(S^*) = 2^{|S|}.$$

Proposition 1.5.3. *Let S be an infinite set. Then:*

(i) *every non-empty G_δ-subset of S^* has a non-empty interior in S^*;*

(ii) *for each $p \in S^*$, the set $\beta S \setminus \{p\}$ is pseudo-compact and non-compact.*

Proof. (i) Let L be a non-empty G_δ-subset of S^*. For each $n \in \mathbb{N}$, take an open set U_n in βS such that $L = \bigcap\{U_n \cap S^* : n \in \mathbb{N}\}$. For an element $p \in L$, choose an infinite subset A_n of S with $p \in A_n^* \subset U_n$; we may suppose that (A_n) is a decreasing sequence of sets in S. There is a sequence (a_n) of distinct points in S such that $a_n \in A_n$ $(n \in \mathbb{N})$; set $A = \{a_n : n \in \mathbb{N}\}$, and take $q \in A^*$. For each $n \in \mathbb{N}$, the set $A \setminus A_n$ is finite, and so $q \in A_n^*$. Thus A^* is a non-empty, clopen subset of L.

(ii) Set $K = \beta S \setminus \{p\}$, so that K is a locally compact space that is not compact. Take $f \in C(K)$. Then f has an extension $\widetilde{f} \in C(\beta S, \mathbb{C}_\infty)$. For $n \in \mathbb{N}$, set

$$F_n = \left\{x \in S^* : \left|\widetilde{f}(x)\right| \geq n\right\}, \quad G_n = \left\{x \in S^* : \left|\widetilde{f}(x)\right| > n\right\}.$$

Assume that $F_n \neq \emptyset$ $(n \in \mathbb{N})$. Then the intersection $\bigcap\{F_n : n \in \mathbb{N}\}$ is non-empty, and so $G := \bigcap\{G_n : n \in \mathbb{N}\}$ is a non-empty G_δ-subset of S^*. By (i), G has a non-empty interior in S^*, and so there exists $x \in K \cap G$, a contradiction. Thus $f \in C^b(K)$, showing that K is pseudo-compact. □

Note that $|\beta\mathbb{N}| = 2^{\mathfrak{c}}$. The following more general result is [112, Theorem 9.2] and [139, Theorem 3.58].

Proposition 1.5.4. *Let S be an infinite set with $|S| = \kappa$. Then*

$$|\beta S| = |S^*| = 2^{2^\kappa}.$$

Proof. Set $\lambda = 2^{2^\kappa}$. Certainly $|\beta S| \leq \lambda$.

Take $\mathscr{F}$ to be the family of all finite subsets of S, and take $\mathscr{G}$ to be the family of all finite subsets of $\mathscr{F}$, so that $|\mathscr{F} \times \mathscr{G}| = \kappa$. To each subset T of S, associate a subset B_T of $\mathscr{F} \times \mathscr{G}$ by setting

$$B_T = \{(F,G) \in \mathscr{F} \times \mathscr{G} : T \cap F \in G\};$$

the complement of B_T in $\mathscr{F} \times \mathscr{G}$ is B_T^c. For each non-empty subset $\mathscr{S}$ of $\mathscr{P}(S)$, define

$$\mathscr{B}_{\mathscr{S}} = \{B_T : T \in \mathscr{S}\} \cup \{B_T^c : T \notin \mathscr{S}\}.$$

We *claim* that each family $\mathscr{B}_{\mathscr{S}}$ has the finite intersection property. Indeed, let $B_{T_1}, \dots, B_{T_m}, B_{T_{m+1}}^c, \dots, B_{T_n}^c$ be distinct members of $\mathscr{B}_{\mathscr{S}}$, where $T_1, \dots, T_m \in \mathscr{S}$ and $T_{m+1}, \dots, T_n \notin \mathscr{S}$. The indices $T_1, \dots, T_n$ are distinct subsets of S, and so, for each $i, j \in \mathbb{N}_n$ with $i < j$, we can choose $t_{i,j}$ that belongs to exactly one of the two sets T_i and T_j. Set $F = \{t_{i,j} : i, j \in \mathbb{N}_n,\, i < j\}$. Then the sets $T_i \cap F$ and $T_j \cap F$ are distinct whenever $i, j \in \mathbb{N}_n$ with $i < j$, and the finite set

$$G = \{T_1 \cap F, \dots, T_m \cap F\}$$

belongs to $\mathscr{G}$. We see that $T_i \cap F \in G$ for $i = 1, \dots, m$ and that $T_j \cap F \notin G$ for $j = m+1, \dots, n$, and this implies that $(F,G) \in B_{T_i}$ for $i = 1, \dots, m$ and $(F,G) \in B_{T_j}^c$ for $j = m+1, \dots, n$. Thus $(F,G) \in B_{T_1} \cap \cdots \cap B_{T_m} \cap B_{T_{m+1}}^c \cap \cdots \cap B_{T_n}^c$. This establishes the claim.

The claim implies that each family $\mathscr{B}_{\mathscr{S}}$ is contained in at least one ultrafilter on $\mathscr{F} \times \mathscr{G}$, and it cannot be that two such families are contained in the same ultrafilter. Since $|\mathscr{P}(\mathscr{P}(S))| = \lambda$, it follows that $|\beta S| \geq \lambda$. □

Proposition 1.5.5. *Let S be an infinite set. Then there are an index set A with $|A| = \mathfrak{c}$ and a family $\{S_\alpha : \alpha \in A\}$ of infinite subsets of S such that $S_\alpha \cap S_\beta$ is finite whenever $\alpha, \beta \in A$ with $\alpha \neq \beta$. Further, $\{S_\alpha^* : \alpha \in A\}$ is a family of non-empty, pairwise disjoint, clopen subsets of S^*, and $w(S^*) \geq \mathfrak{c}$ and $d(C(S^*)) \geq \mathfrak{c}$.*

Proof. We may suppose that S is countable; take S to be $\mathbb{I} \cap \mathbb{Q}$, and take A to be the set of irrationals in $\mathbb{I}$, so that $|A| = \mathfrak{c}$. For each $\alpha \in A$, take S_α to be a sequence of rationals that converges to α. Then each S_α is infinite and $S_\alpha \cap S_\beta$ is finite whenever $\alpha \neq \beta$.

Clearly the sets S_α^* are non-empty, pairwise disjoint, clopen subsets of S^*, and so $w(S^*) \geq |A| = \mathfrak{c}$. Take χ_α to be the characteristic function of S_α^* for $\alpha \in A$. Then $|\chi_\alpha - \chi_\beta|_{S^*} = 1$ whenever $\alpha, \beta \in A$ with $\alpha \neq \beta$, and so $d(C(S^*)) \geq |A| = \mathfrak{c}$. □

Corollary 1.5.6. *Let S be an infinite set. Then S^* contains no non-empty, clopen, Stonean subspace.*

Proof. We first show that $\mathbb{N}^*$ is not Stonean. Indeed, assume that $\mathbb{N}^*$ is Stonean. Then the closure of the union of the sets in every subfamily of the above family

$\{S_\alpha^* : \alpha \in A\}$ would be clopen, and so $\mathbb{N}^*$ would contain $2^{\mathfrak{c}}$ distinct clopen sets. However, there are just $\mathfrak{c}$ such sets, a contradiction.

Now let S be an infinite set. Each non-empty, clopen subspace of S^* has the form $\overline{A} \setminus A$ for some infinite subset A of S. Choose an infinite, countable subset A_0 of A. Then $\overline{A_0} \setminus A_0$ is homeomorphic to $\mathbb{N}^*$, and $\mathbb{N}^*$ is not Stonean. But $\overline{A_0} \setminus A_0$ is a clopen subspace of $\overline{A} \setminus A$ in S^*, and so $\overline{A} \setminus A$ is not Stonean. □

In particular, take $S = \mathbb{N}$. Then $w(\mathbb{N}^*) \geq \mathfrak{c}$ and $d(C(\mathbb{N}^*)) \geq \mathfrak{c}$. Since the family $\{A^* : A \in \mathscr{P}(\mathbb{N})\}$ is a base for the topology of $\mathbb{N}^*$, we have $w(\mathbb{N}^*) \leq 2^{\aleph_0} = \mathfrak{c}$, and so $w(\mathbb{N}^*) = \mathfrak{c}$. Clearly $d(C(\mathbb{N}^*)) \leq |C(\mathbb{N}^*)| \leq |\ell^\infty| = \mathfrak{c}$, and so $d(C(\mathbb{N}^*)) = \mathfrak{c}$. We have shown that

$$w(\mathbb{N}^*) = d(C(\mathbb{N}^*)) = |C(\mathbb{N}^*)| = \mathfrak{c}. \tag{1.6}$$

Lemma 1.5.7. *Let K be a Stonean space, and let U be a subspace of K that is either open or dense in K. Take a compact space L and $f \in C(U,L)$. Then there exists $F \in C(\overline{U},L)$ such that $F \mid U = f$.*

Proof. Take $x \in \overline{U}$, and let (x_α) and (y_β) be nets in U with $\lim_\alpha x_\alpha = \lim_\alpha y_\beta = x$. Then the nets $(f(x_\alpha))$ and $(f(y_\beta))$ have accumulation points, say a and b, respectively, in L. Assume towards a contradiction that $a \neq b$, and take open neighbourhoods N_a and N_b of a and b, respectively, such that $\overline{N_a} \cap \overline{N_b} = \emptyset$. Then the sets

$$\{y \in U : f(y) \in N_a\} \quad \text{and} \quad \{y \in U : f(y) \in N_b\}$$

are disjoint, relatively open subsets of U, and so they have the form $U \cap V$ and $U \cap W$, respectively, for some open subsets V and W in K. In the case where $\overline{U} = K$, we have $V \cap W = \emptyset$, and so $\overline{V} \cap \overline{W} = \emptyset$. In the case where U is open, $(\overline{U \cap V}) \cap (\overline{U \cap W}) = \emptyset$. However $x \in (\overline{U \cap V}) \cap (\overline{U \cap W})$, a contradiction in both cases. Thus $a = b$.

It follows that $(f(x_\alpha))$ converges to a unique limit, say $F(x)$, in L, and that the limit is independent of the net (x_α). It is easy to see that $F : \overline{U} \to L$ is continuous, and so F is the required extension of f. □

Corollary 1.5.8. *Let K be a Stonean space. Then $\beta U = \overline{U}$ for each open subspace U of K. In particular, $\beta D_K = \overline{D_K}$.* □

Proposition 1.5.9. *Let X be a completely regular topological space.*

(i) *Suppose that βX is zero-dimensional. Then X is zero-dimensional.*[1]

(ii) *The space X is extremely disconnected if and only if βX is Stonean.*

(iii) *Suppose that X is extremely disconnected and Y is a dense subspace of X. Then $\beta Y = \beta X$.*

---

[1] An example of C. H. Dowker shows that the converse fails; see [99, 6.2.20] and [112, 16M(Δ_1)]. In fact there is a (non-normal) zero-dimensional, locally compact space X such that βX is not zero-dimensional [236]; this contradicts a sentence in the book [155, p. 169] of Kelley.

Proof. (i) This is obvious.

(ii) Suppose that X is extremely disconnected, and let U and V be disjoint, open subsets of βX. Then $F := \mathrm{cl}_X(U \cap X)$ and $G := \mathrm{cl}_X(V \cap X)$ are disjoint, clopen subsets of X. The function $\chi_F - \chi_G \in C^b(X)$ has a continuous extension, say f, to βX. Since $f = 1$ on $\overline{U}$ and $f = -1$ on $\overline{V}$, it follows that $\overline{U} \cap \overline{V} = \emptyset$, and so βX is extremely disconnected, and hence Stonean.

For the converse, suppose that βX is Stonean. Then X is extremely disconnected because it is dense in βX.

(iii) Take $f \in C^b(Y)$. By (ii), βX is Stonean, and Y is dense in βX. By Lemma 1.5.7, there exists $F \in C(\beta X)$ such that $F \mid Y = f$. Thus $\beta Y = \beta X$. □

Example 1.5.10. By Proposition 1.5.9(ii), the space βS for a discrete space S is always Stonean; in particular, $\beta \mathbb{N}$ is Stonean. However, the closed subspace $\mathbb{N}^*$ of $\beta \mathbb{N}$ is not Stonean, or even basically disconnected: see Example 1.7.14, below.

In Example 1.7.16, on page 45, we shall exhibit a space $G_{\mathbb{I}}$, the Gleason cover of the unit interval, and note that $G_{\mathbb{I}}$ is an infinite, separable, Stonean space without isolated points, and hence $G_{\mathbb{I}}$ is not homeomorphic to either $\beta \mathbb{N}$ or $\mathbb{N}^*$.

One might wonder if any two infinite, separable Stonean spaces without isolated points are homeomorphic. However, this is not the case; see Example 1.7.16. □

The following definition is given in [112, §14.25], for example.

Definition 1.5.11. Let X be a completely regular space. Then X is an *F-space* when every finitely generated ideal in the algebra $C_{\mathbb{R}}(X)$ is principal.

It follows easily from the Tietze extension theorem, Theorem 1.4.30, that each compact subspace of a locally compact F-space is also an F-space. Numerous equivalent conditions for a completely regular space X to be an F-space are listed in [112, §14.25]; a topological condition is that X is an F-space if and only if, given two disjoint, cozero sets U and V, there exists $f \in C(X, \mathbb{I})$ with $f(x) = 0$ $(x \in U)$ and $f(x) = 1$ $(x \in V)$. Indeed, the latter version is the definition of an F-space given in [139, p. 84]. Further, βX is an F-space if and only if X is an F-space.

Let K be a compact space. Then K is an F-space if and only if pairs of disjoint, cozero subsets of K have disjoint closures, and so each Stonean space, together with each of its closed subspaces, is an F-space. In particular, $\mathbb{N}^*$ is a compact F-space as it is a closed subspace of a Stonean space, and so the class of compact F-spaces is strictly larger than the class of Stonean spaces. For a characterization of compact F-spaces, see Theorem 2.3.4.

The following well-known proposition was established in [111, Theorem 2.7]; a proof using ring-theoretic properties of $C(K)$ is given, for example, in the book [68, Proposition 4.2.16]. We give here a short, purely topological, proof from [191, Remark 3.3], which is somewhat less well known.

Proposition 1.5.12. *Let K be a σ-compact, non-compact, locally compact space. Then K^* is a compact F-space.*

Proof. Since K is locally compact, K^* is compact. Let U and V be disjoint, non-empty, cozero sets in K^*; set $W = U \cup V$, and consider the function $f : W \to \mathbb{I}$ defined by $f(U) = \{0\}$ and $f(V) = \{1\}$. It suffices to show that f has an extension in $C(K^*)$. Let $Y = K \cup W$. Then Y is σ-compact, and therefore normal, and W is relatively closed in Y. By the Tietze extension theorem, Theorem 1.4.30, f has an extension, say $\widetilde{f} \in C(Y, \mathbb{I})$; set $g = \widetilde{f} \mid K$. Then g has an extension $\widetilde{g} \in C(\beta K, \mathbb{I})$. The function $F = \widetilde{g} \mid K^*$ is the required extension of f. □

The spaces $(\mathbb{R}^+)^*$ and $(\mathbb{R}^n)^*$ for $n \geq 2$ are connected, compact F-spaces. To see, for example, that the latter spaces are connected, assume that there is a continuous function f that maps this space onto the set $\{0, 1\}$. Then f can be extended to a continuous function, also called f, from $\beta\mathbb{R}^n$ into $\mathbb{I}$. Since the closure in $\beta\mathbb{R}^n$ of the set $\{x \in \mathbb{R}^n : \|x\| > k\}$ is connected for each $k \in \mathbb{N}$, the function f must assume the value $1/2$ somewhere on each of these sets, an obvious contradiction.

Proposition 1.5.13. *Let K be an infinite, compact F-space. Then $\overline{D}$ is homeomorphic to $\beta\mathbb{N}$ for each infinite, countable, discrete subspace D of K. Further, $|K| \geq 2^{\mathfrak{c}}$.*

Proof. Let D be an infinite, countable, discrete subspace of K, say $D = \{x_n : n \in \mathbb{N}\}$, where $x_m \neq x_n$ when $m, n \in \mathbb{N}$ with $m \neq n$.

The map $\varphi : n \mapsto x_n$, $\mathbb{N} \to K$, has a continuous extension to a map $\overline{\varphi} : \beta\mathbb{N} \to K$. Choose (U_n) to be a sequence of pairwise disjoint, cozero subsets of K such that $U_n \cap D = \{x_n\}$ $(n \in \mathbb{N})$, and take two disjoint subsets, A and B, of D. Then

$$\bigcup\{U_n : x_n \in A\} \quad \text{and} \quad \bigcup\{U_n : x_n \in B\}$$

are two disjoint, cozero subsets of K. Since K is an F-space, it follows that $\overline{A} \cap \overline{B} = \emptyset$.

Let p and q be distinct ultrafilters on $\mathbb{N}$. There are disjoint subsets $P \in p$ and $Q \in q$ of $\mathbb{N}$. We have shown that $\overline{\varphi(P)} \cap \overline{\varphi(Q)} = \emptyset$. However $\overline{\varphi}(p) \in \overline{\varphi(P)}$ and $\overline{\varphi}(q) \in \overline{\varphi(Q)}$, and so $\overline{\varphi} : \beta\mathbb{N} \to K$ is an injection, and hence $\overline{\varphi} : \beta\mathbb{N} \to \overline{D}$ is a homeomorphism.

Since the space K contains an infinite, countable, discrete subspace, it follows that $|K| \geq |\beta\mathbb{N}|$. But $|\beta\mathbb{N}| = 2^{\mathfrak{c}}$ by Proposition 1.5.4, and so $|K| \geq 2^{\mathfrak{c}}$. □

The next result first appeared in [212, p. 19].

Proposition 1.5.14. *Let K be a compact F-space satisfying* CCC. *Then K is Stonean.*

Proof. Let U_1 and U_2 be disjoint, open subsets of K. For $j = 1, 2$, take $\mathscr{F}_j$ to be a maximal family of pairwise-disjoint, cozero sets in K contained in U_j, and let V_j be the union of the sets in $\mathscr{F}_j$. Since X satisfies CCC, each $\mathscr{F}_j$ is countable, and so V_j is a cozero set. Clearly $\overline{V_j} = \overline{U_j}$ by the maximality of $\mathscr{F}_j$. Since X is an F-space, $\overline{V_1} \cap \overline{V_2} = \emptyset$, and so $\overline{U_1} \cap \overline{U_2} = \emptyset$. Thus K is Stonean. □

The following result will be used for one later remark; several proofs are given in [185]; a stronger result follows from [139, Theorem 6.38].

Theorem 1.5.15. *An infinite, compact F-space is not homogeneous.* □

We now introduce a possibly new notion.

Definition 1.5.16. Let X be a topological space. A point x of X is an *extremely disconnected point* of X if $x \notin \overline{U} \cap \overline{V}$ for each disjoint pair $\{U, V\}$ of open sets in X.

Thus a topological space X is extremely disconnected if and only if all points of X are extremely disconnected.

As an easy example, consider the space $\mathbb{N}_\infty$: this space is totally disconnected. However, consider the sets $U = \{2n : n \in \mathbb{N}\}$ and $V = \{2n+1 : n \in \mathbb{N}\}$, so that U and V are disjoint open sets in $\mathbb{N}_\infty$. Then $\{\infty\} \in \overline{U} \cap \overline{V}$, so that $\{\infty\}$ is not an extremely disconnected point of $\mathbb{N}_\infty$.

It is proved in [68, Theorem 4.2.23] and [139, Theorem 3.38] that, with (CH), for each separable, σ-compact, locally compact, non-compact space K, and, in particular, for $K = \mathbb{N}$, there is a P-point in K^*. However, it is not provable in ZFC that $\mathbb{N}^*$ has a P-point [241].

Proposition 1.5.17. (i) *Let S be an infinite set. Every extremely disconnected point of S^* is a P-point of S^*.*

(ii) (CH) *Suppose that x is a P-point of $\mathbb{N}^*$. Then x is not an extremely disconnected point of $\mathbb{N}^*$.*

(iii) (CH) *No point of $\mathbb{N}^*$ is an extremely disconnected point of $\mathbb{N}^*$.*

Proof. (i) We shall show the contrapositive of the statement.

Suppose that x is not a P-point of S^*. We shall show that there exist disjoint open subsets U and V of S^* such that $x \in \overline{U} \cap \overline{V}$. Indeed, since x is not a P-point, there is a sequence $(K_n : n \in \mathbb{N})$ of clopen neighbourhoods of x such that $x \notin U$, where

$$U = \operatorname{int}_X\left(\bigcap K_n\right).$$

Then U is an open subset of $\bigcap K_n$, and it follows from Proposition 1.5.3(i) that U is dense in $\bigcap K_n$. Hence $x \in \overline{U}$. Set $V = X \setminus \bigcap K_n$, so that $X \setminus \overline{V} = U$. Then $x \in \overline{V}$, and hence $x \in \overline{U} \cap \overline{V}$ is not an extremely disconnected point of S^*.

(ii) We shall show that there exist disjoint open subsets U and V of $\mathbb{N}^*$ such that $x \in \overline{U} \cap \overline{V}$.

There exists an enumeration $(K_i : i < \mathfrak{c})$ of the clopen neighbourhoods of x in $\mathbb{N}^*$. The set $\bigcap\{V_j : j < i\}$ is a G_δ-set containing x for each $i < \omega_1$; since $\mathfrak{c} = \aleph_1$ (by (CH)), we can inductively choose a sequence $(V_i : i < \omega_1)$ of clopen neighbourhoods of x such that $V_i \subset K_i \cap \bigcap\{V_j : j < i\}$. Set $U_i = V_{i+1} \setminus V_i$ $(i < \omega_1)$, and define

$$U = \bigcup\{U_i : i \text{ is a limit ordinal}\}, \quad V = \bigcup\{U_i : i \text{ is not a limit ordinal}\}.$$

Then the pair $\{U, V\}$ has the required properties.

(iii) This is immediate from (i) and (ii). □

A stronger result than the above, stated in a different language, is given in [29, Theorem 4.16]. In particular, it is shown that clauses (ii) and (iii), above, do not require (CH); they are theorems of ZFC.

Corollary 1.5.18. *There is a dense subset S of $\mathbb{N}^*$ such that, for each $x \in S$, there are disjoint, open subsets U and V of $\mathbb{N}^*$ such that $x \in \overline{(S \cap U)} \cap \overline{(S \cap V)}$.*

Proof. We take S to be the set of points of $\mathbb{N}^*$ which are not P-points. Each non-empty, clopen subset W of $\mathbb{N}^*$ contains an infinite, countable subset that necessarily has a limit point; using Proposition 1.5.3(i), we see that this limit point is not a P-point, and so belongs to $W \cap S$. Thus S is dense in $\mathbb{N}^*$. By Proposition 1.5.17(i), S has the required property. □

Let S be a semigroup, so that S is a non-empty set with an associative binary operation. We shall show that there are two products $\Box$ and $\Diamond$ on the Stone–Čech compactification βS of S such that $(\beta S, \Box)$ and $(\beta S, \Diamond)$ are also semigroups. The theory of the semigroups βS, which has significant applications to combinatorics and topological dynamics, can be found in [139]; see also the text [34] and the memoir [71].

Let S be a non-empty set, and let $*$ be a binary operation on S, so that $*$ is a map from $S \times S$ to S; the image of $(s,t) \in S \times S$ is denoted by $s * t$. For each $s \in S$, the map

$$L_s : t \mapsto s * t, \quad S \to S \subset \beta S,$$

has a unique continuous extension to a map $L_s : \beta S \to \beta S$, and then we define $s \Box u = L_s(u)$ $(u \in \beta S)$. Next, for each $u \in \beta S$, the map

$$R_u : s \mapsto s \Box u, \quad S \to \beta S,$$

has a unique continuous extension to a map $R_u : \beta S \to \beta S$; we define

$$u \Box v = R_v(u) \quad (u, v \in \beta S).$$

Then $\Box$ is a binary operation on βS, and the restriction of $\Box$ to $S \times S$ is the original binary operation $*$. Further, for each $v \in \beta S$, the map R_v is continuous, and, for each $s \in S$, the map L_s is continuous. We see that

$$u \Box v = \lim_{\alpha} \lim_{\beta} s_\alpha * t_\beta \tag{1.7}$$

for $u, v \in \beta S$ whenever (s_α) and (t_β) are two nets in S such that $\lim_\alpha s_\alpha = u$ and $\lim_\beta t_\beta = v$.

Similarly, we can define a binary operation $\Diamond$ on βS such that

$$u \Diamond v = \lim_{\beta} \lim_{\alpha} s_\alpha * t_\beta \tag{1.8}$$

for $u, v \in \beta S$ whenever (s_α) and (t_β) are nets in S with $\lim_\alpha s_\alpha = u$ and $\lim_\beta t_\beta = v$.

Let $(S, \cdot)$ be a semigroup. Then the two extensions of $\cdot$ are the binary operations $\square$ and $\diamond$ on βS; it is immediately checked that both $\square$ and $\diamond$ are associative on βS [139, Theorem 4.4], and so $(\beta S, \square)$ and $(\beta S, \diamond)$ are semigroups.

Definition 1.5.19. A semigroup V which is also a topological space is a *left* (respectively, *right*) *topological semigroup* if L_v (respectively, R_v) is continuous for each $v \in V$.

Theorem 1.5.20. *Let S be a semigroup. Then $(\beta S, \square)$ and $(\beta S, \diamond)$ are right and left, respectively, topological semigroups containing S as a subsemigroup. Further, the maps $L_s : u \mapsto s \square u$ and $R_s : u \mapsto u \diamond s$ are continuous for each $s \in S$.* $\square$

The semigroup maps $\square$ and $\diamond$ that we have described here are defined in a variety of different ways in [139]. It is usual to denote the semigroup operation $\square$ in $\beta \mathbb{N}$ and $\beta \mathbb{Z}$ by '+'.

In fact, $(\beta S, \square)$ is a *maximal semigroup compactification* of the semigroup S. To be precise, let V be a right topological semigroup. Then the *topological centre* of V is the set of elements $v \in V$ for which the map L_v is continuous. A semigroup compactification of a semigroup S is a compact, right topological semigroup V in which S can be densely embedded by a continuous homomorphism which maps S into the topological centre of V. Then $(\beta S, \square)$ is semigroup compactification of S, and it is maximal, in the sense that, for each compact right topological semigroup V, each homomorphism which maps S into the topological centre of V extends to a continuous homomorphism from βS into V.

For details of the above remarks, see [139].

1.6 Projective topological spaces

The notion of a projective object in the category of compact topological spaces and continuous maps is the following.

Definition 1.6.1. A compact space K is *projective* if, for any compact spaces L and M and any continuous surjections $\theta : L \to M$ and $\varphi : K \to M$, there exists a continuous map $\psi : K \to L$ such that $\varphi = \theta \circ \psi$.

We represent this situation with the following commutative diagram:

$$
\begin{array}{ccc}
 & & L \\
 & \psi \nearrow & \downarrow \theta \\
K & \xrightarrow{\;\varphi\;} & M.
\end{array}
$$

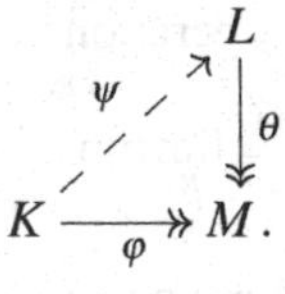

Example 1.6.2. Let S be an infinite set. Then βS is projective. Indeed, take compact spaces L and M and continuous surjections $\theta : L \to M$ and $\varphi : \beta S \to M$. For each $x \in S$, choose $\psi(x) \in L$ with $\theta(\psi(x)) = \varphi(x)$, and then extend ψ to a continuous map $\psi : \beta S \to L$. We have $\varphi = \theta \circ \psi$ because S is dense in βS. □

The following is a famous theorem of *Gleason* [114] from 1958; see also [24, Theorem 7.4] and [84, Theorems D.2.4 and D.2.6].

Theorem 1.6.3. *Let K be a non-empty, compact space. Then the following are equivalent:*

(a) *K is Stonean;*

(b) *K is a retract of the space βK_d;*

(c) *K is a retract of the space βS for some discrete space S;*

(d) *K is projective.*

Proof. (a) ⇒ (b) The pair $(\beta K_d, \pi)$ (as defined on page 26) is a cover of K, and so, by Proposition 1.4.21(iii), there is a closed subspace F_0 of βK_d such that $\pi \mid F_0 : F_0 \to K$ is an irreducible surjection. Since K is Stonean, it follows from Proposition 1.4.24 that $\pi \mid F_0$ is a homeomorphism, and then the map

$$(\pi \mid F_0)^{-1} \circ \pi : \beta K_d \to F_0$$

is a retraction. Hence K is homeomorphic to a retract of βK_d.

(b) ⇒ (c) This is trivial.

(c) ⇒ (d) Let $\nu : \beta S \to K$ be a retraction. Take compact spaces L and M and continuous surjections $\theta : L \to M$ and $\varphi : K \to M$. Since βS is projective, there exists $\mu : \beta S \to L$ with $\varphi \circ \nu = \theta \circ \mu$, and then $\psi := \mu \mid K : K \to L$ is the required map.

(d) ⇒ (a) Let U be a non-empty, open set in K. Take $\{p,q\}$ to be a two-point topological space, and define

$$L = ((K \setminus U) \times \{p\}) \cup (\overline{U} \times \{q\}),$$

so that L is a closed subset of $K \times \{p,q\}$. Take π to be the projection of $K \times \{p,q\}$ onto K, and set $\theta = \pi \mid L$, so that $\theta : L \to K$ is a continuous surjection. Since K is projective, there is a continuous map $\psi : K \to L$ such that $\theta \circ \psi$ is the identity map on K. Since $\theta : U \times \{q\} \to U$ is an injection, we have $\psi(x) = (x,q)$ $(x \in U)$. By the continuity of ψ, it follows that $\psi(x) = (x,q)$ $(x \in \overline{U})$, and hence $\overline{U} = \psi^{-1}(\overline{U} \times \{q\})$. Since $\overline{U} \times \{q\}$ is open in L, it follows that $\overline{U}$ is open in K.

This shows that K is Stonean, and gives (a). □

Lemma 1.6.4. *Let K be a non-empty, compact space, and let $\varphi : K \to K$ be a continuous map which is not the identity map. Then there is a proper, closed subspace F of K such that $K = F \cup \varphi^{-1}(F)$.*

Proof. Choose $x_0 \in K$ with $\varphi(x_0) \neq x_0$, and then choose disjoint, open neighbourhoods U and V of x_0 and $\varphi(x_0)$, respectively. Set

$$F = (U \cap \varphi^{-1}(V))^c = U^c \cup \varphi^{-1}(V)^c .$$

Since $x_0 \in U \cap \varphi^{-1}(V)$, the set F is a proper subset of K. Further,

$$F^c = U \cap \varphi^{-1}(V) \subset \varphi^{-1}(V) \subset \varphi^{-1}(U^c) \subset \varphi^{-1}(F),$$

and so $K = F \cup \varphi^{-1}(F)$. □

Theorem 1.6.5. *Let K be a non-empty, compact space. Then there is a pair (G_K, π_K) such that G_K is a projective, compact space, equivalently, a Stonean space, and the map $\pi_K : G_K \to K$ is an irreducible surjection. Further, G_K is uniquely specified up to homeomorphism.*

Proof. There is a continuous surjection $\varphi : \beta K_d \to K$. Let F be a closed subspace of βK_d such that $\theta := \varphi \mid F : F \to K$ is an irreducible surjection. Since βK_d is projective, there is a continuous map $\beta : \beta K_d \to F$ such that $\varphi = \theta \circ \beta$.

Let $\iota : F \to \beta K_d$ be the identity map. We *claim* that $\beta \circ \iota$ is the identity map on F. First, $\varphi \circ \iota = \theta$, and so $\theta \circ \beta \circ \iota = \varphi \circ \iota = \theta$. Now assume that $\beta \circ \iota$ is not the identity map on F. Then, by Lemma 1.6.4, there is a proper closed subspace C of F such that $F = C \cup (\beta \circ \iota)^{-1}(C)$. But then

$$\theta(F) = \theta(C) \cup (\theta \circ (\beta \circ \iota)^{-1})(C) = \theta(C) \cup (\theta \circ (\beta \circ \iota)(\beta \circ \iota)^{-1})(C) = \theta(C),$$

a contradiction of the irreducibility of θ. Thus $\beta \circ \iota$ is the identity map on F, and so $\beta : \beta K_d \to F$ is a retraction. By Theorem 1.6.3, (b) $\Rightarrow$ (c), F is projective. Set $G_K = F$ and $\pi_K = \theta$. Then the pair (G_K, π_K) has the required properties.

To prove the uniqueness, let (F_1, θ_1) be another pair satisfying the requirements. By the projectivity of F and F_1, there exist two continuous maps $\alpha : F_1 \to F$ and $\alpha_1 : F \to F_1$ such that $\theta_1 = \theta \circ \alpha$ and $\theta = \theta_1 \circ \alpha_1$.

Assume that $\alpha_1 \circ \alpha$ is not the identity map on F_1. Then, by Lemma 1.6.4 again, there is a proper closed subset C of F_1 such that $F_1 = C \cup (\alpha_1 \circ \alpha)^{-1}(C)$. Hence $\theta_1 \circ \alpha_1 \circ \alpha = \theta \circ \alpha = \theta_1$, and so, as before, $\theta_1(F_1) = \theta_1(C)$, contradicting the irreducibility of θ_1. Thus $\alpha_1 \circ \alpha$ is the identity map on F_1. Similarly, $\alpha \circ \alpha_1$ is the identity map on F. Hence α_1 is a homeomorphism. □

Definition 1.6.6. Let K be a compact space. Then the pair (G_K, π_K) is the *Gleason cover* of K.

Usually we say just that 'G_K is the Gleason cover of K'. Note that $G_K = K$ in the case where K is projective; in particular, $G_{G_K} = G_K$. Also note that it follows from Proposition 1.4.19(ii) that π_K is an open map if and only if K is projective.

Let K and L be non-empty, compact spaces such that K can be mapped onto L by an irreducible map. Then K and L clearly have the same Gleason cover. This shows in particular that all compact metric spaces without isolated points have the same Gleason cover because, by Proposition 1.4.23, they are all the irreducible image of the Cantor set, Δ.

Proposition 1.6.7. *Let K be a non-empty, compact space.*

(i) *Suppose that K is separable. Then G_K is separable.*

(ii) *Suppose that K is infinite. Then $|G_K| \geq 2^{\mathfrak{c}}$.*

Proof. (i) This follows from Proposition 1.4.19(i).

(ii) By Theorem 1.6.3, G_K is Stonean, and so G_K is a compact F-space. Since there is a surjection from G_K onto K, the space G_K is infinite. By Proposition 1.5.13, $|G_K| \geq 2^{\mathfrak{c}}$. □

We now give a somewhat different construction of the Gleason cover of a compact space. The following remarks are essentially a summary of [232].

Let (X, τ_X) be a Hausdorff topological space so that τ_X is a complete lattice. As before, an *ultrafilter* on X is a non-empty subset of τ_X which is maximal with respect to the finite intersection property, and then the family of these ultrafilters is the Stone space $St(\tau_X)$. For each $U \in \tau_X$, define

$$\widetilde{U} = \{p \in St(\tau_X) : U \in p\}. \tag{1.9}$$

Then these sets $\widetilde{U}$ form a base for the Stone topology $\widetilde{\tau}_X$ on $St(\tau_X)$, and $(St(\tau_X), \widetilde{\tau}_X)$ is a Stonean space.

Next, denote by E_X the subset of $St(\tau_X)$ consisting of the convergent ultrafilters on X. Then E_X is a dense, extremely disconnected subset of $St(\tau_X)$, and there is a natural mapping $\pi_X : E_X \to X$ that takes a convergent ultrafilter to its limit in X.

Now suppose, further, that X is a regular space. Then π_X is a perfect and irreducible map, and E_X is the unique (up to homeomorphism) extremely disconnected topological space which can be mapped by a perfect, irreducible map onto X. Each extremely disconnected space E 'factors through E_X', in the sense that, for each perfect surjection $\eta : E \to X$, there is a perfect surjection $\overline{\eta} : E \to E_X$ such that $\eta = \overline{\eta} \circ \pi_X$.

Finally, we suppose that K is a compact space. Then $St(\tau_K) = E_K$, and it is clear that $St(\tau_K)$ is exactly the Gleason cover G_K of K. It is easily seen that, for each $U \in \tau_K$, we have $\pi_K(\widetilde{U}) = \overline{U}$.

1.7 Boolean algebras and Boolean rings

At later points in our work, we shall approach some topics through the medium of Boolean rings and Boolean algebras, and so we recall the definition of such objects here. A modern text on Boolean algebras is that of Givant and Halmos [113]; see also [96] and [239, Chapter 2]. For Boolean rings, see [155, Problem 5S, p. 168], and, for a more algebraic approach, [28].

Definition 1.7.1. A *Boolean algebra* is a distributive lattice B with a maximum element 1 and a minimum element 0 and such that, for each $a \in B$, there exists a unique element a', called the *complement* of a, such that $a \vee a' = 1$ and $a \wedge a' = 0$.

A *Boolean ring* is a distributive lattice B with a minimum element 0 such that each order interval $[0,a]$ in B is a Boolean algebra.

For example, the power set, $\mathscr{P}(S)$, of a non-empty set S is a Boolean algebra for the obvious operations. We shall give some further examples of Boolean algebras at the end of this section.

Suppose that B is a Boolean ring. The complement with respect to the Boolean algebra $[0,a]$ of an element $b \in [0,\, a]$ is called the *relative complement* of b with respect to a, and it is denoted by $a - b$. One defines operations $+$ and $\cdot$ on B by setting

$$a + b = (a \vee b) - (a \wedge b), \quad a \cdot b = a \wedge b \quad (a, b \in B).$$

Then $(B, +, \cdot)$ is a ring. Thus a Boolean ring is a ring R (with respect to the maps $+$ and $\cdot$) such that $a \cdot a = a$ for each $a \in R$. This implies that R is commutative and that $a + a = 0$ for each $a \in R$. In fact, a Boolean ring is an algebra over the field $\mathbb{Z}_2$.

Let B be a Boolean ring. Then one recovers the binary relation $\leq$ on B by setting $a \leq b$ for $a, b \in B$ if $a \wedge b = a$; the axioms for $\vee$ and $\wedge$ force $\leq$ to be a partial order on B for which $\vee$ and $\wedge$ are the supremum and infimum.

A Boolean algebra is a Boolean ring with an identity for multiplication; indeed, one sometimes thinks of a Boolean ring as a Boolean algebra with a possibly missing 'top element' (a multiplicative unit). We stress that our Boolean rings do not necessarily have an identity; elsewhere, Boolean rings are implicitly or explicitly assumed to have an identity.

A *Boolean subalgebra* of a Boolean algebra B is a subset of B which contains 0 and 1 and is also a Boolean algebra for the same binary operations; the smallest Boolean subalgebra of B containing a subset S of B is the Boolean subalgebra *generated* by S.

For example, let S be a non-empty set and take $\mathscr{F}$ to be a σ-algebra of subsets of S. Then $\mathscr{F}$ is a Boolean subalgebra of $\mathscr{P}(S)$. The family $\mathfrak{U}_X$ of clopen subsets of a topological space X is a Boolean subalgebra of $\mathscr{P}(X)$; for a Hausdorff space X, the family $\mathfrak{C}_X$ of compact and open subsets of X is a Boolean ring.

A map between two Boolean rings is a *homomorphism* if it is a lattice homomorphism (it necessarily preserves relative complementation); an injective homomorphism is an *embedding* and a bijective homomorphism is an *isomorphism*, and in this case the two Boolean rings are *isomorphic*.

A Boolean ring is *Dedekind complete* (respectively, *Dedekind σ-complete*, *complete*, *σ-complete*) if it has these properties as a lattice. For example, a Boolean algebra is σ-complete if and only if every countable subset has a supremum.

Let I be a subset of a Boolean ring B. Then I is an *ideal* if: $0 \in I$; $a \vee b \in I$ whenever $a, b \in I$; $b \in I$ whenever $b \leq a$ for some $a \in I$. In this case, denote by B/I the quotient $B/\sim$, where $\sim$ is the equivalence relation defined by the condition that $a \sim b$ for $a, b \in B$ if $a + b \in I$. Then the set B/I is also a Boolean ring in a natural way, and the canonical quotient map is a Boolean-ring epimorphism. Thus an ideal in a Boolean ring is just an ideal in the ring $(B,+,\cdot)$, and B/I is the quotient ring. The quotient of a σ-complete Boolean algebra by a σ-complete ideal is σ-complete [113, Theorem 27].

Let B be a Boolean ring. As before, an ultrafilter p on B is a non-empty subset of B which is maximal with respect to the property that

$$b_1 \wedge \cdots \wedge b_n \neq 0 \quad \text{whenever} \quad b_1,\ldots,b_n \in p,$$

and the family of ultrafilters on B is the Stone space of B, denoted by $St(B)$; the topology on $St(B)$ was defined by taking the sets

$$S_b = \{p \in St(B) : b \in p\} = St([0,b]) \subset \mathscr{P}(St(B))$$

for $b \in B$ as a base of the topology of $St(B)$. Note that it follows from Proposition 1.4.9 that the map

$$b \mapsto S_b\,, \quad B \to \mathscr{P}(St(B))\,,$$

is a homomorphism, and so the family $\{S_b : b \in \mathscr{B}\}$ is closed under finite unions and intersections. The Stone space $St(B)$ can also be identified with the space of non-zero homomorphisms from the algebra B onto $\mathbb{Z}_2$, where the space is taken with the topology of pointwise convergence.

The following result is basic and very well known; it is essentially *Stone's representation theorem* from 1936 [231]. See [94, I.12.1], [113, Chapter 22], and [239, §2.6]. However, the proof is usually given just for Boolean algebras, and so we indicate the details in the case of Boolean rings. A full proof is given in [28, Theorem IV.1.12]; see also [155, Problem 5S].

Theorem 1.7.2. *Let B be a Boolean ring. Then:*

(i) *$St(B)$ is a zero-dimensional, Hausdorff space;*

(ii) *$St(B)$ is a locally compact space, S_b is compact and open for each $b \in B$, and, further, $St(B)$ is compact when B is a Boolean algebra;*

(iii) *the subsets of $St(B)$ that are compact and open are precisely those of the form S_b;*

(iv) *$\mathfrak{C}_{St(B)}$ is isomorphic to B;*

(v) $|St(B)| \leq 2^{|B|}$.

Proof. (i) This is a special case of Proposition 1.4.9.

(ii) For $b \in B$, the set S_b is open by definition. Take $b \in B$, and suppose that $\Gamma \subset B$ is such that $\{S_a : a \in \Gamma\}$ is a covering of S_b by basic open sets; we may suppose that the family is closed under finite unions. We *claim* that necessarily $b \in \Gamma$. For otherwise, $b - a \neq 0$ for each $a \in \Gamma$. The family $\{b - a : a \in \Gamma\}$ has the finite intersection property, and so is contained in some $p \in S_b$. But $p \notin \bigcup\{S_a : a \in \Gamma\}$, a contradiction. Thus $b \in \Gamma$, and so S_b is compact.

It follows that $St(B)$ is a locally compact space.

In the case where B is a Boolean algebra with a maximum element 1, the space $St(B) = S_1$ is compact.

(iii) Let U be a compact and open subset of $St(B)$. Then U is a union of sets of the form S_b. Since U is compact, it is a finite union of such sets, and so has the form S_b for some $b \in B$.

(iv) As in Proposition 1.4.9, the map $b \mapsto S_b$, $B \to \mathfrak{C}_{St(B)}$, is a lattice homomorphism; clearly, it is injective.

(v) Clearly $|St(B)| \leq 2^{|B|}$ because $St(B)$ is a subset of $\mathscr{P}(B)$. □

Corollary 1.7.3. *Let B and C be two Boolean rings. Then B and C are isomorphic if and only if $St(B)$ and $St(C)$ are homeomorphic.*

Proof. This follows from clause (iv), above. □

Proposition 1.7.4. *Let B be a Boolean ring. Then:*

(i) *the space $St(B)$ is extremely disconnected if and only if B is Dedekind complete;*

(ii) *the space $St(B)$ is basically disconnected if and only if B is Dedekind σ-complete.*

Proof. (i) Since $B = \bigcup\{S_b : b \in B\}$, it suffices to prove the result when B is a Boolean algebra.

Suppose that B is complete. Each open set U in $St(B)$ has the form $\bigcup\{S_b : b \in \Gamma\}$ for a subset Γ of B. Set $a = \bigvee \Gamma$. We *claim* that $\overline{U} = S_a$. To see this, take $p \in S_a$. For each $c \in p$, we have $c \wedge a \neq 0$, and hence $c \wedge b \neq 0$ for some $b \in \Gamma$, for otherwise, we would have $b \leq c'$ $(b \in \Gamma)$, and hence $a \leq c'$. Thus $S_c \cap U \neq \emptyset$. This shows that $p \in \overline{U}$, and so $S_a \subset \overline{U}$. The reverse inclusion is immediate, and so the claim holds. Hence $\overline{U}$ is open. This shows that $St(B)$ is extremely disconnected.

Conversely, suppose that $St(B)$ is extremely disconnected. Then, for each subset Γ of B, there exists $a \in B$ with $\mathrm{cl}_{St(B)}(\bigcup\{S_b : b \in \Gamma\}) = S_a$. Clearly $a \geq b$ $(b \in \Gamma)$. Suppose that $c \in B$ with $c \geq b$ $(b \in \Gamma)$. Then $S_c \supset S_a$, and so $c \geq a$. Thus $a = \bigvee \Gamma$, showing that B is complete.

(ii) This is similar. □

Corollary 1.7.5. *Let B be a complete Boolean algebra. Then $St(B)$, the Stone space of B, is a Stonean space.* □

Proposition 1.7.6. *Let K be a locally compact, totally disconnected space. Then K is homeomorphic to $St(\mathfrak{C}_K)$.*

Proof. Set $B = \mathfrak{C}_K$. For each $x \in K$, define $p_x = \{U \in B : x \in U\} \in St(B)$. Clearly, the map $f : x \mapsto p_x$, $K \to St(B)$, is a bijection. This map is a homeomorphism because $f(U) = S_U$ and $f^{-1}(S_U) = U$ for each $U \in B$. □

Proposition 1.7.7. *Let X be a completely regular space. Then:*

(i) $\mathfrak{U}_X$ *and* $\mathfrak{C}_{\beta X}$ *are isomorphic Boolean algebras;*

(ii) βX *is totally disconnected if and only if* $St(\mathfrak{U}_X)$ *and* βX *are homeomorphic.*

Proof. (i) Set $B = \mathfrak{U}_X$ and $C = \mathfrak{C}_{\beta X}$, and define $\theta : U \mapsto U \cap X,\ C \to B$. Then θ is a homomorphism. The map θ is injective because $U = \overline{U} = \overline{U \cap X}$ $(U \in C)$. Now take $V \in B$. Then $\chi_V \in C^b(X)$ has a continuous extension $\chi^\beta \in C(\beta X)$. Set $U = \{x \in \beta X : \chi^\beta(x) = 1\}$. Then $U \in C$ and $\theta(U) = V$.

(ii) By Theorem 1.7.2, $St(\mathfrak{U}_X)$ is always totally disconnected, and so βX is totally disconnected whenever $St(\mathfrak{U}_X)$ and βX are homeomorphic.

Now suppose that βX is totally disconnected. By Proposition 1.7.6, βX and $St(\mathfrak{C}_{\beta X})$ are homeomorphic. By (i) and Corollary 1.7.3, $St(\mathfrak{U}_X)$ and $St(\mathfrak{C}_{\beta X})$ are homeomorphic. Hence $St(\mathfrak{U}_X)$ and βX are homeomorphic. □

Definition 1.7.8. Let B be a Boolean ring. A subset S of B is *dense* if, for every $a > 0$ in B, there exists some $s \in S$ with $0 < s \leq a$; B is *separable* if it contains a countable, dense subset. An *atom* in B is an element $a > 0$ in B such that $\{b \in B : b < a\} = \{0\}$; a Boolean ring is *atomless* if it has no atoms.

The locally compact space $St(B)$ is separable as a topological space whenever B is separable as a Boolean algebra (but the converse to this is not true – see Example 1.7.16); isolated points of $St(B)$ correspond to atoms in B.

A *completion* of a Boolean algebra B is a Boolean algebra C such that B is a dense subalgebra of C and every subset of B has a supremum in C. In fact, every Boolean algebra B has a completion, and any two completions are isomorphic via a mapping that is the identity on B. For these results, see [113, Chapter 25].

Lemma 1.7.9. *Let B and C be two Boolean algebras, with C atomless. Suppose that F is a finite Boolean subalgebra of B and that $f : F \to C$ is an embedding. Take $b \in B \setminus F$. Then f can be extended to an embedding of the Boolean subalgebra of B generated by $F \cup \{b\}$ into C.*

Proof. Denote by A the set of atoms of F, and set $A_0 = \{a \in A : 0 < a \wedge b < a\}$. For each $a \in A_0$, choose $c_a \in C$ with $0 < c_a < f(a)$; this is possible because $f(a) > 0$ and C is atomless. The set of atoms of the subalgebra, say F_b, of B generated by $F \cup \{b\}$ is

$$\{a \wedge b : a \in A_0\} \cup \{a - b : a \in A_0\} \cup (A \setminus A_0).$$

We define $f : F_b \to C$ by first setting $f_b(a \wedge b) = c_a$ and $f_b(a - b) = f(a) - c_a$ for $a \in A_0$. Note that $a_1 \wedge a_2 = 0$ and $f(a_1) \wedge f(a_2) = 0$ whenever $a_1, a_2 \in A_0$ with $a_1 \neq a_2$; this implies that f_b is well defined and injective on the atoms of F_b. Clearly f_b extends to an embedding from F_b such that $f_b \mid F = f$. □

Lemma 1.7.10. *Any two countable, infinite, atomless Boolean algebras are isomorphic.*

Proof. Let B and C be two countable, infinite, atomless Boolean algebras, enumerated as sequences (b_n) and (c_n), respectively, where $b_1 = 0_B$ and $c_1 = 0_C$.

Set $B_1 = \{0_B, 1_B\}$ and $C_1 = \{0_C, 1_C\}$, and define $f_1 : B_1 \to C_1$ to be the unique Boolean isomorphism. Now take $n \in \mathbb{N}$, and assume inductively that we have defined finite Boolean subalgebras B_n and C_n of B and C containing $\{1_B, b_1, \dots, b_n\}$ and $\{1_C, c_1, \dots, c_n\}$, respectively, and an isomorphism $f_n : B_n \to C_n$. Suppose that n is odd, and choose $r \in \mathbb{N}$ to be the smallest element of $\mathbb{N}$ with $b_r \notin B_n$. By the above lemma, we can extend f_n to an embedding f_{n+1} of the Boolean subalgebra B_{n+1} of B generated by $B_n \cup \{b_r\}$; set $C_{n+1} = f_{n+1}(B_{n+1})$. Similarly in the case where n is even, we choose $s \in \mathbb{N}$ to be the smallest element of $\mathbb{N}$ with $c_s \notin C_n$, and extend f_n^{-1}.

Continuing this process, we obtain an isomorphism from B onto C. □

Theorem 1.7.11. *Any two infinite, separable, atomless, complete Boolean algebras are isomorphic.*

Proof. Let B and C be infinite, separable, atomless, complete Boolean algebras, and take S and T to be countable, dense subsets of B and C, respectively. We may suppose that, in fact, S and T are Boolean subalgebras of B and C, respectively. By Lemma 1.7.10, there is an isomorphism $f : S \to T$.

We remark that, for subsets A and B of S, we have $\bigvee A = \bigvee B$ if and only if

$$\{s \in S : s \wedge a = 0 \ (a \in A)\} = \{s \in S : s \wedge b = 0 \ (b \in B)\};$$

the corresponding statement also holds for subsets of T. It follows that the map

$$\widetilde{f} : b \mapsto \bigvee \{f(s) : s \in S, s \leq b\}, \quad B \to C,$$

is well defined. It is easy to check that $\widetilde{f} : B \to C$ is an isomorphism that extends the isomorphism $f : S \to T$. □

Theorem 1.7.11 was known in the early 1930's, as explained in [142, p. 483]. For a full proof of more general results, see [203, p. 778].

We now briefly discuss measures on Boolean rings; for a fuller discussion of measures defined on the algebra of Borel sets, see Chapter 4. For a discussion of measures on a Boolean algebra, see [113, Chapter 31].

Definition 1.7.12. Let B be a Boolean ring. A *measure* on B is a map $\mu : B \to \mathbb{C}$ such that $\mu(a \vee b) = \mu(a) + \mu(b)$ whenever $a, b \in B$ with $a \wedge b = 0$. A measure μ on B is: *positive* if $\mu(a) \geq 0 \ (a \in B)$; *normal* if $\lim_\alpha \mu(a_\alpha) = 0$ for each net (a_α) in B such that $a_\alpha \searrow 0$; and *σ-normal* if $\lim_{n\to\infty} \mu(a_n) = 0$ for each sequence (a_n) in B such that $a_n \searrow 0$.

We denote the set of normal measures on a Boolean ring B by $N(B)$, the set of real-valued normal measures by $N_{\mathbb{R}}(B)$, and the set of positive normal measures by $N(B)^+$; we shall relate normal measures on a compact space to normal measures on a Boolean ring in Theorem 4.7.27.

For $\mu, \nu \in N_{\mathbb{R}}(B)$, set $\mu \leq \nu$ if $\mu(b) \leq \nu(b)$ $(b \in B)$. Clearly $(N_{\mathbb{R}}(B), \leq)$ is a partially ordered set. For $a \in B$, set

$$\left.\begin{aligned} (\mu \vee \nu)(a) &= \sup\{\mu(b) + \nu(a-b) : b \in [0,a]\}, \\ (\mu \wedge \nu)(a) &= \inf\{\mu(b) + \nu(a-b) : b \in [0,a]\}. \end{aligned}\right\} \tag{1.10}$$

Then it is easily checked that $\mu \vee \nu, \mu \wedge \nu \in N_{\mathbb{R}}(B)$ and that $(N_{\mathbb{R}}(B), \leq)$ is a Riesz space, and so $N(B)$ is a complex Riesz space.

Suppose that μ is a positive measure on an atomless Boolean ring B and that $\varepsilon > 0$. For each $x > 0$ in B, there exists $y \in B$ with $0 < y < x$, and we may suppose that $\mu(y) \leq \mu(x)/2$ because $\mu(x) = \mu(y) + \mu(x-y)$. Continuing, we see that there exists $z \in B$ with $0 < z < x$ and $\mu(z) < \varepsilon$.

The following fact is proved in many places: the best direct proof is that of Horn–Tarski [142, Theorem 3.2, p. 490], and we now give this proof.

Proposition 1.7.13. *Let B be a separable, atomless Boolean algebra. Then there are no non-zero, σ-normal, positive measures on B.*

Proof. Notice first that a dense set $\{z_\alpha : \alpha \in A\}$ in B has $\bigvee\{z_\alpha : \alpha \in A\} = 1$. To see this, assume that there exists $z \in B$ with $z_\alpha \leq z < 1$ for each $\alpha \in A$. Then, for each $\alpha \in A$, it is not true that $z_\alpha \leq z'$, a contradiction of the fact that $\{z_\alpha : \alpha \in A\}$ is dense.

Assume towards a contradiction that μ is a non-zero, σ-normal, positive measure on B; we may suppose that $\mu(1) = 1$. Let $\{x_n : n \in \mathbb{N}\}$ be a dense subset of B. For each $n \in \mathbb{N}$, choose $y_n \in B$ with $0 < y_n \leq x_n$ and $\mu(y_n) < 1/3^n$. Clearly $\{y_n : n \in \mathbb{N}\}$ is also a dense set, so that $\bigvee\{y_n : n \in \mathbb{N}\} = 1$. Set $z_n = 1 - (y_1 \vee \cdots \vee y_n)$ $(n \in \mathbb{N})$. Then $z_n \searrow 0$, and so $\mu(z_n) \searrow 0$, whence $\mu(y_1 \vee \cdots \vee y_n) \nearrow 1$. But

$$\mu(y_1 \vee \cdots \vee y_n) \leq \sum_{i=1}^{n} \mu(y_i) \leq \sum_{i=1}^{n} \frac{1}{3^i} < \frac{1}{2} \quad (n \in \mathbb{N}),$$

a contradiction. The result follows. □

We conclude this section by giving some examples of Boolean algebras that will be relevant for us.

Example 1.7.14. The power set Let S be a non-empty set, with power set $\mathscr{P}(S)$. Then $\mathscr{P}(S)$ is a complete Boolean algebra. An ultrafilter with respect to the Boolean algebra $\mathscr{P}(S)$ is exactly an ultrafilter on the set S, and so the Stone space, $St(\mathscr{P}(S))$, is immediately identified with the Stone–Čech compactification βS.

Let $\mathscr{F}$ be the family of finite subsets of S, so that $\mathscr{F}$ is an ideal in the Boolean algebra $\mathscr{P}(S)$; the quotient Boolean algebra is $\mathscr{P}(S)/\mathscr{F}$, and it is easy to see that the Stone space $St(\mathscr{P}(S)/\mathscr{F})$ is identified with the growth S^* of S in βS.

We note that the Boolean algebra $\mathscr{P}(S)/\mathscr{F}$ is not σ-complete whenever S is infinite. For let $\{S_n : n \in \mathbb{N}\}$ be a family of pairwise-disjoint, infinite subsets of S, and suppose that $T \subset S$ is such that $T/\mathscr{F}$ is an upper bound for $\{S_n/\mathscr{F} : n \in \mathbb{N}\}$ in $\mathscr{P}(S)/\mathscr{F}$. For each $n \in \mathbb{N}$, the set T contains all but finitely many points of S_n; form a new set U in S by deleting one further point in $S_n \cap T$ for each $n \in \mathbb{N}$. Then $U/\mathscr{F}$ is also an upper bound for $\{S_n/\mathscr{F} : n \in \mathbb{N}\}$ in $\mathscr{P}(S)/\mathscr{F}$. However $U/\mathscr{F} \leq T/\mathscr{F}$ and $U/\mathscr{F} \neq T/\mathscr{F}$ in $\mathscr{P}(S)/\mathscr{F}$ because $T \setminus U$ is infinite, and so $T/\mathscr{F}$ is not a supremum of $\{S_n/\mathscr{F} : n \in \mathbb{N}\}$ in $\mathscr{P}(S)/\mathscr{F}$.

It follows from Proposition 1.7.4(ii) that S^* is not basically disconnected, and hence is not a Stonean space. □

Example 1.7.15. The algebra of complemented faces of a simplex Let L be a convex set in a real-linear space E. The family, called $\mathfrak{F}(L)$, of all faces (see page 8) of L, partially ordered by inclusion, clearly forms a complete lattice with $0 = \emptyset$ and $1 = L$ because an arbitrary intersection of faces is a face.

In general, this lattice is not distributive, as is easily seen from the example of a square with sides A, B, C and D, where A and B meet at a corner point x. Then $A \wedge (B \vee C) = A$, but $(A \wedge B) \vee (A \wedge C) = \{x\} \vee \emptyset = \{x\}$, and $A \not\subset \{x\}$, violating the distributivity law. On the other hand, it is easy to verify that the lattice of faces of a triangle is distributive. This example is a special case of the following concept.

An important class of convex sets L is obtained when the ambient space E is a Riesz space, L is a subset of the positive cone E^+ of E, and every non-zero element of E^+ is uniquely represented as a positive multiple of some element of L. Such an L, and each of its affine isomorphs, is called a *simplex*; see [118, p. 156]. This requirement has far-reaching consequences. In particular, the convex hull of any union of faces of a simplex is a face [118, Proposition 10.10], and every complemented face of a simplex is a split face (as defined on page 8) [118, Proposition 10.12].

We *claim* that, for a simplex L, the family $\mathfrak{F}(L)$ is indeed a distributive lattice, that the complemented elements of this lattice are precisely the complemented faces of L as specified in our definition on page 8, and that the partially ordered set, Comp_L, of complemented faces is a Boolean algebra.

To establish the distributivity of $\mathfrak{F}(L)$ in the above situation, we use the fact that

$$A \wedge B = A \cap B \quad \text{and} \quad A \vee B = \mathrm{co}(A \cup B)$$

for faces A and B. Indeed, we need to show only that, for faces A, B, and C in L, we have

$$A \cap \mathrm{co}(B \cup C) \subset \mathrm{co}((A \cap B) \cup (A \cap C))$$

(the reverse inclusion being obvious). For this, we suppose that $a \in A$ is given by $a = tb + (1-t)c$, where $b \in B$, $c \in C$, and $t \in \mathbb{I}$. If $t = 0$ (respectively, $t = 1$), then $a \in C$ (respectively, $a \in B$), and, if $0 < t < 1$, then, because A is a face, $b, c \in A$, and hence $b \in A \cap B$ and $c \in A \cap C$, so that $a \in \mathrm{co}((A \cap B) \cup (A \cap C))$, establishing the above formula. Hence, the lattice $\mathfrak{F}(L)$ is distributive. Clearly, the complemented elements of the lattice $\mathfrak{F}(L)$ are the complemented faces of L, so that Comp_L is a Boolean algebra (as is the case for the complemented elements in any distributive lattice with top and bottom elements).

So far, all the discussion of convex sets and faces has been algebraic and geometric, with no reference to any possible topology. Let us now suppose, further, that our ambient Riesz space E is a locally convex space (see page 48) and that L is a *compact* simplex in E. Then L is called a *Choquet simplex* [118, p. 163]. In this case, it is known from [117] that a face $F \in \mathfrak{F}(L)$ is a complemented face if and only if F is *σ-convex*, in the sense that, whenever $x_1, x_2, \ldots \in F$ and $\alpha_1, \alpha_2, \ldots \geq 0$ with $\sum_{i=1}^{\infty} \alpha_i = 1$, then $\sum_{i=1}^{\infty} \alpha_i x_i \in F$. Clearly an intersection of an arbitrary family of σ-convex sets is σ-convex, which implies that Comp_L is closed under arbitrary meets. Since we know from the preceding discussion that Comp_L is a Boolean algebra, it is therefore a complete Boolean algebra (as shown in [117]).

See Theorem 5.4.5, below, for an application of this Boolean algebra. □

Example 1.7.16. The regular-open algebra Let X be a non-empty topological space. The collection $\mathfrak{R}_X$ of regular–open subsets of X is a Boolean algebra with respect to the operations $\wedge$ and $\vee$, where $\wedge$ and $\vee$ are defined by

$$U \wedge V = U \cap V \quad \text{and} \quad U \vee V = \mathrm{int}_X \overline{U \cup V}$$

for $U, V \in \mathfrak{R}_X$. Now $U' = \mathrm{int}\,(X \setminus U)$ for $U \in \mathfrak{R}_X$. It is routine, but not entirely trivial, to check that $\mathfrak{R}_X$ is a Boolean algebra for these operations; there is an account in [239, §2.3]. The Boolean algebra $\mathfrak{R}_X$ is the *regular-open algebra* of X.

The regular-open algebra is always complete: the supremum and infimum of a family, $\mathscr{F}$, of regular–open sets are given respectively by

$$\bigvee \mathscr{F} = \mathrm{int}_X \left(\overline{\bigcup \{U : U \in \mathscr{F}\}} \right) \quad \text{and} \quad \bigwedge \mathscr{F} = \mathrm{int}_X \left(\bigcap \{U : U \in \mathscr{F}\} \right).$$

In the case where K is a compact space, the Stone space of the Boolean algebra $\mathfrak{R}_K$ is exactly the Gleason cover, G_K, of K. To see this, one simply checks that $St(\mathfrak{R}_K)$ has the defining properties of the Gleason cover. Specifically, observe first that $St(\mathfrak{R}_K)$ is extremely disconnected, and hence a Stonean space, because $\mathfrak{R}_K$ is a complete Boolean algebra. Second, define the canonical map

$$\pi : p \mapsto \bigcap \{\overline{V} : V \in p\}, \quad St(\mathfrak{R}_K) \to K.$$

The map π is an irreducible surjection, so that, by the uniqueness assertion of Theorem 1.6.5, $G_K = St(\mathfrak{R}_K)$. The Boolean algebra $\mathfrak{R}_K$ is isomorphic to $\mathfrak{R}_{G_K}$.

It follows from Proposition 1.6.7(ii) that $|St(\mathfrak{R}_K)| \geq 2^{\mathfrak{c}}$ whenever K is infinite.

Let K and L be two compact spaces. Then $\mathfrak{R}_K$ and $\mathfrak{R}_L$ are isomorphic as Boolean algebras if and only if the Gleason covers G_K and G_L are homeomorphic.

The family $\mathfrak{C}_K$ of compact and open subsets of K is a Boolean subalgebra of $\mathfrak{R}_K$; in the case where K is Stonean, $\mathfrak{C}_K = \mathfrak{R}_K$, and so $\mathfrak{C}_K$ is a complete Boolean algebra. See also [239, p. 288] for similar results.

The algebra $\mathfrak{R}_{\mathbb{I}}$ is an infinite, separable, atomless, complete Boolean algebra. (To see that $\mathfrak{R}_{\mathbb{I}}$ is separable, take the family of open intervals with rational endpoints as the subset S of Definition 1.7.8.) By Theorem 1.7.11, each infinite, separable, atomless, complete Boolean algebra is isomorphic to $\mathfrak{R}_{\mathbb{I}}$. Further, by Proposition 1.6.7(i), $G_{\mathbb{I}}$ is an infinite, separable Stonean space without isolated points.

Set $L = \mathbb{Z}_2^{\mathfrak{c}}$, a compact space. By Proposition 1.4.7, L is separable, and $\mathfrak{R}_L$ is also atomless and complete. However, we shall show that $\mathfrak{R}_{\mathbb{I}}$ and $\mathfrak{R}_L$ are not isomorphic as Boolean algebras, and hence that $G_{\mathbb{I}} = St(\mathfrak{R}_{\mathbb{I}})$ and $G_L = St(\mathfrak{R}_L)$ are separable Stonean spaces without isolated points which are not mutually homeomorphic.

To see that $\mathfrak{R}_{\mathbb{I}}$ and $\mathfrak{R}_L$ are not isomorphic, we shall show that the Boolean algebra $\mathfrak{R}_L$ is not separable. Indeed, let S be a dense subset of $\mathfrak{R}_L$; we may suppose that S consists of basic open sets of the form specified in equation (1.4). Then each clopen set of the form $\{(\varepsilon_\tau) \in L : \varepsilon_\sigma = 0\}$, where $\sigma < \mathfrak{c}$, contains a member of S, and so $|S| = \mathfrak{c}$. Hence $\mathfrak{R}_L$ is not separable.

The regular–open algebra plays an important role in the theory of forcing within mathematical logic; see the introduction to this subject for analysts given by Dales and Woodin in [74], for example. □

Section 9 of Bade's 1957 Notes [23], based on earlier work by Dilworth [87], established the representation of the 'normal completion' of the lattice $C_{\mathbb{R}}(K)$ for a compact space K as $C_{\mathbb{R}}(St(\mathfrak{R}_K))$. Later, after Gleason [114] appeared, this representation became recognized as $C_{\mathbb{R}}(G_K)$.

Example 1.7.17. The Borel sets Let X be a non-empty topological space. Then $\mathfrak{B}_X$, the family of Borel subsets of X, is a Boolean algebra with respect to the operations of union and intersection, and $\mathfrak{U}_X$ is a Boolean subalgebra of $\mathfrak{B}_X$. The Boolean algebra $\mathfrak{B}_X$ is always σ-complete, and so, by Proposition 1.7.4(ii), $St(\mathfrak{B}_X)$ is always basically disconnected. However, suppose that X is an uncountable Polish space. Then, by Corollary 1.4.15, X has a subset, say S, that is not Borel; the family of finite subsets of S is a net in $\mathfrak{B}_X$ with respect to inclusion, and this net has no supremum in $\mathfrak{B}_X$. Hence $\mathfrak{B}_X$ is not complete.

Let $\mathfrak{M}_X$ be the subset of $\mathfrak{B}_X$ consisting of the meagre Borel subsets of X. Then $\mathfrak{M}_X$ is a σ-complete ideal in $\mathfrak{B}_X$, and $\mathfrak{B}_X/\mathfrak{M}_X$ is a σ-complete Boolean algebra.

Suppose that the topological space X is completely metrizable or locally compact. Then we have noted on page 13 that, for each Borel set B in X, there is a unique regular–open set U with $B \equiv U$; further, each open set is equivalent to a regular–open set, and so $\mathfrak{B}_X/\mathfrak{M}_X$ is isomorphic to $\mathfrak{R}_X$, and hence $\mathfrak{B}_X/\mathfrak{M}_X$ is a complete Boolean algebra. (This is also true whenever X is a 'Baire space'; see [159, Proposition 12.9].) Thus, for compact K, $St(\mathfrak{B}_K/\mathfrak{M}_K)$ is G_K. This is also explained in [113, Chapter 29] and [239, p. 288]; it was first stated by Tarski.

Let X be an infinite, Hausdorff topological space, and take an infinite, countable subset, say S. Every ultrafilter p on S defines an ultrafilter $\widetilde{p}$ on $\mathfrak{B}_X$ by setting

$$\widetilde{p} = \{B \in \mathfrak{B}_X : B \cap S \in p\}.$$

The mapping $p \mapsto \widetilde{p}$ is injective because every subset of S belongs to $\mathfrak{B}_X$, and so $|St(\mathfrak{B}_X)| \geq |\beta S_d| = 2^{\mathfrak{c}}$. □

Chapter 2
Banach Spaces and Banach Lattices

We shall now give some background in the theory of normed and Banach spaces, including the key definitions of dual and bidual spaces and of an isomorphism and an isometric isomorphism between two normed spaces. In particular, we shall show how certain bidual spaces can be embedded in other Banach spaces. In §2.3, we shall also recall some basic results and theorems concerning Banach lattices. We shall define complemented subspaces of a Banach space in §2.4, and also we shall discuss, in §2.5, the projective and injective objects in the category of Banach spaces and bounded operators. We shall conclude the chapter by discussing dentability and the Krein–Milman property for Banach spaces in §2.6.

2.1 Banach spaces

We now recall the basics of the Banach-space theory that we shall use.

There is a huge literature on the theory of normed and Banach spaces; for example, see [3, 6, 30, 82, 85, 94, 100, 166, 175, 176, 183, 218, 225]. There is a collection of instructive essays on topics in Banach-space theory in [147]. We shall regard the texts of Albiac and Kalton [3], Allan [6], and Rudin [218] as accessible and elementary accounts and shall rarely repeat proofs from those sources.

Let E be a linear space or a real-linear space, with underlying field $\mathbb{K}$, still always $\mathbb{C}$ or $\mathbb{R}$. A *semi-norm* on E is a map $p : E \to \mathbb{R}^+$ such that

$$p(x+y) \leq p(x) + p(y) \quad (x,y \in E), \quad p(\alpha x) = |\alpha|\, p(x) \quad (\alpha \in \mathbb{K}, x \in E);$$

the semi-norm is a *norm* if, further, $p(x) = 0$ if and only if $x = 0$. Then $(E, \|\cdot\|)$ is a *normed space* if $\|\cdot\|$ is a norm on E; $(E, \|\cdot\|)$ is a *Banach space* if it is complete with respect to the metric d_E defined by

$$d_E(x,y) = \|x - y\| \quad (x,y \in E).$$

H.G. Dales et al., *Banach Spaces of Continuous Functions as Dual Spaces*,
CMS Books in Mathematics, DOI 10.1007/978-3-319-32349-7_2

For example, $(C^b(X), |\cdot|_X)$ is a Banach space for each non-empty topological space X.

Let E be a normed space. Then there is a Banach space containing E as a dense subspace (with the same norm); the latter space is the *completion* of E. Let p be a semi-norm on a linear space E, and set $F = \{x \in E : p(x) = 0\}$. Then we can regard p as a norm on the quotient space E/F and on the completion of E/F.

Two norms $\|\cdot\|_1$ and $\|\cdot\|_2$ on a linear space E are *equivalent* it there exist constants $m, M > 0$ such that

$$m\|x\|_1 \le \|x\|_2 \le M\|x\|_1 \quad (x \in E),$$

and so the two norms define equivalent metrics and the same topology on E. For example, any two norms on the linear space $\mathbb{C}^n$, where $n \in \mathbb{N}$, are equivalent.

Let F be a real linear space, with complexification $E = F \oplus \mathrm{i}F$, so that E is a (complex) linear space. Suppose that F is a normed space. Then E is a normed space for the norm specified by

$$\|x + \mathrm{i}y\| = \sup\{\|x\cos\theta - y\sin\theta\| : \theta \in [0, 2\pi]\},$$

and F is a closed real-linear subspace of E; E is a Banach space whenever F is a Banach space. But the above is not always the most appropriate choice of a norm on F; indeed, various choices for various different purposes can be made. For a discussion of this point, see [187]. For example, for the norm on the complexification of a Banach lattice, see equation (2.7).

Let $(E, \|\cdot\|)$ be a normed space. We denote by $E_{[1]}$ the *closed unit ball* of E; more generally,

$$E_{[r]} = \{x \in E : \|x\| \le r\} \quad \text{and} \quad B_r(x) = \{y \in E : \|y - x\| < r\}$$

for $r \ge 0$ and $x \in E$; the *unit sphere* of E is

$$S_E = \{x \in E : \|x\| = 1\}.$$

A *barrel* in E is a closed, bounded, absolutely convex, absorbent set; in the case where E is a Banach space, each of these is the closed unit ball of E with respect to a norm on E that is equivalent to the given norm.

Let F be a closed subspace of a normed space $(E, \|\cdot\|)$, with quotient map $\pi : E \to E/F$. Set

$$\|x + F\| = \inf\{\|x + y\| : y \in F\} = \inf\{\|z\| : z \in E,\, \pi(z) = x + F\} \quad (x \in E).$$

Then $\|\cdot\|$ is the *quotient norm* on E/F; always $(E/F, \|\cdot\|)$ is a normed space, called the *quotient space*, and the quotient map π is continuous and open; $(E/F, \|\cdot\|)$ is a Banach space whenever E is a Banach space.

A (Hausdorff) *locally convex space* is a linear space E (over $\mathbb{R}$ or $\mathbb{C}$) with a collection $\mathscr{P}$ of semi-norms on E such that $\mathscr{P}$ separates the points of E, in the sense that, for each $x \in E$ with $x \ne 0$, there exists $p \in \mathscr{P}$ with $p(x) \ne 0$. We define a topology on E by saying that a subset U of E is open if, for each $x \in U$, there are $p_1, \dots, p_n \in \mathscr{P}$ and $\varepsilon > 0$ such that

$$\{y \in E : p_j(y - x) < \varepsilon \ (j \in \mathbb{N}_n)\} \subset U.$$

A *topological linear space* is a linear space with a Hausdorff topology such that addition and scalar multiplication are continuous. A topological linear space E is a locally convex space if and only if there is a base of neighbourhoods of 0_E consisting of convex sets.

Let E be a locally convex space. We denote by E' the *dual* space of E, so that E' is the space of all continuous linear functionals on E. The action of $\lambda \in E'$ on $x \in E$ gives the complex number $\lambda(x)$ that we shall usually denote by $\langle x, \lambda \rangle$.

In the case where E is a normed space, the dual space E' is itself a Banach space for the norm specified by

$$\|\lambda\| = \sup\{|\langle x, \lambda \rangle| : x \in E_{[1]}\} \quad (\lambda \in E').$$

The dual space $(E')'$ of $(E', \|\cdot\|)$ is denoted by E''; it is called the second dual or *bidual space* of E. Occasionally, we shall refer to the third dual of E; this is $E''' = (E'')'$.

For examples of locally convex spaces, let E be a normed space, and define

$$p_\lambda(x) = |\langle x, \lambda \rangle| \quad (x \in E)$$

for each $\lambda \in E'$. Then each p_λ is a semi-norm on E, and the family $\{p_\lambda : \lambda \in E'\}$ defines a topology, called $\sigma(E, E')$, with respect to which E is a locally convex space; this topology is the *weak topology* on E. Let (x_γ) be a net in E, and take $x \in E$. Then $\lim_\gamma x_\gamma = x$ weakly (i.e., with respect to the weak topology) if and only if $\lim_\gamma \langle x_\gamma, \lambda \rangle = \langle x, \lambda \rangle$ $(\lambda \in E')$. The closure of a set S in E with respect to the weak topology is called the *weak closure*, etc.

Now define

$$p_x(\lambda) = |\langle x, \lambda \rangle| \quad (\lambda \in E')$$

for each $x \in E$. Then each p_x is a semi-norm on E', and the family $\{p_x : x \in E\}$ defines a topology, called $\sigma(E', E)$, with respect to which E' is a locally convex space. The topology $\sigma(E', E)$ is the *weak* topology* on E'. Clearly $\sigma(E', E) \subset \sigma(E', E'')$; every weakly convergent net in E' is weak*-convergent. We have $(E, \sigma(E, E'))' = E'$ and $(E', \sigma(E', E))' = E$, for example. Later we shall use the weak* topology $\sigma(E'', E')$ on E''.

For a discussion of locally convex spaces and these topologies, see [6, 68, 94, 144, 183, 218], for example.

We shall mention the following class of spaces at a few later points; in particular, see §4.5.

Definition 2.1.1. Let E be a Banach space. Then E is a *Grothendieck space*, or E has the *Grothendieck property*, if every weak*-convergent sequence in E' is weakly convergent.

The proto-typical example of a Grothendieck space is the space $C(K)$, where K is a Stonean space [124, Théorème 9, p. 168], as we shall show in Theorem 4.5.6. This class of examples includes the spaces $\ell^\infty(S)$ for each set S as particular instances; a generalization of these examples will be noted in Example 6.7.1. Many characterizations of Grothendieck space are listed, without proofs, in [85, Theorem p. 179]; some of these are proved in [184, Proposition 5.3.10].

The following is a form of the *Hahn–Banach theorem*; see [6, Corollaries 3.4 and 3.27], for example.

Theorem 2.1.2. (i) *Let E be a normed space, and suppose that F is a linear subspace of E. Take $\lambda \in F'$. Then there exists $\Lambda \in E'$ with $\Lambda \mid F = \lambda$ and $\|\Lambda\| = \|\lambda\|$.*

(ii) *Let E be a real locally convex space, and let A and B be non-empty, convex subsets of E with A compact, B closed, and $A \cap B = \emptyset$. Then there exists $\lambda \in E'$ with*

$$\sup_{x\in A}\langle x,\lambda\rangle < \inf_{x\in B}\langle x,\lambda\rangle .$$

(iii) *Let E be a complex locally convex space, and suppose that B is an absolutely convex, closed subset of E and that $x_0 \in E \setminus B$. Then there exists $\lambda \in E'$ with $|\langle x,\lambda\rangle| \le 1$ $(x \in B)$ and $\langle x_0,\lambda\rangle > 1$.* □

The functional Λ in clause (i), above, is a *norm-preserving extension* of λ.

Corollary 2.1.3. *Let E be a normed space, and let S be a circled subset of S_E. Then $\overline{\mathrm{co}}\, S = E_{[1]}$ if and only if*

$$\|\lambda\| \le \sup\{|\langle x,\lambda\rangle| : x \in S\} \quad (\lambda \in E'). \tag{2.1}$$

Proof. Suppose that $\overline{\mathrm{co}}\, S = E_{[1]}$, and take $\lambda \in E'$. For each $\varepsilon > 0$, there exist $n \in \mathbb{N}$, $x_1,\dots,x_n \in S$, and $\alpha_1,\dots \alpha_n \in \mathbb{I}$ with $\sum_{i=1}^n \alpha_i = 1$ such that

$$\left|\left\langle \sum_{i=1}^n \alpha_i x_i, \lambda \right\rangle\right| > \|\lambda\| - \varepsilon ,$$

and so $|\langle x_i,\lambda\rangle| > \|\lambda\| - \varepsilon$ for some $i \in \mathbb{N}_n$. Hence (2.1) follows.

Conversely, assume that there exists $x_0 \in E_{[1]}$ with $x_0 \notin \overline{\mathrm{co}}\, S$. Since S is circled, the set $\overline{\mathrm{co}}\, S$ is absolutely convex, and so, by Theorem 2.1.2(iii), there exists $\lambda \in E'$ with $|\langle x,\lambda\rangle| \le 1$ $(x \in S)$, but with $\langle x_0,\lambda\rangle > 1$, a contradiction of equation (2.1). Thus $\overline{\mathrm{co}}\, S = E_{[1]}$. □

Let E be a normed space. It follows from Theorem 2.1.2(i) that, for each $x \in E$, there exists $\lambda \in S_{E'}$ with $\|x\| = \langle x, \lambda\rangle$. The action of $\Phi \in E''$ on $\lambda \in E'$ gives the complex number $\langle \Phi, \lambda\rangle$, and we define the *canonical embedding* $\kappa_E : E \to E''$ by

$$\langle \kappa_E(x), \lambda\rangle = \langle x, \lambda\rangle \quad (x \in E, \lambda \in E') .$$

Clearly κ_E is a linear map; by our remark, $\|\kappa_E(x)\| = \|x\|$ $(x \in E)$, and so κ_E identifies E as a closed subspace of E''; the space E is *reflexive* if $E = E''$ under this identification. For example, ℓ^p and $L^p(\mathbb{I})$ are both reflexive whenever $1 < p < \infty$.

Here are some further standard theorems; the weak* topology on E' is $\sigma(E', E)$.

Theorem 2.1.4. *Let E be a Banach space.*

(i) *The closed unit ball $E'_{[1]}$ is weak*-compact and convex.*

(ii) *The space $\kappa_E(E_{[1]})$ is weak*-dense in $E''_{[1]}$.*

(iii) *The weak* topology on $E'_{[1]}$ is metrizable if and only if $(E, \|\cdot\|)$ is separable.*

(iv) *The following conditions on a linear functional* M *on E' are equivalent:*

(a) $\mathrm{M} \in \kappa_E(E)$;

(b) M *is weak*-continuous on E';*

(c) M *is weak*-continuous on $E'_{[1]}$.*

(v) *The weak and norm closures of a convex subset of E coincide.*

(vi) *A convex set C in E' is weak*-closed if and only if $C \cap E'_{[r]}$ is weak*-closed for each $r > 0$.*

(vii) *A subset S of E is relatively weakly compact if and only if each countable, infinite subset of S has a weak limit point in E if and only if each sequence in S has a subsequence converging weakly in E.*

Proof. Clause (i) is the *Banach–Alaoglu theorem*; see [6, Theorem 3.21]. Clause (ii) is *Goldstine's theorem*; see [6, Corollary 3.30]. For (iii) and (iv), see [94, Theorems V.5.1, V.5.6], for example. Clause (v) is *Mazur's theorem* [6, Corollary 3.28], clause (vi) is the *Krein–Šmulian theorem* [94, V.5.7], and clause (vii) is the *Eberlein–Šmulian theorem* [3, Theorem 1.6.3]. □

Proposition 2.1.5. *Let X be a completely regular topological space. Then there exists a compactification K of X such that each $f \in C^b(X)$ has an extension to a function $f^\beta \in C(K)$.*

Proof. Set $E = (C^b(X), |\cdot|_X)$, a Banach space. The weak* topology on E' is denoted by σ, so that $(E'_{[1]}, \sigma)$ is compact by Theorem 2.1.4(i). For $x \in X$, define $\varepsilon(x)$ on E by

$$\varepsilon(x)(f) = f(x) \quad (f \in E).$$

Then $\varepsilon(x) \in E'_{[1]}$ $(x \in X)$, and the map $\varepsilon : X \to (E'_{[1]}, \sigma)$ is a continuous injection; since X is completely regular, it is easily seen that ε is a homeomorphism onto its range, and so we can regard X as a subspace of $(E'_{[1]}, \sigma)$. Take K to be the closure of X in $(E'_{[1]}, \sigma)$, so that K is a compactification of X, and, for $f \in C^b(X)$, define f^β on K by

$$f^\beta(\lambda) = \langle f, \lambda \rangle \quad (\lambda \in K).$$

Then $f^\beta \in C(K)$ and f^β extends f, identified with $f \circ \varepsilon$. □

Thus $K = \beta X$ is the Stone–Čech compactification of X, as discussed in §1.5.

Recall that $w(X)$ and $d(X)$ are the weight and density character, respectively, of a topological space X.

Proposition 2.1.6. *Let E be a normed space. Then $d(E) \leq d(E')$. In particular, E is separable whenever E' is separable.*

Proof. We may suppose that $d(E')$ is infinite.

Let S be the unit sphere of E'. Then there is a dense subset, say $\{\lambda_\alpha : \alpha \in A\}$, of S with $|A| = d(E')$. For each $\alpha \in A$, choose $x_\alpha \in E_{[1]}$ with $|\langle x_\alpha, \lambda_\alpha \rangle| > 1/2$, and set $F = \mathrm{lin}\,\{x_\alpha : \alpha \in A\}$.

Assume towards a contradiction that F is not dense in E. By the Hahn–Banach theorem, there exists $\lambda \in S$ with $\lambda \mid F = 0$. There exists $\alpha \in A$ with $\|\lambda_\alpha - \lambda\| < 1/2$, and so

$$\frac{1}{2} < |\langle x_\alpha, \lambda_\alpha \rangle| \leq |\langle x_\alpha, \lambda_\alpha - \lambda \rangle| + |\langle x_\alpha, \lambda \rangle| = |\langle x_\alpha, \lambda_\alpha - \lambda \rangle| \leq \|\lambda_\alpha - \lambda\| < \frac{1}{2},$$

a contradiction. Thus $\overline{F} = E$.

It follows that linear combinations with coefficients in $\mathbb{Q} + \mathrm{i}\mathbb{Q}$ of the elements x_α constitute a dense subset of E with cardinality $|A|$. Hence $d(E) \leq d(E')$. □

Theorem 2.1.7. *Let K be a non-empty, compact space. Then:*

(i) $C(K)$ is separable if and only if K is metrizable;

(ii) $w(K) = d(C(K))$.

Proof. For (i) and (ii), it is clearly sufficient to prove the analogous results for the real Banach space $C_{\mathbb{R}}(K)$; set $E = C_{\mathbb{R}}(K)$.

We regard K as a subset of $(E'_{[1]}, \sigma(E', E))$ by identifying $x \in K$ with $\varepsilon_x \in E'_{[1]}$, where $\varepsilon_x(f) = f(x)$ $(f \in E)$, as above. The restriction of the topology $\sigma(E', E)$ to K is the original topology on K.

(i) Suppose that E is separable. By Theorem 2.1.4(iii), $(E'_{[1]}, \sigma(E', E))$ is metrizable, and so K is metrizable.

Conversely, suppose that d is a metric that defines the topology of K. Then (K, d) is separable, say $\{x_n : n \in \mathbb{N}\}$ is a dense subset of K. For $n \in \mathbb{N}$, define $f_n \in E$ by setting $f_n(x) = d(x, x_n)$ $(x \in K)$. Let A and B be the subsets of E formed by taking all the elements $p(1, f_1, \dots, f_n)$, where p is a polynomial in $n+1$ variables, $n \in \mathbb{N}$, and p has coefficients in $\mathbb{Q}$ and $\mathbb{R}$, respectively. Then A is countable and dense in B, and B is a subalgebra of E. Further, B contains the constants and separates strongly the points of K. By the Stone–Weierstrass theorem, Theorem 1.4.26(i), B is dense in $(E, |\cdot|_K)$, and so A is also dense in this space. Thus E is separable.

(ii) Set $\kappa = d(E)$; necessarily κ is infinite.

Let $\{f_\alpha : \alpha < \kappa\}$ be a dense subset of E, and define $U_\alpha = \{x \in K : f_\alpha(x) > 0\}$ for $\alpha < \kappa$. It is easy to check that $\{U_\alpha : \alpha < \kappa\}$ is a subbase for the topology of K, and so $w(K) \leq \kappa$.

Conversely, let $\mathscr{B}$ be a base for the topology of K with $|\mathscr{B}| = w(K)$, and let $\mathscr{A}$ be the family of all pairs $(U,V) \in \mathscr{B} \times \mathscr{B}$ such that $\overline{U} \cap \overline{V} = \emptyset$, so that $|\mathscr{A}| = w(K)$. By Urysohn's lemma, Theorem 1.4.25, for each $(U,V) \in \mathscr{A}$, there exists $f \in C(K,\mathbb{I})$ with $f \mid U = 1$ and $f \mid V = 0$. Form the sets A and B with respect to these functions f as in (i). Then $|A| = w(K)$, and again A is dense in E by the Stone–Weierstrass theorem. Thus $\kappa \leq w(K)$. □

Corollary 2.1.8. *Let E be a normed space, and set $B = (E'_{[1]}, \sigma(E',E))$. Then $w(B) = d(E)$.*

Proof. The space B is compact. We regard E as a linear subspace of $C(B)$ that separates the points of B. By Corollary 1.4.27, $d(E) = d(C(B))$, and, by Theorem 2.1.7(ii), $d(C(B)) = w(B)$. Hence $d(E) = w(B)$. □

The following result was originally proved by Choquet [54, p. 7] by a rather indirect and complicated argument; our simple proof is taken from [104, Proposition 2.9] and [201, Proposition 1.3].

Proposition 2.1.9. *Let K be a compact, convex set in a locally convex space E. Suppose that the relative topology on K is metrizable. Then* $\operatorname{ex} K$ *is a G_δ-set in E. Further,* $\operatorname{ex} K$ *is either countable or has cardinality* $\mathfrak{c}$.

Proof. Let d be a metric that gives the relative topology on K from E. For $n \in \mathbb{N}$, take K_n to be the set of points x in K such that $2x = y + z$ for some $y, z \in K$ for which $d(y,z) \geq 1/n$. Then each K_n is closed in K, and the complement of the union of the sets K_n is a G_δ-set. But this set is exactly $\operatorname{ex} K$.

By Proposition 1.4.14, $\operatorname{ex} K$ is either countable or has cardinality $\mathfrak{c}$. □

Let E be a normed space, and let $* : E \to E$ be an isometric linear involution on E. For $\lambda \in E'$, define $\lambda^* \in E'$ by

$$\langle x, \lambda^* \rangle = \overline{\langle x^*, \lambda \rangle} \quad (x \in E).$$

Then the map $* : \lambda \mapsto \lambda^*$, $E' \to E'$, is an isometric linear involution; this map is clearly also continuous with respect to the topology $\sigma(E',E)$. Continuing, we obtain an isometric linear involution $*$ on E''; the restriction of this linear involution to the subspace E of E'' is the original linear involution.

Let $\{(E_\alpha, \|\cdot\|_\alpha) : \alpha \in A\}$ be a family of normed spaces, defined for each α in a non-empty index set A (perhaps finite). Then we shall consider the following spaces.

First set

$$\bigoplus_\infty E_\alpha = \left\{ (x_\alpha : \alpha \in A) : \|(x_\alpha)\| = \sup_\alpha \|x_\alpha\|_\alpha < \infty \right\}.$$

Similarly, for p with $1 \leq p < \infty$, we define

$$\bigoplus_p E_\alpha = \left\{ (x_\alpha : \alpha \in A) : \|(x_\alpha)\| = \left(\sum_\alpha \|x_\alpha\|_\alpha^p \right)^{1/p} < \infty \right\}.$$

Clearly $\bigoplus_\infty E_\alpha$ and $\bigoplus_p E_\alpha$ are normed spaces; they are Banach spaces if each of the spaces E_α is a Banach space. We write

$$E \oplus_\infty F \quad \text{and} \quad E \oplus_p F$$

for the sum of two normed spaces E and F with the appropriate norms, etc., and we write $\ell_n^p(E)$ for E^n with the norm given by

$$\|(x_1, \dots, x_n)\| = \left(\sum_{i=1}^n \|x_i\|^p \right)^{1/p} \qquad (x_1, \dots, x_n \in E).$$

We write $\ell^p(E)$ and $\ell^\infty(E)$ for $\bigoplus_p \{E_n : n \in \mathbb{N}\}$ and $\bigoplus_\infty \{E_n : n \in \mathbb{N}\}$, respectively, when each E_n is equal to E.

Take p with $1 \leq p < \infty$. Then it is easy to see that we can identify the dual space of the normed space $\bigoplus_p E_\alpha$ with $\bigoplus_q E'_\alpha$, where q is the conjugate index to p, with the obvious duality, so that

$$\left(\bigoplus_p E_\alpha \right)' = \bigoplus_q E'_\alpha. \tag{2.2}$$

We shall sometimes refer to $E \times F$, the *product* of two Banach spaces E and F; there are many equivalent norms on $E \times F$ making it into a Banach space. For example, we could take the norm to be given by

$$\|(x,y)\| = \|x\| + \|y\| \quad (x \in E, y \in F).$$

Thus we can identify this space with $E \oplus_1 F$ in an obvious way.

Proposition 2.1.10. *Let E be a normed space, and suppose that F and G are subspaces such that $E = F \oplus_1 G$. Then* $\operatorname{ex} E_{[1]} = \operatorname{ex} F_{[1]} \cup \operatorname{ex} G_{[1]}$.

Proof. First take $x \in E$ with $\|x\| = 1$. We *claim* that there exist $y \in F$ and $z \in G$ with $\|y\| = \|z\| = 1$ and $\alpha, \beta \in \mathbb{I}$ with $\alpha + \beta = 1$ such that $x = \alpha y + \beta z$. Indeed, set $x = y_1 + z_1$ with $y_1 \in F$ and $z_1 \in G$ and $\|y_1\| + \|z_1\| = 1$. We may suppose that $y_1 \neq 0$ and $z_1 \neq 0$, for otherwise the claim is trivial. Set $\alpha = \|y_1\|$ and $\beta = \|z_1\|$, so that $\alpha, \beta > 0$ and $\alpha + \beta = 1$, and set $y = \alpha^{-1} y_1$ and $z = \beta^{-1} z_1$. Then the requirements are satisfied.

To show that $\operatorname{ex} E_{[1]} \subset \operatorname{ex} F_{[1]} \cup \operatorname{ex} G_{[1]}$, take $x \in \operatorname{ex} X_{[1]}$. Then $\|x\| = 1$. If $x \in F$, then trivially $x \in \operatorname{ex} F_{[1]}$, and similarly if $x \in G$. Assume that $x \notin F \cup G$. By the claim, x is a convex combination of two norm 1 elements from F and G, with coefficients in $(0,1)$, a contradiction.

The reverse inclusion is trivial. □

2.2 Isomorphisms and isometric isomorphisms

Let E and F be normed spaces over $\mathbb{R}$ or $\mathbb{C}$. Then a linear operator T from E to F is bounded if and only if it is continuous on E if and only if it is continuous at 0_E, and then

$$\|T\| = \sup\{\|Tx\| : x \in E_{[1]}\}$$

defines the *operator norm* $\|\cdot\|$ of T. The linear space $\mathscr{B}(E,F)$ of all bounded linear operators from E to F is itself a normed space with respect to the operator norm; we write $\mathscr{B}(E)$ for $\mathscr{B}(E,E)$, so that $\mathscr{B}(E)$ is a normed algebra with respect to the composition of operators; see §3.1. The space $\mathscr{B}(E,F)$ is a Banach space whenever F is a Banach space. Of course the basic inequality is that

$$\|Tx\| \leq \|T\| \, \|x\| \quad (x \in E, T \in \mathscr{B}(E,F)).$$

An operator $T \in \mathscr{B}(E,F)$ is a *contraction* if $\|T\| \leq 1$; a projection in $\mathscr{B}(E)$ is a *bounded projections* on E.

The following is the famous *uniform boundedness theorem*.

Theorem 2.2.1. *Let E be a Banach space, let $\{E_\alpha : \alpha \in A\}$ be a family of normed spaces, and let $T_\alpha : E \to E_\alpha$ be a bounded operator for each $\alpha \in A$. Suppose that $\sup\{\|T_\alpha x\| : \alpha \in A\} < \infty$ for each $x \in E$. Then $\sup\{\|T_\alpha\| : \alpha \in A\} < \infty$.* □

Corollary 2.2.2. *Let E be a normed space. Then a subset of E is bounded if and only if it is weakly bounded.* □

The following is a form of the *open mapping theorem*, together with *Banach's isomorphism theorem*.

Theorem 2.2.3. *Let E and F be Banach spaces, and suppose that $T \in \mathscr{B}(E,F)$ is a surjection. Then T is an open mapping. In particular, in the case where T is a bijection, $T^{-1} \in \mathscr{B}(F,E)$.* □

Definition 2.2.4. Let E and F be normed spaces. A bijection T in $\mathscr{L}(E,F)$ is an *isomorphism* or a *linear homeomorphism* if both T and T^{-1} are bounded. Two normed spaces E and F are *isomorphic* if there is an isomorphism from E onto F, and in this case we write

$$E \sim F.$$

An operator $T \in \mathscr{B}(E,F)$ is an *embedding* if it is an isomorphism onto a subspace of F, and E *embeds* in F if there is such an embedding.

Of course, in the case where E and F are Banach spaces, each bijection in $\mathscr{B}(E,F)$ is an isomorphism. An operator $T \in \mathscr{B}(E,F)$ is an embedding if and only if there exists $\delta > 0$ with $\|Tx\| \geq \delta \|x\| \ (x \in E)$. Indeed, when we consider an embedding $T : E \to F$ as an isomorphism onto its range, we see that T has a bounded inverse $T^{-1} : T(E) \to E$; clearly, for $x \in E$, we have $\|x\| \leq \|T^{-1}\| \|Tx\|$, and so, when $E \neq \{0\}$, we can take $\delta = \|T^{-1}\|^{-1}$. (The constant δ is sometimes called the *embedding constant* of T.)

Definition 2.2.5. Let $\mathscr{C}$ be a class of Banach spaces. Then a property is an *isomorphic invariant* for the class $\mathscr{C}$ if each Banach space E in $\mathscr{C}$ has the property whenever E is isomorphic to another Banach space in $\mathscr{C}$ that has the property.

For example, it is clear that 'separability' and 'having a separable dual space' are isomorphic invariants of the class of all Banach spaces. Also, the Grothendieck property of Definition 2.1.1 is an isomorphic invariant of this class.

Let E and F be normed spaces, and take $T \in \mathscr{B}(E,F)$. Then, as on page 9, T induces a linear map

$$\overline{T} : x + \ker T \mapsto Tx\,, \quad E/\ker T \to F\,,$$

such that $\overline{T} : E/\ker T \to T(E)$ is a linear isomorphism from $E/\ker T$ onto $T(E)$. In our present setting, $\ker T$ is closed in E, and $\overline{T}$ is bounded with $\|\overline{T}\| \leq \|T\|$ when $E/\ker T$ has the quotient norm; in the case where E and F are Banach spaces and T has closed range, the map $\overline{T} : E/\ker T \to T(E)$ is an embedding.

Many, but not all, Banach spaces E have the property that $E \sim E \times E$. (This is not true, for example, for the James space, J, described in [3, p. 233].) In particular, the following is clear.

Proposition 2.2.6. *Take p with $1 \leq p \leq \infty$, and let E be either of the two Banach spaces ℓ^p and $L^p(\mathbb{I})$. Then $E \sim E \times E$.* □

We caution that it is possible that two (complex) Banach spaces E and F can fail to be isomorphic (as complex Banach spaces), but to be such that their underlying real spaces are isomorphic (as real Banach spaces): see [43] and, for a more elementary example, [151].

The following definition is given in [3, Definition 7.4.5], for example.

Definition 2.2.7. Let E and F be isomorphic normed spaces. Then the *Banach-Mazur distance*, $d(E,F)$, from E to F is given by

$$d(E,F) = \inf\{\|T\| \|T^{-1}\| : T \in \mathscr{B}(E,F) \text{ is an isomorphism}\}\,.$$

Definition 2.2.8. Let E and F be normed spaces. A map $T \in \mathscr{B}(E,F)$ is *isometric* if $\|Tx\| = \|x\| \ (x \in E)$, and then T is a *linear isometry*; T is an *isometric isomorphism* if it is a surjective linear isometry from E onto F. When there is such an isometric isomorphism, we say that E and F are *isometrically isomorphic* and write

$$E \cong F\,.$$

A linear isometry from E onto a subspace of F is an *isometric embedding*.

Thus $d(E,F) = 1$ whenever $E \cong F$; the converse is true when E and F are both finite-dimensional spaces, but it is not true in general. An isomorphism $T \in \mathscr{B}(E,F)$ is isometric if and only if T and T^{-1} are both contractions.

Sometimes, with a slight abuse of language, we say that '$E = F$' or 'E is F' when, strictly, we mean that '$E \cong F$'. For example, in the case where $1 \leq p < \infty$, we say that the duals of the Banach spaces ℓ^p and $L^p(\mathbb{I})$ are ℓ^q and $L^q(\mathbb{I})$, respectively, where q is the conjugate index to p. Also, the dual of c_0 is ℓ^1, so that $c_0'' = (\ell^1)' = \ell^\infty = C(\beta\mathbb{N})$.

The difference between the corresponding 'isomorphic' and 'isometric' theories of Banach spaces is of great significance, as we shall see shortly. For example, here is a result that applies in the isometric, but not necessarily in the isomorphic, theory: Let E and F be Banach spaces, and suppose that $T : E \to F$ is an isometric isomorphism. Then $T(\mathrm{ex}\, E_{[1]}) = \mathrm{ex}\, F_{[1]}$.

Since we shall be concerned with linear isometries, we give a gem, the *Mazur–Ulam theorem* from 1932; see [30, Chapitre 6, §3].

Lemma 2.2.9. *Let E and F be two real Banach spaces. Suppose that a map $\Psi : (E, d_E) \to (F, d_F)$ is isometric and that $\Psi(0_E) = 0_F$. Then Ψ is real-linear.*

Proof. Take $x_1, x_2 \in E$ with $x_1 \neq x_2$, and set $x_0 = (x_1 + x_2)/2$. We inductively define subsets E_n of E for $n \in \mathbb{N}$ by setting

$$E_1 = \{x \in E : 2d_E(x, x_1) = 2d_E(x, x_2) = d_E(x_1, x_2)\}$$

and

$$E_{n+1} = \{x \in E_n : 2d_E(x, y) \leq \operatorname{diam} E_n \ (y \in E_n)\}\,.$$

Then $x_0 \in E_1$ and $\operatorname{diam} E_1 < \infty$. Further $\operatorname{diam} E_{n+1} \leq (\operatorname{diam} E_n)/2$ $(n \in \mathbb{N})$, and so $\lim_{n\to\infty} \operatorname{diam} E_n = 0$.

We *claim* that, for each $n \in \mathbb{N}$, the point $\overline{y} := x_1 + x_2 - y$ belongs to E_n whenever $y \in E_n$. First suppose that $y \in E_1$. Then

$$d_E(\overline{y}, x_1) = d_E(x_2, y) \quad \text{and} \quad d_E(\overline{y}, x_2) = d_E(x_1, y)\,,$$

and so $\overline{y} \in E_1$. Now assume that the claim holds for $n \in \mathbb{N}$, and take $y \in E_{n+1}$. For each $z \in E_n$, we have $\overline{z} \in E_n$, and so

$$2d_E(\overline{y}, z) = 2d_E(y, \overline{z}) \leq \operatorname{diam} E_n\,,$$

and hence $\overline{y} \in E_{n+1}$. Thus the claim follows by induction on $n \in \mathbb{N}$.

We next *claim* that $x_0 \in E_n$ $(n \in \mathbb{N})$. Clearly $x_0 \in E_1$. Take $n \in \mathbb{N}$ and $y \in E_n$. Then $d_F(y, \overline{y}) \leq \operatorname{diam} E_n$. But $2d_E(x_0, y) = d_E(y, \overline{y})$, and so $x_0 \in E_{n+1}$. This gives the claim.

Since $\lim_{n\to\infty} \operatorname{diam} E_n = 0$ and the metric space (E, d_E) is complete, it now follows that $\bigcap_{n=1}^{\infty} E_n = \{x_0\}$.

Set $y_1 = \Psi(x_1)$, $y_2 = \Psi(x_2)$, and $y_0 = (y_1 + y_2)/2$. We now define subsets F_n of F in an analogous way to the above with respect to y_1 and y_2, so that $\bigcap_{n=1}^{\infty} F_n = \{y_0\}$. However $\Psi(E_n) = F_n$ $(n \in \mathbb{N})$ because $\Psi : (E, d_E) \to (F, d_F)$ is an isometry, and so $\Psi(x_0) = y_0$. It follows that

$$\Psi\left(\frac{1}{2}(x_1 + x_2)\right) = \frac{1}{2}(\Psi(x_1) + \Psi(x_2)) \quad (x_1, x_2 \in E).$$

Further, $\Psi(x/2) = \Psi(x)/2$ $(x \in E)$ because $\Psi(0_E) = 0_F$, and so

$$\Psi(x_1 + x_2) = \Psi(x_1) + \Psi(x_2) \quad (x_1, x_2 \in E).$$

It follows that $\Psi(\alpha x) = \alpha \Psi(x)$ $(\alpha \in \mathbb{Q}, x \in E)$, and so Ψ is linear over $\mathbb{Q}$. Since Ψ is continuous, it follows that Ψ is real-linear. □

We cannot say that an isometric map Ψ between two Banach spaces is (complex) linear in the above situation: indeed, the map $\Psi : z \mapsto \bar{z}$, $\mathbb{C} \to \mathbb{C}$, is an isometry with $\Psi(0) = 0$.

Theorem 2.2.10. *Let E and F be two real Banach spaces, and suppose that there is an isometry from E onto F. Then E and F are isometrically isomorphic.*

Proof. Let $\Phi : E \to F$ be an isometry, and set $\Psi(x) = \Phi(x) - \Phi(0_E)$ $(x \in E)$. Then $\Psi : E \to F$ is also an isometry with $\Psi(0_E) = 0_F$, and so, by Lemma 2.2.9, Ψ is a real-linear isometry; it is a surjection whenever Φ is a surjection. □

Let F be a normed space, and let M and N be closed subspaces of F and F', respectively. Define

$$M^\circ = \{\lambda \in F' : \langle x, \lambda \rangle = 0 \ (x \in M)\} \quad {}^\circ N = \{x \in F : \langle x, \lambda \rangle = 0 \ (\lambda \in N)\}.$$

Then M° and ${}^\circ N$ are closed linear subspaces of $(F', \sigma(F', F))$ and $(F, \sigma(F, F'))$, respectively; M° is the *annihilator* of M and ${}^\circ N$ is the *pre-annihilator* of N. Clearly, $({}^\circ N)^\circ$ is the $\sigma(F', F)$-closure of N in F', so that $({}^\circ N)^\circ = N$ whenever N is $\sigma(F', F)$-closed.

Now suppose that $F' \cong E$. Then it is standard that $M' \cong E/M^\circ$ and $(F/M)' \cong M^\circ$, and so, in the case where N is $\sigma(E, F)$-closed in E, we obtain the following result by setting $M = {}^\circ N$.

Proposition 2.2.11. *Let E be a Banach space with $E \cong F'$ for a normed space F. Suppose that N is a $\sigma(E, F)$-closed linear subspace of E. Then $N \cong (F/{}^\circ N)'$ and $E/N \cong ({}^\circ N)'$.* □

Let E and F be normed spaces. The *dual* (or *adjoint*) of $T \in \mathscr{B}(E, F)$ is the operator $T' \in \mathscr{B}(F', E')$, defined by the formula

$$\langle x, T'\lambda \rangle = \langle Tx, \lambda \rangle \quad (x \in E, \lambda \in F');$$

of course, $\|T'\| = \|T\|$ and $T' : F' \to E'$ is weak*-weak*-continuous. Using dual maps, it is easy to see that $E' \sim F'$ and $E' \cong F'$ whenever $E \sim F$ and $E \cong F$, respectively. Suppose that there is a bounded linear surjection T from a Grothendieck space onto a Banach space E. Then, by consideration of T' and T'', it is easily seen that E is also a Grothendieck space.

The following standard results are given in [6, §3.16] and [183, pp. 287–293].

Proposition 2.2.12. *Let E and F be Banach spaces, and take $T \in \mathscr{B}(E,F)$. Then:*

(i) *T is a surjection if and only if $T' : F' \to E'$ is an embedding if and only if there exists $c > 0$ such that $\|T'\lambda\| \ge c\|\lambda\|$ $(\lambda \in E)$;*

(ii) *T is an injection if and only if $T'(F')$ is weak*-dense in E';*

(iii) *T is an injection with closed range if and only if T' is a surjection;*

(iv) *T is a bijection if and only if T' is a bijection.* □

Proposition 2.2.13. *Let E and F be Banach spaces. Then each weak*-weak*-continuous operator from F' to E' has the form T' for some operator $T \in \mathscr{B}(E,F)$.* □

Proposition 2.2.14. *Let E be a normed space, and take B to be the weak*-compact space $E'_{[1]}$.*

(i) *The map*

$$J : x \mapsto \kappa_E(x) \mid B, \quad E \to C(B) \subset \ell^\infty(B),$$

is an isometric embedding.

(ii) *Suppose that S is a weak*-closed, circled subspace of B with $\overline{\mathrm{co}}(S) = B$. Then the map*

$$J : x \mapsto \kappa_E(x) \mid S, \quad E \to C(S),$$

is an isometric embedding.

Proof. (i) Clearly $Jx \in C(B)$ $(x \in E)$, and the map $J : E \to C(B)$ is linear. Further,

$$|Jx|_B = \sup\{|\langle \kappa_E(x), \lambda\rangle| : \lambda \in B\} = \|x\| \quad (x \in E),$$

and so T is an isometric embedding of E into $C(B)$.

(ii) This follows from Corollary 2.1.3. □

Definition 2.2.15. Let E and F be normed spaces, and take $T \in \mathscr{B}(E,F)$. Then T is a *quotient operator* if T maps the open unit ball in E onto the open unit ball in F.

Proposition 2.2.16. *Let E and F be normed spaces, and take $T \in \mathscr{B}(E,F)$. Then T is a quotient operator if and only if the induced operator $\overline{T} : E/\ker T \to T(E)$ is an isometric embedding into F.*

Proof. Certainly $\overline{T}$ is always a bounded operator with $\|\overline{T}\| \le \|T\|$.

It is clear that $\overline{T}$ is an isometric isomorphism onto $T(E)$ whenever T is a quotient operator.

Now suppose that $\overline{T}$ is an isometric isomorphism onto $T(E)$. Then $\|Tx\| < 1$ when $\|x\| < 1$ in E. Take $y \in T(E)$ with $\|y\| < 1$. Then there exists $z \in E/F$ with $\|z\| < 1$ and $\overline{T}z = y$, and there exists $x \in E$ with $\|x\| < 1$ and $x + F = z$. We have $Tx = y$, and so T is a quotient operator. □

Each Banach space E is a quotient of a space $\ell^1(\Gamma)$ for some index set Γ. Indeed, we can take $\Gamma = E_{[1]}$ and define the map

$$\sum \alpha_\gamma \delta_\gamma \mapsto \sum \alpha_\gamma \gamma, \quad \ell^1(\Gamma) \to E.$$

Proposition 2.2.17. *Let E be a separable Banach space. Then:*

(i) *there is an isometric embedding of E into* ℓ^∞*;*

(ii) *there is a quotient operator from* ℓ^1 *onto E;*

(iii) *there is an isometric embedding of* E' *into* ℓ^∞.

Proof. (i) Let $S = \{x_n : n \in \mathbb{N}\}$ be a dense subset of S_E. For each $n \in \mathbb{N}$, choose $\lambda_n \in E'$ with $\langle x_n, \lambda_n \rangle = \|\lambda_n\| = 1$. Then the map

$$T : x \mapsto (\langle x, \lambda_n \rangle), \quad E \to \ell^\infty.$$

is an isometric embedding.

(ii) Let $S = \{x_n : n \in \mathbb{N}\}$ be a dense subset of $E_{[1]}$. We define

$$T : (\alpha_n) \mapsto \sum_{n=1}^{\infty} \alpha_n x_n, \quad \ell^1 \to E.$$

Then clearly T is a linear contraction with $T\delta_n = x_n$ $(n \in \mathbb{N})$.

Now take $x \in E$ with $0 < \|x\| < 1$, say $\|x\| = \eta$, choose $\varepsilon > 0$ with $\varepsilon < 1 - \eta$, and set $y = x/\eta$, so that $y \in S_E$. First choose $n_1 \in \mathbb{N}$ such that $\|y - x_{n_1}\| < \varepsilon$, and then inductively choose a strictly increasing sequence (n_k) in $\mathbb{N}$ such that

$$\left\| y - \left(\sum_{j=1}^{k} \varepsilon^{j-1} x_{n_j} \right) \right\| < \varepsilon^k \quad (k \in \mathbb{N}).$$

Set $\alpha = \sum_{j=1}^{\infty} \varepsilon^{j-1} \delta_{n_j}$, so that $\alpha \in \ell^1$ with $\|\alpha\|_1 = (1-\varepsilon)^{-1}$ and

$$y = \sum_{j=1}^{\infty} \varepsilon^{j-1} x_{n_j} = T\alpha \in T(\ell^1).$$

Thus $x = T(\eta\alpha)$ with $\|\eta\alpha\|_1 < 1$. This shows that $T : \ell^1 \to E$ is a quotient operator.

(iii) By Proposition 2.2.16, E is isometrically isomorphic to $\ell^1/\ker T$, and then E' is isometrically isomorphic to $(\ker T)^\circ$, a closed subspace of ℓ^∞. □

The following notion, which may be new, will be useful in answering questions about when a certain Banach space is not embedded in another.

Definition 2.2.18. Let E be a normed space. Then a subset S of E has *bounded finite sums* if there is a constant $M > 0$ such that

$$\left\|\sum\{x : x \in F\}\right\| \leq M$$

for each finite subset F of S.

The following is clear.

Proposition 2.2.19. *Let E and F be normed spaces. Suppose that E has a subset of cardinality κ that has bounded finite sums, and suppose that there is an embedding of E into F. Then F has a subset of cardinality κ that has bounded finite sums.* □

Example 2.2.20. Let K be a non-empty, separable, locally compact space. Then each subset of $C_0(K)$ that has bounded finite sums is countable.

Indeed, take $\{x_n : n \in \mathbb{N}\}$ to be a dense subset of K, and take S to be a subset of $C_0(K)$ that has bounded finite sums. For each $n \in \mathbb{N}$, the set $\{f \in S : f(x_n) \neq 0\}$ is countable, and so

$$\bigcup\{\{f \in S : f(x_n) \neq 0\} : n \in \mathbb{N}\}$$

is countable. But the above set is $S \setminus \{0\}$, and so S is countable. □

Example 2.2.21. Let S be an infinite set. By Proposition 1.5.5, there is a family $\{S_\alpha^* : \alpha \in A\}$ of non-empty, pairwise-disjoint, clopen subsets of S^*, where $|A| = \mathfrak{c}$. The family of characteristic functions in $C(S^*)$ of the sets S_α^* has cardinality $\mathfrak{c}$ and has bounded finite sums. □

Example 2.2.22. It is immediate from the above two examples and Proposition 2.2.19 that there is no embedding of $C(\mathbb{N}^*) \cong \ell^\infty/c_0$ into $C(\beta\mathbb{N}) \cong \ell^\infty$. A stronger result will be given in Corollary 2.2.25. □

Theorem 2.2.23. *Let K be a non-empty, locally compact space. Then the following conditions on K are equivalent:*

(a) *K does not satisfy* CCC;

(b) *there is an uncountable subset of $C_0(K)$ with bounded finite sums;*

(c) *there is an uncountable set Γ such that $c_0(\Gamma)$ embeds into $C_0(K)$.*

Proof. (a) $\Rightarrow$ (b), (c) Let $\{U_\gamma : \gamma \in \Gamma\}$ be a pairwise-disjoint family of non-empty, open subsets of K such that each $\overline{U_\gamma}$ is compact. For each $\gamma \in \Gamma$, choose $f_\gamma \in C_0(K, \mathbb{I})$ such that $|f_\gamma|_K = 1$ and $\operatorname{supp} f_\gamma \subset U_\gamma$. Then $\{f_\gamma : \gamma \in \Gamma\}$ has bounded finite sums, and, by (a), the set Γ is uncountable, giving (b).

Let $\alpha = (\alpha_\gamma : \gamma \in \Gamma)$ be an element of $c_0(\Gamma)$, and set

$$T\alpha = \sum \{\alpha_\gamma f_\gamma : \gamma \in \Gamma\}.$$

Then $T\alpha \in C_0(K)$, and $T : c_0(\Gamma) \to C_0(K)$, is a linear isometry that identifies $c_0(\Gamma)$ with the closed subspace $\overline{\operatorname{lin}}\{f_\gamma : \gamma \in \Gamma\}$ of $C_0(K)$, giving (c).

(c) $\Rightarrow$ (b) The family $\{\chi_{\{\gamma\}} : \gamma \in \Gamma\}$ has bounded finite sums, and so (b) follows from Proposition 2.2.19.

(b) $\Rightarrow$ (a) Let $S := \{f_\gamma : \gamma \in \Gamma\}$ be a family with bounded finite sums, where Γ is an uncountable index set and $f_\gamma \neq f_\delta$ when $\gamma, \delta \in \Gamma$ with $\gamma \neq \delta$. By replacing the set S by $\{\Re f_\gamma : \gamma \in \Gamma\}$ or $\{\Im f_\gamma : \gamma \in \Gamma\}$, we may suppose that $S \subset C_{0,\mathbb{R}}(K)$. There exists $\eta > 0$ such that $\{f \in S : |f|_K > \eta\}$ is uncountable, and so we may suppose, by passing to a subset of Γ and scaling, that $|f|_K > 1$, and in fact that $\sup\{f(x) : x \in K\} > 1$, for each $f \in S$.

For each $\gamma \in \Gamma$, set

$$U_\gamma = \{x \in K : f_\gamma(x) > 1\},$$

so that U_γ is a non-empty, open set in K. The assumption that S has bounded finite sums implies that there exists $M \in \mathbb{N}$ such that the intersection of any family of M of the sets U_γ is empty. [1]

We shall inductively define a certain family $\{W_\alpha : \alpha < \omega_1\}$ of pairwise-disjoint, non-empty, open subsets of K to satisfy the following properties for each $\alpha < \omega_1$:

(i) for each $\beta < \alpha$, there is a finite subset Γ_β of Γ such that $W_\beta = \bigcap\{U_\gamma : \gamma \in \Gamma_\beta\}$;

(ii) for each $\beta < \alpha$, we have $W_\beta \cap U_\gamma = \emptyset$ $(\gamma \in \Gamma \setminus \Gamma_\beta)$;

(iii) for each $\beta_1, \beta_2 < \alpha$ with $\beta_1 \neq \beta_2$, we have $\Gamma_{\beta_1} \cap \Gamma_{\beta_2} = \emptyset$.

First choose a subset Γ_1 of Γ to be maximal with respect to the property that $\bigcap\{U_\gamma : \gamma \in \Gamma_1\} \neq \emptyset$, and set $W_1 = \bigcap\{U_\gamma : \gamma \in \Gamma_1\}$. We observe that $|\Gamma_1| < M$ and that $W_1 \cap U_\gamma = \emptyset$ $(\gamma \in \Gamma \setminus \Gamma_1)$.

Now take $\alpha < \omega_1$, and assume that we have defined W_β for each $\beta < \alpha$ such that (i), (ii), and (iii) hold. We observe that $\Gamma \setminus \bigcup\{\Gamma_\beta : \beta < \alpha\}$ is uncountable; we then choose $\Gamma_\alpha \subset \Gamma \setminus \bigcup\{\Gamma_\beta : \beta < \alpha\}$ to be maximal with respect to the property that $\bigcap\{U_\gamma : \gamma \in \Gamma_\alpha\} \neq \emptyset$, and set $W_\alpha = \bigcap\{U_\gamma : \gamma \in \Gamma_\alpha\}$. This continues the inductive construction.

In this way, we obtain a family of cardinality $\aleph_1$ of non-empty, pairwise-disjoint, open subsets of K. Thus (a) holds. □

Corollary 2.2.24. *Let K and L be two non-empty, locally compact spaces. Suppose that $C_0(K)$ embeds in $C_0(L)$ and that L satisfies* CCC. *Then K satisfies* CCC.

[1] An immediate contradiction can be obtained at this point by an appeal to a lemma of Rosenthal (see [131, Proposition 7.21]); we provide a somewhat simpler, self-contained argument here.

Proof. This follows from Proposition 2.2.19 and Theorem 2.2.23. □

For more comprehensive versions of Corollary 2.2.24, see [131, Theorem 7.22] and [211, Theorem 4.6].

Corollary 2.2.25. *Let S be an infinite set, and let L be a compact space that satisfies* CCC. *Then there is no embedding of $C(S^*)$ into $C(L)$.*

Proof. By Proposition 1.5.5, S^* does not satisfy CCC. Thus the claim follows immediately from Corollary 2.2.24. □

Definition 2.2.26. Let X be a topological space. Then the *Souslin number* of X is the minimum cardinal number κ such that every family of non-empty, pairwise-disjoint, open subsets of X has cardinality at most κ; it is denoted by $c(X)$.

Thus X satisfies CCC if and only if $c(X) \leq \aleph_0$.

An easy modification of the above argument shows that $c(K) = c(L)$ whenever K and L are two non-empty, locally compact spaces with $C_0(K) \sim C_0(L)$, and so $c(K)$ is an isomorphic invariant of the spaces $C_0(K)$; for further isomorphic invariants of these spaces, see §6.1.

We now introduce a definition that encapsulates a key theme of this work.

Definition 2.2.27. Let E be a Banach space. Then a Banach space F is an *isometric predual* of E if $E \cong F'$ and an *isomorphic predual* of E if $E \sim F'$. Similarly, a Banach space F is an *isometric pre-bidual* of E if $E \cong F''$ and an *isomorphic pre-bidual* of E if $E \sim F''$. We say that E is *isomorphically/isometrically a (bi) dual space* if E has an isomorphic/isometric pre-(bi)dual.

It will be apparent through several later examples that a Banach space E might have many isomorphic preduals, but no isometric preduals. In fact, there is a general result of this nature, due to Davis and Johnson [79]. Let $(E, \|\cdot\|)$ be a Banach space that is not reflexive. Then there is a norm $|||\cdot|||$ on E that is equivalent to $\|\cdot\|$ and such that $(E, |||\cdot|||)$ is not isometrically a dual space. Thus, let F be a non-reflexive Banach space, and set $E = F'$. Then there is a norm $|||\cdot|||$ on E such that F is an isomorphic predual, but not an isometric predual, of $(E, |||\cdot|||)$.

Theorem 2.2.28. *Let E be a Banach space.*

(i) *The space E is isometrically a dual space if and only if there is a topology τ on E such that (E, τ) is a locally convex space and $(E_{[1]}, \tau)$ is compact.*

(ii) *The space E is isomorphically a dual space if and only if there is a topology τ on E such that (E, τ) is a locally convex space and for which (B, τ) is compact for some barrel B in E.*

Proof. (i) Suppose that $E \cong F'$ for some Banach space F. Then we take τ to be the topology $\sigma(E,F)$.

Conversely, suppose that there is a topology τ as specified, and let $F \subset E'$ consist of the τ-continuous functionals on E, so that F is a closed subspace of E'. There is a natural mapping $j : E \to F'$ defined by

$$\langle \lambda, j(x) \rangle = \langle x, \lambda \rangle \quad (x \in E, \lambda \in F).$$

Clearly j is injective and continuous, with $\|j(x)\| \leq \|x\|$ $(x \in E)$. Furthermore, j is continuous from (E,τ) to $(F', \sigma(F',F))$, and so $j(E_{[1]})$ is a $\sigma(F',F)$-compact subset of $F'_{[1]}$. It follows immediately from Theorem 2.1.2(iii) that $j(E_{[1]}) = F'_{[1]}$. We now see that $j : E \to F'$ is an isometry.

(ii) This follows easily from (i). □

We now consider the uniqueness of isometric preduals.

Definition 2.2.29. Let E be a Banach space with an isometric predual F. Then F is *unique* if, whenever G is also an isometric predual of E, it follows that $F \cong G$. The unique predual of a Banach space E is denoted by E_* when it exists. Further, F is *strongly unique* if, whenever G is also a Banach space and $T : E \to G'$ is an isometric isomorphism, the map $T' : G'' \to F'' = E'$ carries $\kappa_G(G)$ onto $\kappa_F(F)$.

A predual F of a Banach space E is strongly unique if and only if the above map $T : F' \to G'$ is weak*-weak*-continuous. Thus E has a unique predual whenever it has a strongly unique predual. All known examples of Banach spaces with a unique predual actually have a strongly unique predual.

For a fine survey concerning the existence and uniqueness of isometric preduals of Banach spaces, including a discussion of strongly unique preduals, see [115]; see also [50]. The definition of 'E has a strongly unique predual' in [115] is that there is a unique bounded projection $\pi : E''' \to E'$ with $\|\pi\| = 1$ and such that $\ker \pi$ is weak*-closed; as noted in [50], this is equivalent to our definition.

It is certainly not the case that every Banach space that is isometrically a dual space has a unique predual; for example, we shall discuss the many isometric preduals of the Banach space ℓ^1 in §6.3.

We continue this section with a representation theorem for the bidual E'' of a Banach space E that we shall use later.

Let E be a Banach space. We shall suppose that there is a subset S of the unit sphere $S_{E'}$ of E' such that, for each $\mu \in S$, there is a closed subspace F_μ of E' and that the family $\{F_\mu : \mu \in S\}$ of these subspaces has the property that

$$\|\Lambda\| = \sup_{\mu \in S}\{|\langle \Lambda, y \rangle| : y \in (F_\mu)_{[1]}\} \quad (\Lambda \in E''). \tag{2.3}$$

Set

$$F = \bigoplus_1 \{F_\mu : \mu \in S\}, \quad \text{so that} \quad F' = \bigoplus_\infty \{F'_\mu : \mu \in S\}. \tag{2.4}$$

Thus

$$F' = \{(\lambda_\mu) = (\lambda_\mu : \mu \in S) : \lambda_\mu \in F'_\mu,\ \sup\|\lambda_\mu\| < \infty\},$$

with $\|(\lambda_\mu)\|_\infty = \sup\{\|\lambda_\mu\| : \mu \in S\}$. For each $x \in E$ and $\mu \in S$, define $x_\mu \in F'_\mu$ by

$$\langle y, x_\mu \rangle = \langle y, x \rangle \quad (y \in F_\mu).$$

Then it is easy to see that the map $x \mapsto (x_\mu),\ E \to F'$, is a linear isometry. We shall extend this map to a representation of E''.

Suppose, further, that E has an isometric linear involution $*$, so that $*$ induces an isometric linear involution on E', and that each F_μ is a $*$-closed subspace of E'. Then we define a linear involution on F coordinatewise, and hence obtain an isometric linear involution on F; in turn we obtain an isometric linear involution on F'. Clearly, the map $x \mapsto (x_\mu),\ E \to F'$, is $*$-linear and each F'_μ is a $*$-closed subspace of F'.

Theorem 2.2.30. *Let E be a Banach space, and let S, $\{F_\mu : \mu \in S\}$, and F be as above. For each $\Lambda \in E''$ and $\mu \in S$, define Λ_μ on F_μ by*

$$\Lambda_\mu(y) = \langle \Lambda, y \rangle \quad (y \in F_\mu).$$

Then $\Lambda_\mu \in F'_\mu$ with $\|\Lambda_\mu\| \leq \|\Lambda\|$. Further, $(\Lambda_\mu : \mu \in S) \in F'$ with $\|(\Lambda_\mu)\|_\infty = \|\Lambda\|$. The map

$$T : \Lambda \mapsto (\Lambda_\mu : \mu \in S), \quad E'' \to F',$$

is a linear isometry, and $T : (E'', \sigma(E'', E')) \to (F', \sigma(F', F))$ is continuous.

Suppose, further, that E has an isometric linear involution $*$ and that F_μ is a $*$-closed subspace of E'. Then $T : E'' \to F'$ is $*$-linear.

Proof. It is clear that, for each $\mu \in S$, we have $\Lambda_\mu \in F_\mu$ with $\|\Lambda_\mu\| \leq \|\Lambda\|$, and so $(\Lambda_\mu : \mu \in S) \in F'$ with $\|(\Lambda_\mu)\|_\infty \leq \|\Lambda\|$. By (2.3), for each $\varepsilon > 0$, there exist $\nu \in S$ and $y \in (F_\nu)_{[1]}$ with $|\langle \Lambda, y \rangle| \geq \|\Lambda\| - \varepsilon$, and so

$$\|(\Lambda_\mu)\|_\infty \geq \|\Lambda_\nu\| \geq \|\Lambda\| - \varepsilon.$$

This holds for each $\varepsilon > 0$, and hence $\|(\Lambda_\mu)\|_\infty \geq \|\Lambda\|$. Thus $T : E'' \to F'$ is a linear isometry.

Let $\Lambda_\alpha \to 0$ in $(E'', \sigma(E'', E'))$. Then

$$\langle (\Lambda_\alpha)_\mu, y \rangle = \langle \Lambda_\alpha, y \rangle \to 0 \quad (y \in F_\mu)$$

for each $\mu \in S$, and so $T(\Lambda_\alpha) \to 0$ in $(F', \sigma(F', F))$. This shows that the linear map $T : (E'', \sigma(E'', E')) \to (F', \sigma(F', F))$ is continuous.

We check immediately that T is $*$-linear in the case where the further hypotheses hold. □

Finally in this section we give a technical result that will be used later; see Theorem 4.6.8.

Proposition 2.2.31. *Let G be a Banach space, and let F be a separable Banach space. Suppose that J is an uncountable set and that, for each $j \in J$, L_j is a Banach space which does not embed into G. Then there is no embedding of the Banach space*

$$\bigoplus_1 \{L_j : j \in J\}$$

into $F \oplus_1 G$.

Proof. Set $E = \bigoplus_1 \{L_j : j \in J\}$ and regard each L_j as a closed subspace of E. Set $H = F \oplus_1 G$, with F and G as stated in the theorem, and take $\pi_F : H \to F$ and $\pi_G : H \to G$ to be the associated bounded projections, so that

$$y = \pi_F(y) + \pi_G(y) \quad (y \in H).$$

Assume towards a contradiction that there is an embedding $T : E \to H$. Then there exists $\delta > 0$ such that $\|Tx\| \geq \delta \|x\|$ $(x \in E)$. For each $j \in J$, there exists $x_j \in L_j$ with $\|x_j\| = 1$ such that $\|\pi_G(Tx_j)\| < \delta/2$. Indeed, otherwise there exists $j \in J$ such that $\|\pi_G(Tx)\| \geq \delta \|x\|/2$ $(x \in L_j)$, so that $(\pi_G \circ T)(L_j)$ is a subspace of G that is isomorphic to L_j, contradicting the assumption on L_j.

For each $i, j \in J$ with $i \neq j$, we have

$$\|Tx_i - Tx_j\| \geq \delta \|x_i - x_j\| = \delta(\|x_i\| + \|x_j\|) = 2\delta.$$

Now

$$\|y\| = \|\pi_F(y) + \pi_G(y)\| \leq \|\pi_F(y)\| + \|\pi_G(y)\| \quad (y \in H),$$

and so (taking $y = Tx_i - Tx_j$), we have

$$\|\pi_F(Tx_i) - \pi_F(Tx_j)\| \geq \|Tx_i - Tx_j\| - \|\pi_G(Tx_i) - \pi_G(Tx_j)\| > 2\delta - 2(\delta/2) = \delta.$$

Thus there is an uncountable family of mutually disjoint balls in F, contradicting the hypothesis that F is separable. □

2.3 Banach lattices

We shall require some basic notions in the theory of Banach lattices; for much more on Banach lattices, see [1, 174, 184, 223], for example.

Definition 2.3.1. Let $(E, \leq)$ be a Riesz space. A norm $\|\cdot\|$ on E is a *lattice norm* if $\|x\| \leq \|y\|$ whenever $x, y \in E$ with $|x| \leq |y|$. A *normed Riesz space* is a Riesz space equipped with a lattice norm. A *real Banach lattice* is a normed Riesz space which is a Banach space with respect to the norm.

For example, the spaces $L^p_{\mathbb{R}}(\mathbb{I})$ for $1 \le p \le \infty$ and the spaces $C_{0,\mathbb{R}}(K)$ for a non-empty, locally compact space K are real Banach lattices with respect to the pointwise lattice operations and the specified norm.

We recall that a linear subspace F of a real Banach lattice E is a sublattice if $x \vee y, x \wedge y \in F$ whenever $x, y \in F$ and a lattice ideal if $x \in F$ whenever $x \in E$ and $|x| \le |y|$ for some $y \in F$.

Suppose that E is a linear space such that $E = E_{\mathbb{R}} \oplus \mathrm{i}E_{\mathbb{R}}$ for a real Banach lattice $(E_{\mathbb{R}}, \|\cdot\|)$, so that E, a linear space over the field $\mathbb{C}$, is a complex Riesz space. Then we make the following definitions. First, set $E^+ = E^+_{\mathbb{R}}$ and

$$E^+_{[r]} = E_{[r]} \cap E^+ \quad (r > 0).$$

Take $z \in E$, say $z = x + \mathrm{i}y$, where $x, y \in E_{\mathbb{R}}$, and define the *modulus* $|z| \in E^+$ of z by

$$|z| = \left(|x|^2 + |y|^2\right)^{1/2} \tag{2.5}$$

(the right-hand side of (2.5) is well defined in $E_{\mathbb{R}}$ by the 'Youdine–Krivine functional calculus'). Alternatively, we can set

$$|z| = |x + \mathrm{i}y| = \sup\{x \cos\theta + y \sin\theta : 0 \le \theta \le 2\pi\}; \tag{2.6}$$

the supremum always exists in E^+ and the two definitions of $|z|$ are consistent. We then define

$$\|z\| = \| \, |z| \, \| \quad (z \in E). \tag{2.7}$$

We see that $\|\cdot\|$ is a norm on E and that $(E, \|\cdot\|)$ is a Banach space. This complexification of a real Banach lattice is defined to be a (complex) *Banach lattice*.

For example, the spaces $L^p(\mathbb{I})$ for $1 \le p \le \infty$ and the spaces $C_0(K)$ for a non-empty, locally compact space K are Banach lattices which are the complexifications of the analogous real Banach lattices.

Again, a linear subspace F of a Banach lattice E is a *lattice ideal* if $x \in F$ whenever $x \in E$ and $|x| \le |y|$ for some $y \in F$.

Let $\{E_\alpha : \alpha \in A\}$ be a family of Banach lattices, and take p with $1 \le p \le \infty$. Then the Banach space $\bigoplus_p \{E_\alpha : \alpha \in A\}$ is also a Banach lattice for the obvious operations.

For details of these remarks, including a discussion of the Youdine–Krivine functional calculus, see [1, §3.2], [73], [176, §1.d], [180], [184, §2.2], [223, Chapter II, §11], and [245, §13].

Definition 2.3.2. A Banach lattice is *Dedekind complete* (respectively, *Dedekind σ-complete*) if it is Dedekind complete (respectively, Dedekind σ-complete) as a complex Riesz space.

Clearly, to show that a Banach lattice E is Dedekind complete, it suffices to show that each increasing net in E^+ that is bounded above has a supremum.

The following well-known theorem is proved in [68, Proposition 4.2.29(i)].

Theorem 2.3.3. *Let K be a non-empty, compact space. Then K is Stonean if and only if $C(K)$ is Dedekind complete, and K is basically disconnected if and only if $C(K)$ is Dedekind σ-complete.*

Proof. Suppose that $C_{\mathbb{R}}(K)$ is Dedekind complete, and let U be an open set in K. Take $\mathscr{F}$ to be the family of functions $f \in C_{\mathbb{R}}(K)$ such that $f(x) = 0$ $(x \in K \setminus U)$ and $0 \le f \le 1$. Then $\mathscr{F}$ has a supremum, say $f_0 \in C_{\mathbb{R}}(K)$. Clearly $f_0(x) = 1$ $(x \in U)$ and $f_0(x) = 0$ $(x \in K \setminus \overline{U})$, and so $f_0 = \chi_{\overline{U}}$. Thus $\overline{U}$ is open. This shows that K is Stonean.

Conversely, suppose that K is Stonean, and let $\mathscr{F}$ be a family in $C(K)^+$ which is bounded above, say by 1. For $r \in \mathbb{I}$, define

$$U_r = \bigcup\{\{x \in K : f(x) > r\} : f \in \mathscr{F}\}.$$

Then U_r is open in K, and so $V_r := \overline{U_r}$ is also open in K. Clearly $V_1 = \emptyset$. Define

$$g(x) = \sup\{r \in \mathbb{I} : x \in U_r\} \in \mathbb{I}.$$

If $g(x) \in (r,s)$, then $x \in V_r \setminus V_s$, and, if $x \in V_r \setminus V_s$, then $g(x) \in [r,s]$.

Take $x_0 \in K$, and take a neighbourhood V of $g(x_0)$. Then there exist $r,s \in \mathbb{R}$ with $g(x_0) \in (r,s) \subset [r,s] \subset V$. Since $V_r \setminus V_s$ is an open set and

$$x_0 \in V_r \setminus V_s \subset g^{-1}([r,s]) \subset g^{-1}(V),$$

we see that g is continuous at x_0. Thus $g \in C_{\mathbb{R}}(K)$.

Now take $h \in C_{\mathbb{R}}(K)$ with $h \ge f$ $(f \in \mathscr{F})$. Assume that there exists $x_0 \in K$ with $h(x_0) < g(x_0)$. Then $h(x_0) < r$ for some r with $x_0 \in V_r$. Let W be a neighbourhood of x_0 with $h(x) < r$ $(x \in W)$. Then there exists $x \in W$ with $f(x) > r$ for some $f \in \mathscr{F}$, a contradiction. Thus $h \ge g$, and so $g = \sup \mathscr{F}$. We have shown that $C_{\mathbb{R}}(K)$ is Dedekind complete.

The proof that K is basically disconnected if and only if $C_{\mathbb{R}}(K)$ is Dedekind σ-complete is a small variation of the above. □

In fact, the term 'Stonean' was used first by Dixmier in the seminal work [91], where a Stonean space was defined to be a compact space K such that $(C_{\mathbb{R}}(K), \le)$ is Dedekind complete.

The following, related theorem was proved by Seever in [224]; this paper is based on his thesis written under the direction of William Bade. In the proof, we shall use the notation $\prec$ from page 24. See also [225, Theorem 24.7.5].

Theorem 2.3.4. *Let K be a non-empty, compact space. Then K is an F-space if and only if, whenever (f_n) and (g_n) are sequences in $C_{\mathbb{R}}(K)$ with $f_m \le g_n$ $(m,n \in \mathbb{N})$, there exists $f \in C_{\mathbb{R}}(K)$ with $f_m \le f \le g_n$ $(m,n \in \mathbb{N})$.*

Proof. Suppose that $C_{\mathbb{R}}(K)$ has the stated property, and take disjoint cozero sets U and V, say

$$U = \{x \in K : f(x) > 0\} \quad \text{and} \quad V = \{x \in K : g(x) > 0\},$$

where $f, g \in C(K, \mathbb{I})$. For $n \in \mathbb{N}$, set $f_n = 1 \wedge nf$ and $g_n = (1-g)^n$. For $m, n \in \mathbb{N}$, we have $f_m \leq g_n$ in $C(K)^+$, and so there exists $h \in C(K)^+$ with $f_m \leq h \leq g_n$ $(m, n \in \mathbb{N})$. Clearly $h(x) = 1$ $(x \in U)$ and $h(x) = 0$ $(x \in V)$, and so K is an F-space.

Conversely, suppose that K is an F-space, and take (f_n) and (g_n) to be as specified; we may suppose that

$$0 < f_n(x) \leq f_{n+1}(x) \leq g_{n+1}(x) \leq g_n(x) < 1 \quad (x \in K, n \in \mathbb{N}).$$

Let D be the set of dyadic rationals in $[0, 1]$, and, for $r \in D$, define

$$U(r) = \bigcup \{\{x \in K : f_n(x) > r\} : n \in \mathbb{N}\}, \quad V(r) = \bigcup \{\{x \in K : g_n(x) < r\} : n \in \mathbb{N}\},$$

so that $U(r)$ and $V(r)$ are disjoint cozero sets, and so $\overline{U(r)} \cap \overline{V(r)} = \emptyset$ because K is an F-space. Further, $U(r) \supset U(s)$ and $V(r) \subset V(s)$ when $r, s \in D$ with $r < s$, and $U(1) = V(0) = \emptyset$.

We *claim* that there exist cozero sets $W(r)$ in K for $r \in D$ such that:

(i) $V(r) \prec W(r) \prec K \setminus U(r)$ for $r \in D$;

(ii) $W(r) \prec W(s)$ for $r, s \in D$ with $r < s$.

Indeed, start with $W(0) = \emptyset$ and $W(1) = K$. Now take $n \in \mathbb{N}$, and assume inductively that the sets $W(k/2^n)$ have been defined for $k = 0, \dots, 2^n$. Take $k \in \{0, \dots, 2^n - 1\}$. We have

$$V((2k+1)/2^{n+1}) \subset V((k+1)/2^n) \prec W((k+1)/2^n)$$

and $W(k/2^n) \prec W((k+1)/2^n)$, and so

$$V((2k+1)/2^{n+1}) \cup W(k/2^n) \prec W((k+1)/2^n).$$

Also, $V((2k+1)/2^{n+1}) \prec K \setminus U((2k+1)/2^{n+1})$ and

$$W(k/2^n) \prec K \setminus U(k/2^n) \subset K \setminus U((2k+1)/2^{n+1}),$$

and so

$$V((2k+1)/2^{n+1}) \cup W(k/2^n) \prec K \setminus U((2k+1)/2^{n+1}).$$

Thus

$$V((2k+1)/2^{n+1}) \cup W(k/2^n) \prec W((k+1)/2^n) \cap (K \setminus U((2k+1)/2^{n+1})).$$

By a remark on page 24, there is a cozero set U with

$$V((2k+1)/2^{n+1}) \cup W(k/2^n) \prec U \prec W((k+1)/2^n) \cap (K \setminus U((2k+1)/2^{n+1}));$$

we take $W((2k+1)/2^{n+1})$ to be this set U. This completes the definition of the sets $W(k/2^{n+1})$ for $k = 0, \dots, 2^{n+1}$. We see that the recursion continues.

Now define $f(x) = \inf\{r \in D : x \in W_r\}$ $(x \in K)$. As in Theorem 2.3.3, $f \in C(K)^+$.

Fix $n \in \mathbb{N}$. For each $x \in K$ and $\varepsilon > 0$, choose $r, s \in D$ with

$$f(x) - \varepsilon < r < f(x) < s < f(x) + \varepsilon .$$

Then $x \notin W(r)$. By (i), $x \notin V(r)$, and so $g_n(x) \geq r$. Hence $g_n(x) \geq f(x) - \varepsilon$. This holds true for each $\varepsilon > 0$, and so $g_n(x) \geq f(x)$, whence $f \leq g_n$. Similarly, $f \geq f_n$. Thus f has the required properties. □

Let E and F be Banach lattices that are the complexifications of the real Banach lattices $E_{\mathbb{R}}$ and $F_{\mathbb{R}}$, respectively. An operator $T \in \mathscr{B}(E,F)$ is a *Banach-lattice homomorphism* or *Banach-lattice isomorphism* if $T \mid E_{\mathbb{R}} : E_{\mathbb{R}} \to F_{\mathbb{R}}$ is a Riesz homomorphism or a Riesz isomorphism, respectively; it is a *Banach-lattice isometry* if, further, T is a linear isometry. The Banach lattices E and F are *Banach-lattice isomorphic* or *Banach-lattice isometric* if there is a Banach-lattice isomorphism or isometry, respectively, between them; an *isometric lattice embedding* is an isometric embedding that is a lattice homomorphism.

Let E and F be Banach lattices, and take $T \in \mathscr{B}(E,F)$. Then T is *positive* if $T(E^+) \subset F^+$. It is clear that an isomorphism $T \in \mathscr{B}(E,F)$ such that T and T^{-1} are positive operators is a lattice isomorphism from $E_{\mathbb{R}}$ onto $F_{\mathbb{R}}$, and so is a Banach-lattice isomorphism.

The following remark establishes some consistency in our terminology.

Proposition 2.3.5. *Let E and F be Banach lattices, and suppose that $T \in \mathscr{B}(E,F)$ is a Banach-lattice isomorphism such that $\|Tx\| = \|x\|$ $(x \in E^+)$. Then*

$$|Tz| = T(|z|) \quad (z \in E),$$

and $T : E \to F$ is a Banach-lattice isometry.

Proof. Take $z = x + \mathrm{i}y \in E$, where $x, y \in E_{\mathbb{R}}$, and set

$$S = \{x \cos\theta + y \sin\theta : 0 \leq \theta \leq 2\pi\} .$$

Then $T(S) = \{(Tx)\cos\theta + (Ty)\sin\theta : 0 \leq \theta \leq 2\pi\}$, and, by (2.6), $\sup S = |z|$ and $\sup T(S) = |Tz|$. Since T is a lattice isomorphism, $T(\sup S) = \sup T(S)$, i.e., $T(|z|) = |Tz|$. Hence $\|Tz\| = \|\,|Tz|\,\| = \|T(|z|)\| = \|\,|z|\,\| = \|z\|$, and so $T : E \to F$ is a linear isometry. □

Let E be a real Banach lattice, with dual space E'. Then E' is ordered by the requirement that $\lambda \in E'$ belongs to $(E')^+$ if and only if $\langle x, \lambda \rangle \geq 0$ $(x \in E^+)$ (*cf.* page 9). One checks easily that this ordering gives a lattice ordering, and so E' becomes a real Banach lattice. The equations that define the lattice operations are the following; they are called the *Riesz–Kantorovich formulae*. Take $\lambda, \mu \in E'$. Then $\langle x, \lambda \vee \mu \rangle$ and $\langle x, \lambda \wedge \mu \rangle$ are defined for $x \in E^+$ by

$$\begin{cases} \langle x, \lambda \vee \mu \rangle = \sup\{\langle y, \lambda \rangle + \langle z, \mu \rangle : y, z \in E^+, y + z = x\} , \\ \langle x, \lambda \wedge \mu \rangle = \inf\{\langle y, \lambda \rangle + \langle z, \mu \rangle : y, z \in E^+, y + z = x\} , \end{cases} \tag{2.8}$$

and then $\lambda \vee \mu$ and $\lambda \wedge \mu$ are extended linearly to all of E'. The dual of a real Banach lattice E is also a real Banach lattice for these operations; this is the *dual Banach lattice* of E.

It is standard that a dual Banach lattice is Dedekind complete. Indeed, let E be a Banach lattice, and take $\mathscr{F}$ to be a non-empty, bounded subset of $(E')^+$. Consider the net $\mathscr{G}$ of finite subsets of $\mathscr{F}$, and, for each $x \in E^+$, set

$$\lambda(x) = \lim \left\{ \left\langle x, \bigvee S \right\rangle : S \in \mathscr{G} \right\}.$$

Then λ is additive, homogeneous, and positive on E^+, and thus λ extends uniquely to an element, also λ, of E'. Clearly $\lambda = \bigvee\{F : F \in \mathscr{F}\}$, and so E' is Dedekind complete.

Let F be a real Banach lattice, and set $E = F \oplus \mathrm{i}F$, its complexification. Let λ be a continuous, real-linear functional on F. Then λ extends to a continuous, complex-linear functional on E: indeed, we define

$$\lambda(x + \mathrm{i}y) = \lambda(x) + \mathrm{i}\lambda(y) \quad (x, y \in F),$$

and so we may regard F' as a real-linear subspace of E'. For each λ in E', there exist λ_1 and λ_2 in F' such that

$$\lambda(x) = \lambda_1(x) + \mathrm{i}\lambda_2(x) \quad (x \in F),$$

and so E' is isomorphic as a complex Banach space to the complexification $F' \oplus \mathrm{i}F'$. In fact, this identification is isometric; the details of this are given in [1, Corollary 3.26] and [184, Proposition 2.2.6], for example. Thus we obtain the *dual Banach lattice* of a Banach lattice.

Let E be a Banach lattice, and take $\lambda \in E'$. Then clearly $(E')^+$ is weak*-closed in E'. We have

$$\|\lambda\| = \sup\{\langle x, \lambda \rangle : x \in E^+_{[1]}\} \quad (\lambda \in (E')^+).$$

Further, take $x \in E$. Then $x \in E^+$ if and only if $\langle x, \lambda \rangle \geq 0$ $(\lambda \in (E')^+)$; this follows from the Hahn–Banach theorem.

The bidual E'' of a Banach lattice E is also a Banach lattice, and the embedding $\kappa_E : E \to E''$ is an isometric lattice embedding. It also follows from the Hahn–Banach theorem that $E^+_{[1]}$ is weak*-dense in $(E'')^+_{[1]}$.

Proposition 2.3.6. *Let F be a real Banach lattice which is isometrically the dual of a real Banach space. Then the complexification of F is also isometrically a dual space.*

Proof. Set $E = F \oplus \mathrm{i}F$, the complexification of F; we recall that $E' = F' \oplus \mathrm{i}F'$.

Suppose that $F \cong G'$ for a real Banach space G, and regard G as a closed subspace of F'; set $H = G \oplus \mathrm{i}G$, so that H is a closed subspace of $E' \cong G'' \oplus \mathrm{i}G'' = H''$, and hence H is a Banach space. We shall show that $H' \cong E$, which will give the result.

Take $z \in E$, and set $\lambda(h) = \langle z, h \rangle$ $(h \in H)$. Then $\lambda \in H'$ with $\|\lambda\| \leq \|z\|$, and the map $S : z \mapsto \lambda$, $E \to H'$, is a linear contraction.

Take $\lambda \in H'$, and set $\lambda_1 = \Re\lambda \mid G$ and $\lambda_2 = \Im\lambda \mid G$, so that λ_1 and λ_2 are bounded, real-linear functionals on G with $\lambda = \lambda_1 + \mathrm{i}\lambda_2$. Thus there exist unique elements x and y in F such that $\lambda_1(g) = \langle x, g \rangle$ and $\lambda_2(g) = \langle y, g \rangle$ for $g \in G$. Set $z = x + \mathrm{i}y \in E$. Then, for each $g_1, g_2 \in G$, we have

$$\begin{aligned}\lambda(g_1 + \mathrm{i}g_2) &= (\lambda_1 + \mathrm{i}\lambda_2)(g_1 + \mathrm{i}g_2) = \langle x, g_1 \rangle - \langle y, g_2 \rangle + \mathrm{i}(\langle y, g_1 \rangle + \langle x, g_2 \rangle) \\ &= \langle x + \mathrm{i}y, g_1 + \mathrm{i}g_2 \rangle = \langle z, g_1 + \mathrm{i}g_2 \rangle\,,\end{aligned}$$

and so $\lambda = Sz$. Thus $S : E \to H'$ is a surjection.

Now fix $\varepsilon > 0$. Since H is weak*-dense in H'', there exists $h \in H$ with $\|h\| = 1$ and $|\langle z, h \rangle| > \|z\| - \varepsilon$, and hence $\|\lambda\| > \|z\| - \varepsilon$. This holds for each $\varepsilon > 0$, and so $\|\lambda\| \geq \|z\|$. We have shown that $S : E \to H'$ is an isometric isomorphism. □

A somewhat more general version of the above result is given in [187, Proposition 7]. We shall prove the converse of the above theorem in the special case where $F = C_{\mathbb{R}}(K)$ in Proposition 6.2.5; we do not know whether the converse holds in general.

Finally, we define some special types of Banach lattices.

Definition 2.3.7. A (real or complex) Banach lattice $(E, \|\cdot\|)$ is an *AL-space* (or *abstract L-space*) if

$$\|x + y\| = \|x\| + \|y\| \quad \text{whenever} \quad x, y \in E^+ \quad \text{with} \quad x \wedge y = 0\,,$$

and an *AM-space* (or *abstract M-space*) if

$$\|x \vee y\| = \max\{\|x\|, \|y\|\} \quad \text{whenever} \quad x, y \in E^+ \quad \text{with} \quad x \wedge y = 0\,.$$

For example, each space of the form $L^1(\Omega, \mu)$, where (Ω, μ) is a measure space, is an *AL*-space, and each space $C_0(K)$, where K is a non-empty, locally compact space, is an *AM*-space.

Let E be a Banach lattice. Then it is standard that E is an *AL*-space if and only if

$$\|x + y\| = \|x\| + \|y\| \quad (x, y \in E^+)\,,$$

and an *AM*-space if and only if

$$\|x \vee y\| = \max\{\|x\|, \|y\|\} \quad (x, y \in E^+)\,.$$

The following duality result is [5, Theorem 4.23] or [184, Proposition 1.4.7], for example.

Theorem 2.3.8. *Let E be a Banach lattice, with dual Banach lattice E'. Then E is an AL-space if and only if E' is an AM-space, and E is an AM-space if and only if E' is an AL-space.* □

The following central representation theorem is proved in [1, Theorems 3.5 and 3.6], [5, Theorems 4.27 and 4.29], [174, II. §1.b], and [184, Theorems 2.1.3 and 2.7.1], for example. The proofs are usually given for real Banach lattices, but the complex versions are valid; the technique for the complex version is illustrated in [1, Theorem 3.20]. We shall call this result '*Kakutani's theorem*'; detailed attributions for the various statements are given in [1].

An *AM-unit* in a Banach lattice E is an element $e \in E$ with $e > 0$ such that $E_{[1]} = \{x \in E : |x| \leq e\}$. Thus $\|x\| = \inf\{r \in \mathbb{R} : |x| \leq re\}$ for $x \in E$, and so E is an AM-space. An AM-unit is an order unit in the ordered linear space $(E_{\mathbb{R}}, \leq)$.

Theorem 2.3.9. (i) *A Banach lattice is an AL-space if and only if it is Banach-lattice isometric to a Banach lattice of the form $L^1(\Omega,\mu)$ for some measure space (Ω,μ).*

(ii) *A Banach lattice is an AM-space if and only if it is Banach-lattice isometric to a closed sublattice of a space $C(K)$ for some non-empty, compact space K.*

(iii) *A Banach lattice with an AM-unit is Banach-lattice isometric to a space $C(K)$ for some non-empty, compact space K.* □

2.4 Complemented subspaces of Banach spaces

We first define complemented subspaces of a normed space; earlier we defined complemented subspaces of a linear space.

Definition 2.4.1. Let E be a normed space. A closed subspace F of E is *complemented* in E if there is a closed subspace G of E such that $E = F \oplus G$.

In the case that a Banach space E is such that $E = F \oplus G$ for closed subspaces F and G, we have $E \sim F \times G$ and $E/F \sim G$.

It is elementary that finite-dimensional subspaces and subspaces of finite codimension in a normed space E are complemented in E, but we shall see soon that there are closed subspaces of a Banach space that are not complemented. It is remarkable that there is an infinite-dimensional Banach space E such that the *only* closed subspaces that are complemented in E are those that are either of finite dimension or of finite codimension; see page 195.

The following result is a standard consequence of the closed graph theorem.

Proposition 2.4.2. *Let E be a Banach space, and suppose that F and G are closed subspaces of E such that $E = F \oplus G$. Then there is a unique projection $P \in \mathscr{B}(E)$ with $P(E) = F$ and $(I_E - P)(E) = G$.* □

It follows immediately from the preservation of the Grothendieck property by bounded linear surjections (see page 59) that a closed, complemented subspace of a Grothendieck space is also a Grothendieck space.

Definition 2.4.3. A closed, complemented subspace F of a Banach space E is *λ-complemented* (for $\lambda \geq 1$) if there is a projection $P \in \mathscr{B}(E)$ with $P(E) = F$ and $\|P\| \leq \lambda$.

Thus a closed, complemented subspace of a Banach space is λ-complemented for some $\lambda \geq 1$.

Proposition 2.4.4. *Let E be a normed space. Then E' is 1-complemented in E'''.*

Proof. The required bounded projection from E''' to E' is the dual of the canonical injection $\kappa_E : E \to E''$; it is called the *Dixmier projection.* □

Corollary 2.4.5. *Let E be a Banach space such that E is isomorphically a dual space. Then E is complemented in E''.* □

Proposition 2.4.6. *Let K be an infinite, locally compact space.*

(i) *The space $C_0(K)$ contains a subspace that is isometrically isomorphic to c_0.*

(ii) *Suppose that K contains a convergent sequence of distinct points. Then $C_0(K)$ contains a 2-complemented subspace that is isometrically isomorphic to c_0.*

Proof. (i) The space K contains sequences (x_n) of distinct points and (U_n) of pairwise-disjoint, open subsets such that $x_n \in U_n$ and $\overline{U_n}$ is compact for each $n \in \mathbb{N}$. For each $n \in \mathbb{N}$, there exists $f_n \in C_0(K)^+$ with $f_n(x_n) = |f_n|_K = 1$ and $\operatorname{supp} f_n \subset U_n$. Essentially as in the proof of Theorem 2.2.23, (a) $\Rightarrow$ (c), set

$$T\alpha = \sum_{n=1}^{\infty} \alpha_n f_n \quad (\alpha = (\alpha_n) \in c_0).$$

Then $T : c_0 \to C_0(K)$ is a linear isometry that identifies c_0 with the closed subspace $\overline{\operatorname{lin}}\{f_n : n \in \mathbb{N}\}$ of $C_0(K)$.

(ii) Let (x_n) be a convergent sequence of distinct points in K, say $x_n \to x_0$ as $n \to \infty$; we may suppose that $x_n \neq x_0$ $(n \in \mathbb{N})$. Choose neighbourhoods of each x_n as in (i) such that $x_0 \notin U_n$ $(n \in \mathbb{N})$, and let (f_n) and $T : c_0 \to C_0(K)$ be as in (i), so that T is an isometric embedding.

For $g \in C_0(K)$, set

$$Pg = \sum_{n=1}^{\infty} (g(x_n) - g(x_0)) f_n .$$

Then $Pg \in C_0(K)$ $(g \in C_0(K))$ and $P \in \mathscr{B}(C_0(K))$ is a projection onto $T(c_0)$ with $\|P\| = 2$. Hence $T(c_0)$ is 2-complemented in $C_0(K)$. □

In fact, the following result concerning complemented copies of c_0 in $C(K)$–spaces is given in [184, Corollary 5.3.12 and Proposition 5.3.6], for example; Grothendieck spaces were defined in Definition 2.1.1.

Proposition 2.4.7. *Let K be a non-empty, compact space. Then the Banach space $C(K)$ is a Grothendieck space if and only if $C(K)$ contains no complemented subspace that is isomorphic to c_0.* □

Proposition 2.4.8. *The Banach space ℓ^1 is isometrically isomorphic to a 1-complemented subspace of $L^1(\mathbb{I})$.*

Proof. Let $\{I_n : n \in \mathbb{N}\}$ be a family of pairwise-disjoint, closed intervals in $\mathbb{I}$, and, for each $n \in \mathbb{N}$, let χ_n be the characteristic function of I_n, ℓ_n the length of I_n, and $f_n = \chi_n/\ell_n$, so that $\|f_n\|_1 = 1$. Then take $E = \overline{\mathrm{lin}}\{f_n : n \in \mathbb{N}\}$, so that E is a closed subspace of $L^1(\mathbb{I})$.

Take $\alpha = (\alpha_n) \in \ell^1$. Then it is clear that the map

$$\alpha \mapsto \sum_{n=1}^{\infty} \alpha_n f_n, \quad \ell^1 \to E,$$

is an isometric embedding, and so $E \cong \ell^1$.

Define the map

$$P : f \mapsto \sum_{n=1}^{\infty} \left(\int_{I_n} f \right) f_n, \quad L^1(\mathbb{I}) \to L^1(\mathbb{I}).$$

Clearly P is a linear map with

$$P f_n = f_n \quad (n \in \mathbb{N}) \quad \text{and} \quad \|Pf\|_1 \le \|f\|_1 \quad (f \in L^1(\mathbb{I})).$$

Thus P is a bounded projection onto E with $\|P\| = 1$, and so E is a 1-complemented subspace in $L^1(\mathbb{I})$. □

A similar argument [3, Proposition 6.4.1] shows that, for each p with $1 < p < \infty$, the Banach space ℓ^p is isometrically isomorphic to a 1-complemented subspace of $L^p(\mathbb{I})$.

We also remark that, for $r, p > 1$, the Banach space ℓ^r is isomorphic to a complemented subspace of $L^p(\mathbb{I})$ if and only if $r = p$ or $r = 2$ [3, Proposition 6.4.21]. Now take $r \ge 1$. For p with $1 \le p \le 2$, the Banach space ℓ^r is isomorphic to a closed subspace of $L^p(\mathbb{I})$ if and only if $p \le r \le 2$, and, for $2 < p < \infty$, the space ℓ^r is isomorphic to a closed subspace of $L^p(\mathbb{I})$ if and only if $r = 2$ or $r = p$ [3, Proposition 6.4.19].

We now present a beautiful result of Pełczyński from [196]; it will be used later. It is called the *Pełczyński decomposition method.* Our proof is taken from [3, Theorem 2.2.3].

Theorem 2.4.9. *Let E and F be normed spaces such that both E and F are isomorphic to complemented subspaces of the other. Further, suppose that either $E \sim E \times E$ and $F \sim F \times F$ or that $E \sim \ell^\infty(E)$. Then $E \sim F$.*

Proof. There exist normed spaces G and H such that $F \sim E \oplus G$ and $E \sim F \oplus H$, so that $F \sim E \times G$ and $E \sim F \times H$.

In the first case, we have

$$E \sim F \times H \sim (F \times F) \times H \sim F \times (F \times H) \sim F \times E$$

and, similarly, $F \sim E \times F$. But $E \times F \cong F \times E$, and so $E \sim F$.

In the second case, we have $E \sim E \times E$, and so $F \sim E \times F$, as before. But now

$$E \sim \ell^\infty(E) \sim \ell^\infty(F \times H) \sim \ell^\infty(F) \times \ell^\infty(H) \sim F \times \ell^\infty(F) \times \ell^\infty(H) \sim F \times E,$$

and so we again see that $E \sim F$. □

Since we shall discuss complemented subspaces of Banach spaces of the form $C(K)$, it is important to note that not all such closed subspaces are complemented; indeed, the most famous counter-example to this possibility is given by *Phillips' theorem* that c_0 is not complemented in ℓ^∞. A slightly stronger version of this theorem already follows easily from a previous result. Indeed, assume towards a contradiction that c_0 is complemented in ℓ^∞. Then there is an embedding of ℓ^∞/c_0 into $\ell^\infty \cong C(\beta\mathbb{N})$. However it follows from Example 2.2.22 that there is no such embedding. See also [148, p. 19].

Nevertheless, we wish to give the classical, elementary proof of Phillips' theorem; it is taken from [240]. See also [3, Theorem 2.5.5] and [183, Theorem 3.2.20].

Definition 2.4.10. Let E be a Banach space. A subset T of E' is *total* if $x = 0$ whenever $x \in E$ and $\langle x, \lambda \rangle = 0$ $(\lambda \in T)$; a Banach space E has *property* (T) if E' contains a countable, total subset.

Note that property (T) is preserved under isomorphisms and under the passage to closed subspaces.

Theorem 2.4.11. *The subspaces c_0 and c are not complemented in ℓ^∞.*

Proof. First, assume towards a contradiction that there is a closed subspace F of ℓ^∞ such that $\ell^\infty = c_0 \oplus F$. We regard F as a Banach space by setting

$$\|x\| = d(x, c_0) \quad (x \in F),$$

the distance from x to c_0 in ℓ^∞, thus identifying F with the quotient space ℓ^∞/c_0.

Clearly $\{\delta_n : n \in \mathbb{N}\}$ is a countable, total subset of $(\ell^\infty)'$, and so $(F, \|\cdot\|)$ has property (T).

Let $\{S_\alpha : \alpha \in A\}$ be a family of subsets of $\mathbb{N}$ as specified in Proposition 1.5.5, and, for $\alpha \in A$, let f_α be the coset in F that corresponds to χ_{S_α}, so that $\|f_\alpha\| = 1$ $(\alpha \in A)$.

Take $\lambda \in F'$. We *claim* that the set $\{\alpha \in A : \langle f_\alpha, \lambda \rangle \neq 0\}$ is countable. For this, it suffices to show that, for each $n \in \mathbb{N}$, the set

$$C_n := \{\alpha \in A : |\langle f_\alpha, \lambda \rangle| \geq 1/n\}$$

is finite. Indeed, fix $n \in \mathbb{N}$, and then, for $m \in \mathbb{N}$ with $m \leq |C_n|$, choose distinct elements $\alpha_1, \ldots, \alpha_m \in C_n$; set $g_i = f_{\alpha_i}$ $(i \in \mathbb{N}_m)$ and

$$\beta_i = \operatorname{sgn}\langle g_i, \lambda \rangle \quad (i \in \mathbb{N}_m) \quad \text{and} \quad g = \sum_{i=1}^{m} \beta_i g_i .$$

Then there exists a number $N \in \mathbb{N}$ such that $S_{\alpha_i} \cap S_{\alpha_j} \subset \mathbb{N}_N$ for $i, j \in \mathbb{N}_m$ with $i \neq j$, and so $\|g\| = 1$, regarding g as an element of F. Thus

$$\|\lambda\| \geq |\langle g, \lambda \rangle| = \sum_{i=1}^{m} |\langle g_i, \lambda \rangle| \geq \frac{m}{n},$$

and so $|C_n| \leq n \|\lambda\| + 1$. Hence C_n is finite, and the claim follows.

Now suppose that Λ is a countable set in $(F, \|\cdot\|)'$. Then there are only countably many values of $\alpha \in A$ such that $\langle f_\alpha, \lambda \rangle \neq 0$ for some $\lambda \in \Lambda$, and so there exists an index $\alpha \in A$ with $\langle f_\alpha, \lambda \rangle = 0$ for all $\lambda \in \Lambda$. Thus the set Λ is not total in $(F, \|\cdot\|)'$, a contradiction of the fact that F has property (T).

It follows that c_0 is not complemented in ℓ^∞. Clearly c is not complemented in ℓ^∞: if c were so complemented, then c_0 would be complemented in ℓ^∞ because it is complemented in c. □

The following generalization by Conway of Phillips' theorem is taken from [64].

Theorem 2.4.12. *Let K be a non-empty, locally compact space that is not pseudo-compact. Then $C_0(K)$ is not complemented in $C^b(K)$.*

Proof. There is a function $f \in C(K, \mathbb{R}^+) \setminus C^b(K)$. Choose $x_1 \in K$ with $f(x_1) > 1$, and then inductively choose (x_n) in K such that $f(x_{n+1}) > f(x_n) + 4$ for each $n \in \mathbb{N}$; set

$$U_n = \{x \in K : |f(x) - f(x_n)| < 1\} \quad (n \in \mathbb{N}).$$

For each $n \in \mathbb{N}$, choose $f_n \in C_0(K, \mathbb{I})$ with $f_n(x_n) = 1$ and $\operatorname{supp} f_n \subset U_n$, and define

$$T\alpha = \sum_{n=1}^{\infty} \alpha_n f_n \quad (\alpha = (\alpha_n) \in \ell^\infty).$$

For each $x \in K$, the neighbourhood $\{y \in K : |f(y) - f(x)| < 1\}$ of x has non-empty intersection with at most one set U_n, and it follows easily from this that $T\alpha \in C^b(K)$ for each $\alpha \in \ell^\infty$. We see that $T(c_0) \subset C_0(K)$ and that $T : \ell^\infty \to C^b(K)$ is a linear isometry.

Define $Sg = (g(x_n))$ $(g \in C_0(K))$. Since the sequence (x_n) has no accumulation point in K, each compact subset of K contains at most finitely many points of this set, and so $Sg \in c_0$ $(g \in C_0(K))$. Clearly $S : C_0(K) \to c_0$ is a linear isometry and $(S \circ T)(\alpha) = \alpha$ $(\alpha \in c_0)$.

Assume to the contrary that there is a bounded projection $P : C^b(K) \to C_0(K)$. Then the map $S \circ P \circ T : \ell^\infty \to c_0$ is a bounded projection. But this is a contradiction of Theorem 2.4.11. Thus $C_0(K)$ is not complemented in $C^b(K)$. □

An elementary special case of the above is the following.

Corollary 2.4.13. *Let Γ be an infinite set. Then $c_0(\Gamma)$ is not complemented in $\ell^\infty(\Gamma)$.* □

The space $[0, \omega_1)$ is pseudo-compact. Here $C_0([0, \omega_1))$ has codimension 1, and so is complemented, in $C^b([0, \omega_1)) \cong C([0, \omega_1])$.

The following result, called *Sobczyk's theorem*, is taken from [3, Theorem 2.5.8], [20, Theorem 2.3], and [175, Theorem 2.f.5]; the elegant proof is due to Veech [238].

Theorem 2.4.14. *Let E be a separable Banach space containing c_0 as a closed subspace. Then c_0 is 2-complemented in E.*

Proof. Since E is separable, it follows from Theorem 2.1.4(iii) that there is a metric, say d, giving the weak* topology on $E'_{[1]}$.

Let $n \in \mathbb{N}$. Then the map $\delta_n : (\alpha_m) \mapsto \alpha_n$ is a continuous linear functional on c_0 with $\|\delta_n\| = 1$. Let $\lambda_n \in E'$ be a norm-preserving extension of δ_n, and set

$$S = \{\lambda \in E'_{[1]} : \lambda \mid c_0 = 0\}.$$

Since each weak*-limit point of $\{\lambda_n : n \in \mathbb{N}\}$ belongs to S, $\lim_{n\to\infty} d(\lambda_n, S) = 0$, and so there is sequence (μ_n) in S with $\lim_{n\to\infty} d(\lambda_n, \mu_n) = 0$. Since $\lim_{n\to\infty}(\lambda_n - \mu_n) = 0$ in $(E'_{[1]}, \sigma(E', E))$, the map $P : x \mapsto (\langle x, \lambda_n - \mu_n\rangle)$, $E \to c_0$, is a bounded projection onto c_0, and clearly $\|P\| \leq 2$. □

For interesting extensions of Sobczyk's theorem, see [14]. In fact, it is a theorem of Zippin that a Banach space that is complemented in every separable Banach space that contains the space as a closed subspace is isomorphic to c_0 [246, 247]. For an entertaining essay on Sobczyk's theorem and Phillips' theorem, see [48].

Theorem 2.4.15. *Let E be a Banach space containing c_0 as a closed, complemented subspace. Then E is not complemented in E'' and E is not isomorphically a dual space. In particular, c_0 is not isomorphically a dual space.*

Proof. There is a bounded projection P of E onto c_0. Assume that there is a bounded projection Q of E'' onto E. We may regard the spaces c_0 and $\ell^\infty = c_0''$ as closed subspaces of E'', and then $(P \circ Q) \mid \ell^\infty$ is a bounded projection of ℓ^∞ onto c_0, a contradiction of Theorem 2.4.11. Thus E is not complemented in E''. By Corollary 2.4.5, E is not isomorphically a dual space. □

Corollary 2.4.16. *Let E be a separable Banach space containing c_0 as a closed subspace. Then E is not complemented in E'' and E is not isomorphically a dual space.*

Proof. By Theorem 2.4.14, c_0 is complemented in E, and so this follows from the theorem. □

Corollary 2.4.17. *Let K be a locally compact space that contains a convergent sequence of distinct points. Then $C_0(K)$ is not complemented in $C_0(K)''$ and $C_0(K)$ is not isomorphically a dual space.*

Proof. By Proposition 2.4.6(ii), $C_0(K)$ contains c_0 as a closed, complemented subspace, and so the result follows from Theorem 2.4.15. □

In particular, the above corollary covers the cases where K is an infinite, compact, metrizable space, where $K = [0, \alpha]$ for an ordinal $\alpha \geq \omega$, and where $K = \mathbb{Z}_2^\kappa$, the Cantor cube of weight κ: in each of these cases, it is easy to see that the space contains a convergent sequence of distinct points.

Definition 2.4.18. Let E be a Banach space. Then E is *prime* if every complemented, infinite-dimensional, closed subspace of E is isomorphic to E.

Clause (i) of the following theorem is a famous result of Pełczyński [3, Theorem 2.2.4]; clause (ii) is a theorem of Lindenstrauss [3, Theorem 5.6.5].

Theorem 2.4.19. (i) *The spaces c_0 and ℓ^p, for $1 \leq p < \infty$, are prime Banach spaces.*

(ii) *The space ℓ^∞ is a prime Banach space.* □

Definition 2.4.20. Let E be a Banach space. Then E is *primary* if, whenever E is isomorphic to the direct sum of two Banach spaces, E is isomorphic to one of the two summands.

As stated in [3, p. 122], $L^1(\mathbb{I})$ and $C(\mathbb{I})$ are not prime, but both are primary. In fact, each space $L^p(\mathbb{I})$ for $1 \leq p \leq \infty$ is primary [176, Theorem 2.d.11].

It is easily seen that $C(\mathbb{N}^*)$ is isomorphic to $C(\mathbb{N}^*) \oplus \ell^\infty$, and so we can regard (a copy of) ℓ^∞ as a complemented, infinite-dimensional, closed subspace of $C(\mathbb{N}^*)$. However, by Example 2.2.22, ℓ^∞ is not isomorphic to $C(\mathbb{N}^*)$, and so $C(\mathbb{N}^*)$ is not prime. It is known that, with CH, $C(\mathbb{N}^*)$ is primary [92], but it is not known whether this is a theorem of ZFC. Incidentally, we note that it is proved in [92] that, with CH, $C(\mathbb{N}^*) \sim \ell^\infty(C(\mathbb{N}^*))$ and in [46] that it is consistent with ZFC that $C(\mathbb{N}^*)$ is not isomorphic to $\ell^\infty(E)$ for any Banach space E.

A major result in this area is the following solution of the *complemented subspace problem*, due to Lindenstrauss and Tzafriri [173]. For a proof of this theorem, see [3, §12.4].

Theorem 2.4.21. *Let E be an infinite-dimensional Banach space such that every closed subspace of E is complemented in E. Then E is isomorphic to a Hilbert space.* □

2.5 Projection properties and injective Banach spaces

We now consider the appropriate versions of projectivity and injectivity in the category of Banach spaces and bounded operators that we are considering.

Definition 2.5.1. A Banach space E has the *projection property* if, whenever F is a closed subspace of a Banach space G that is isometrically isomorphic to E, the space F is complemented in G. More generally, a Banach space E is a P_λ*-space* (for $\lambda \geq 1$) if such a space F is λ-complemented in G.

Suppose that E is a P_λ-space for some $\lambda \geq 1$. Then the *projection constant* of E is the infimum of the numbers λ such that E is a P_λ-space.

We represent the above situation with the following commutative diagram:

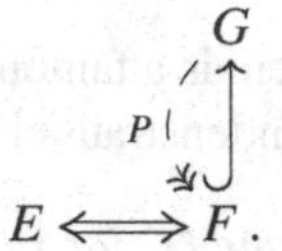

The following is an immediate property of P_λ-spaces. Let E be a P_λ-space, and suppose that E is a closed subspace of a Banach space G, that F is a Banach space, and that $T \in \mathscr{B}(E,F)$. Then there is an extension $\widetilde{T}$ of T in $\mathscr{B}(G,F)$ such that $\left\|\widetilde{T}\right\| \leq \lambda \|T\|$. Indeed, let $P : G \to E$ be a bounded projection with $\|P\| \leq \lambda$, and set $\widetilde{T} = T \circ P$.

We represent the above situation with the following commutative diagram:

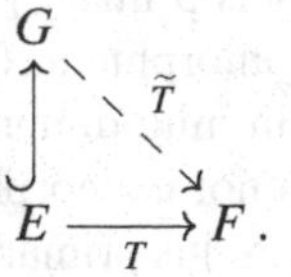

It is proved in [171, Theorem 6.10] that a real Banach space which is a $P_{1+\varepsilon}$-space for each $\varepsilon > 0$ is already a P_1-space. It seems to be unknown whether the same result holds for complex Banach spaces. However an example in [143] shows that a (real) Banach space which is a $P_{2+\varepsilon}$-space for each $\varepsilon > 0$ is not necessarily a P_2-space.

The next definition gives a similar concept with the spaces E and F 'the other way round'.

Definition 2.5.2. A Banach space E is *injective* if, for every Banach space G, every closed subspace F of G, and every $T \in \mathscr{B}(F,E)$, there is an extension $\widetilde{T} \in \mathscr{B}(G,E)$ of T; the space E is λ*-injective* if, further, we can always find such a $\widetilde{T}$ such that $\left\|\widetilde{T}\right\| \leq \lambda \|T\|$.

We represent this situation with the following commutative diagram:

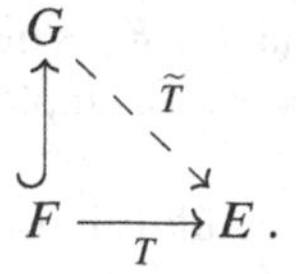

For a discussion of injective spaces, see [20, Chapter 1].

Clearly an injective space is complemented in any Banach space that contains it as a closed subspace, and injectivity is an isomorphic invariant for the class of all Banach spaces. For example, by Theorem 2.4.12 and Corollary 2.4.17, respectively, $C_0(K)$ is not injective whenever K is a non-empty, locally compact space that is not pseudo-compact and whenever K is a compact space that contains a convergent sequence of distinct points.

We see that a real Banach space is injective if and only if its complexification is injective.

We shall use the following obvious remark.

Proposition 2.5.3. *A complemented subspace of an injective space is injective; a 1-complemented subspace of a 1-injective space is 1-injective.* □

The next proposition is immediate from Theorem 2.4.9.

Proposition 2.5.4. *Let E and F be injective Banach spaces such that $E \sim E \times E$ and $F \sim F \times F$ and such that both E and F are isomorphic to closed subspaces of the other. Then $E \sim F$.* □

The following result was first noted by Phillips in [202, Corollary 7.2].

Proposition 2.5.5. *The space $\ell^\infty(S) = C(\beta S)$ is 1-injective for each non-empty set S.*

Proof. Take a Banach space G, a closed subspace F, and $T \in \mathscr{B}(F, \ell^\infty(S))$. For each $s \in S$, the functional $\lambda_s : x \mapsto (Tx)(s)$ on F is continuous with $\|\lambda_s\| \leq \|T\|$. By the Hahn–Banach theorem, Theorem 2.1.2(i), each λ_s has a norm-preserving extension $\widetilde{\lambda}_s$ to G. Set

$$(\widetilde{T}x)(s) = \langle x, \widetilde{\lambda}_s \rangle \quad (s \in S, x \in G).$$

Then $\widetilde{T} \in \mathscr{B}(G, \ell^\infty(S))$ is an extension of T with $\left\|\widetilde{T}\right\| = \|T\|$. □

Corollary 2.5.6. *Let E be a Banach space. Then E is isometrically isomorphic to a subspace of a 1-injective space.*

Proof. By Proposition 2.2.14(i), E is isometrically isomorphic to a closed subspace of a space of the form $\ell^\infty(S)$. □

Take $p \in \mathbb{N}^*$, and set $M_p = \{f \in C(\beta\mathbb{N}) : f(p) = 0\}$. Then M_p is a complemented subspace of $C(\beta\mathbb{N})$, and so M_p is injective. This gives an example of a non-compact space $K = \beta\mathbb{N} \setminus \{p\}$ such that $C_0(K)$ is injective; see also Example 6.9.1 for a slightly stronger fact. Of course, as in Example 1.5.3(ii), K is a pseudo-compact space.

On the other hand, the following result is immediate from Theorem 2.4.15.

Proposition 2.5.7. *Let E be a Banach space containing c_0 as a closed, complemented subspace. Then E is not injective.* □

Proposition 2.5.8. *Let E be a separable, infinite-dimensional Banach space. Then E is not injective.*

Proof. By Proposition 2.2.17(i), there is an isometric embedding of E into ℓ^∞.

Assume to the contrary that E is injective. Then E is complemented in ℓ^∞. But, by Theorem 2.4.19(ii), ℓ^∞ is prime, and so E is isomorphic to ℓ^∞. But ℓ^∞ is not separable, a contradiction. □

It follows from Theorem 2.1.7(i) that $C(K)$ is not injective whenever K is an infinite, compact, metrizable space; a stronger result was given in Corollary 2.4.17.

Proposition 2.5.9. *Take $\lambda \geq 1$. Then a Banach space is λ-injective if and only if it is a P_λ-space, and it is injective if and only if it has the projection property.*

Proof. Suppose that the Banach space E is λ-injective. Take F to be a closed subspace of a Banach space G such that $E \cong F$, and let $T : F \to E$ be a linear isometry. Then there is an extension $S \in \mathscr{B}(G,E)$ of T with $\|S\| \leq \lambda$. Set $P = T^{-1} \circ S : G \to F$. Then P is a bounded projection with $\|P\| \leq \lambda$, and so E is a P_λ-space.

Now suppose that E is a P_λ-space. Take F to be a closed subspace of a Banach space G, and take $T \in \mathscr{B}(F,E)$. By Corollary 2.5.6 , we can identify E as a closed subspace of a 1-injective space, say H. There is a bounded projection P from H onto E with $\|P\| \leq \lambda$, and, since $T \in \mathscr{B}(F,H)$, there is a norm-preserving extension, say $L \in \mathscr{B}(G,H)$, of T. Set $\widetilde{T} = P \circ L$ to obtain the required extension of T.

Similarly, E is injective if and only if it has the projection property. □

Proposition 2.5.10. *A Banach space with the projection property is a P_λ-space for some $\lambda \geq 1$.*

Proof. It is easy to see that F_0 is a $P_{\lambda\mu}$-space whenever F is a P_λ-space, and there is a bounded projection of norm μ from F onto the subspace F_0. By Corollary 2.5.6, each Banach space E is a closed subspace of a 1-injective space F. In the case where E has the projection property, there is a bounded projection $P : F \to E$, and so E is P_λ-space with $\lambda = \|P\|$. □

It follows that an injective Banach space is λ-injective for some $\lambda \geq 1$.

Let K and L be two non-empty, compact spaces. First, let $\eta : K \to L$ be a continuous map, and define

$$\eta^\circ : f \mapsto f \circ \eta, \quad C(L) \to C(K). \tag{2.9}$$

Then η° is a bounded operator with $\|\eta^\circ\| = 1$. Further, η° is a surjection if and only if η is an injection, and η° is an injection if and only if η is a surjection if and only if η° is isometric. In particular, let (G_L, π_L) be the Gleason cover of L, as in Theorem 1.6.5. Then the map $\pi_L^\circ : C(L) \to C(G_L)$ is an isometric embedding.

We first generalize Proposition 2.5.5.

Theorem 2.5.11. *Let K be a non-empty, Stonean space. Then $C(K)$ is* 1*-injective. Further, $C(K)$ is isometrically isomorphic to a complemented subspace of $C(\beta K_d)$, which is isometrically a bidual space.*

Proof. By Theorem 1.6.3, (a) $\Rightarrow$ (b), there is a retraction $\theta : \beta K_d \to K$.

Let $\eta : K \to \beta K_d$ be the natural embedding, so that $\theta \circ \eta$ is the identity on K. Then the map $\theta^\circ : C(K) \to C(\beta K_d)$ is an isometry and $\eta^\circ : C(\beta K_d) \to \theta^\circ(C(K))$ is a linear surjection with $\|\eta^\circ\| = 1$. Since $\eta^\circ \circ \theta^\circ$ is the identity on $C(K)$, the map η° is a bounded projection. By Proposition 2.5.5, $C(\beta K_d)$ is 1-injective, and so $C(K)$ is 1-injective.

Of course, $C(\beta K_d)$ is isometrically the bidual of $C_0(K_d)$. □

We shall see in Theorem 6.8.3 that, conversely, K is Stonean whenever $C(K)$ is 1-injective. Indeed, Question 3 on page 212 will raise the possibility that the only injective Banach spaces are those isomorphic to $C(K)$ for K a Stonean space.

Corollary 2.5.12. *Let K be a non-empty, compact space. Then $C(G_K)$ is* 1*-injective.*

Proof. By Theorems 1.6.5, G_K is a Stonean space. □

There is a closely related theory of extensions of Banach spaces. Some of these results will be used in the characterization of 1-injective Banach spaces to be given in Theorem 6.8.6. The next few results are based on Bade's notes [23, 24]; see also [166, §11].

Definition 2.5.13. Let E be a closed subspace of a Banach space F. Then:

(i) F is an *essential extension* of E if, for each Banach space G and each contraction $T \in \mathscr{B}(F, G)$ such that $T \mid E$ is an isometry, T is also an isometry;

(ii) F is a *rigid extension* of E if, for each contraction $T \in \mathscr{B}(F)$ such that $T \mid E = I_E$, necessarily $T = I_F$.

Proposition 2.5.14. *Let E be a Banach space. Then each essential extension of E is rigid.*

Proof. Let F be an essential extension of E, and assume towards a contradiction that F is not rigid. Then there are a contraction $T \in \mathscr{B}(F)$ and $y \in F$ such that $T \mid E = I_E$ and $Ty \neq y$, say $z = y - Ty$, so that $z \neq 0$; we may suppose that $\|y\| = 1$. Set $M = \mathbb{C}z$. Then the quotient map $q : F \to F/M$ is a contraction that is not an isometry because $q(z) = 0$.

We *claim* that $q \mid E$ is an isometry; this will give the required contradiction. Indeed, assume that $q \mid E$ is not an isometry. Then there exist $x \in E$ with $\|x\| = 1$ and $\delta > 0$ such that $\|x + \delta z\| < 1$. There exists $\eta > 0$ such that $\|w + \delta z\| < \|w\|$ whenever $\|w - x\| \leq \eta$. For such an element w, we have

$$\|w\| < \|w - \eta z\| , \tag{2.10}$$

for otherwise

$$\|w\| \leq \frac{\eta}{\delta + \eta} \|w + \delta z\| + \frac{\delta}{\delta + \eta} \|w - \eta z\| < \|w\| .$$

We apply (2.10) with $w = x + \eta y$, so that $w - \eta z = x + \eta Ty = T(x + \eta y)$, to see that $\|x + \eta y\| < \|T(x + \eta y)\|$; this is a contradiction of the fact that T is a contraction. Thus $q \mid E$ is an isometry. □

Proposition 2.5.15. *Let E be a closed subspace of a Banach space $(F, \|\cdot\|)$. Then the following are equivalent:*

(a) *for each semi-norm p on F with $p(x) = \|x\|$ $(x \in E)$ and $p(y) \leq \|y\|$ $(y \in F)$, necessarily $p(y) = \|y\|$ $(y \in F)$;*

(b) *F is an essential extension of E.*

Proof. (a) $\Rightarrow$ (b) Let G be a Banach space, and suppose that $T : F \to G$ is a contraction such that $T \mid E$ is an isometry. Set $p(y) = \|Ty\|$ $(y \in F)$. Then p is a semi-norm on F satisfying the conditions in (a), and so $p(y) = \|y\|$ $(y \in F)$, whence T is an isometry.

(b) $\Rightarrow$ (a) Let p be a semi-norm on F satisfying the conditions in (a), and set $K = \{y \in F : p(y) = 0\}$. Take q to be the quotient map from F onto the space F/K, let F/K have the norm induced by p, and take G to be the completion of this space. Then $q : F \to G$ is a contraction and $q \mid E$ is an isometry, and so, by (b), q is an isometry. It follows that $p(y) = \|y\|$ $(y \in F)$. □

Theorem 2.5.16. *Let E be a closed subspace of a Banach space F, and suppose that F is a 1-injective space. Then there is a closed subspace G of F containing E such that G is a 1-injective space and G is a rigid extension of E.*

Proof. Let $\mathscr{F}$ be the family of semi-norms p on F such that $p(x) = \|x\|$ $(x \in E)$ and $p(y) \leq \|y\|$ $(y \in F)$. For $p, q \in \mathscr{F}$, set $p \leq q$ if $p(y) \leq q(y)$ $(y \in F)$. Then $(\mathscr{F}, \leq)$

is a partially ordered space. Clearly each chain in $(\mathscr{F},\leq)$ has a lower bound, and so $(\mathscr{F},\leq)$ has a minimal element, say p_0. Let H be the completion of $F/\ker p_0$, and let $\pi : F \to F/\ker p_0$ be the quotient map.

We can regard E as a closed subspace of H. Since F is 1-injective, there is a contraction $T : H \to F$ with $T \mid E = I_E$. Set $P = T \circ \pi$, so that $Px = x$ $(x \in E)$ and $\|P\| = 1$.

Set $p_1(y) = \|Py\|$ $(y \in F)$. Then $p_1 \leq p_0$ in $(\mathscr{F},\leq)$, and so $p_1 = p_0$ by the minimality of p_0. Next define

$$p_2(y) = \limsup_{n\to\infty}\left\|\frac{1}{n}\sum_{i=1}^{n}P^i y\right\| \quad (y \in F).$$

Then $p_2 \leq p_1$ in $(\mathscr{F},\leq)$, and so $p_2 = p_0$. Further,

$$p_2(y - Py) = \limsup_{n\to\infty}\left\|\frac{1}{n}(Py - P^{n+1}y)\right\| = 0 \quad (y \in F),$$

and so $\|Py - P^2y\| = p_1(y - Py) = p_2(y - Py) = 0$ $(y \in F)$. This shows that $P^2 = P$ in $\mathscr{B}(F)$.

Set $G = P(F)$. Then G is a closed subspace of F containing E and G is a 1-injective space.

Finally, we show that G is an essential extension of E; for this, we verify clause (a) of Proposition 2.5.15. Indeed, let p be a semi-norm on G such that $p(x) = \|x\|$ $(x \in E)$ and $p(y) \leq \|y\|$ $(y \in G)$. Then $p \circ P \in \mathscr{F}$, and

$$(p \circ P)(y) \leq \|Py\| = p_1(y) = p_0(y) \quad (y \in F),$$

and so $p \circ P = p_0$ and $p(y) = \|y\|$ $(y \in G)$, as required.

By Proposition 2.5.14, G is a rigid extension of E. □

The rigid extension G of E clearly has the property that, for each 1-injective subspace H of G with $E \subset H$, necessarily $H = G$. Further, suppose that H has the same properties as G. Then H is isometrically isomorphic to G by a map that is the identity on E. The space G is the *injective envelope* of E; we shall see in Theorem 6.8.6 that an injective envelope of a Banach space has the form $C(K)$ for a certain Stonean space K.

Recall from page 15 that Δ denotes the Cantor set.

Proposition 2.5.17. *Let E be a separable Banach space. Then there is an isometric embedding of E into $C(\Delta)$.*

Proof. By Proposition 2.2.14(i), there is a non-empty, compact, metrizable space B and an isometric isomorphism $T : E \to C(B)$. By Proposition 1.4.6(i), there is a continuous surjection $\eta : \Delta \to B$. Thus $\eta^\circ : C(B) \to C(\Delta)$ is an isometric embedding. The map $\eta^\circ \circ T : E \to C(\Delta)$ is also an isometric embedding. □

The above results say that $C(\Delta)$ is *universal* in the class of separable Banach spaces. It follows easily that $C(\mathbb{I})$ is also universal in the class of separable Banach spaces. This is the *Banach–Mazur theorem*, already given in [30, Chapitre XI, §8]; see also [3, Theorem 1.4.3] and [225, Theorem 8.7.2]. These results are contained in [23, Chapter 4]; early texts in which they appeared are [82, p. 123] and [225], and the standard account is [175, §2f]. For example, it is proved in [175, Theorem 2.f.3] that every infinite-dimensional injective Banach space contains a closed subspace that is isomorphic to ℓ^∞. For a more recent discussion of these properties, see [247]. It is stated in [172, p. 337] that a Banach space is injective if and only if it is a so-called $\mathscr{L}_\infty$ space and is isomorphic to a complemented subspace of a dual space.

We have noted in equation (1.6) that $w(\mathbb{N}^*) = d(C(\mathbb{N}^*)) = |C(\mathbb{N}^*)| = \mathfrak{c}$. By a famous theorem of Parovichenko (see [99, p. 236] and [239, p. 81]), every compact (Hausdorff) space of weight at most $\aleph_1$ is a continuous image of $\mathbb{N}^*$. Recall from Proposition 2.2.14(i) that each Banach space E is isometrically embedded in the space $C(B)$, where $B = E'_{[1]}$ and that $d(E) = w(B)$ by Corollary 2.1.8. Hence every Banach space of density at most $\aleph_1$ can be isometrically embedded in $C(\mathbb{N}^*)$, and so, with CH, $C(\mathbb{N}^*)$ is universal in the class of Banach spaces of density $\mathfrak{c}$. However this is not a result of the theory ZFC: it is consistent with ZFC that there is no isometrically universal Banach space of density $\mathfrak{c}$ [226]. For further related and stronger results, see [45, 46]. For example, it is consistent with ZFC that the Banach space $C([0,\mathfrak{c}])$ does not embed into $C(\mathbb{N}^*)$.

There is an extension of the notion of an injective space. A Banach space E is *separably injective* if, for every separable Banach space G, every closed subspace F of G, and every $T \in \mathscr{B}(F,E)$, there is an extension $\widetilde{T} \in \mathscr{B}(G,E)$ of T. Obviously, every injective space is separably injective. By Zippin's theorem, mentioned above, the only separable and separably injective Banach space is c_0. The idea of extending the notion of separably injective spaces to non-separable spaces was introduced by Rosenthal in [214]. Examples of non-separable spaces which are separably injective but not injective are certain Banach spaces $\ell^\infty_c(\Gamma)$, to be discussed below at Example 6.7.1, and $C(\mathbb{N}^*)$ (due to Lindenstrauss). For accounts of separably injective Banach spaces, including these examples, see [19, 20] and [247, p. 1722].

We shall discuss the injectivity of $C(K)$-spaces further in §6.8.

Although it is not strictly relevant to our work, we briefly introduce the dual concept to that of an injective space.

Definition 2.5.18. A Banach space E is *projective* if, for every Banach space G, every quotient Banach space F of G with quotient map $q : G \to F$, and every operator $T \in \mathscr{B}(E,F)$, there is a *lifting* $\widetilde{T} \in \mathscr{B}(E,G)$ of T, in the sense that $T = q \circ \widetilde{T}$; the space E is λ*-projective* (for $\lambda \geq 1$) if, further, we can always find such a $\widetilde{T}$ with $\left\|\widetilde{T}\right\| \leq \lambda \|T\|$.

We represent the above situation with the following commutative diagram:

$$\begin{array}{ccc} & & G \\ & \overset{\widetilde{T}}{\nearrow} & \big\downarrow q \\ E & \xrightarrow[T]{} & F. \end{array}$$

Each projective Banach space is λ-projective for some $\lambda \geq 1$. The following results give characterizations of projective Banach spaces.

Theorem 2.5.19. *A Banach space is $(1+\varepsilon)$-projective for each $\varepsilon > 0$ if and only if it is isometrically isomorphic to a Banach space of the form $\ell^1(\Gamma)$ for a non-empty set Γ.*

Proof. This is proved in [166, Theorem 9, p. 178] and in [225, Theorem 27.4.2]; that a 1-projective space has the form $\ell^1(\Gamma)$ is due to Grothendieck [125]. □

Theorem 2.5.20. *A Banach space is 1-projective if and only if it is isometrically isomorphic to a Banach space of the form $L^1(\Omega,\mu)$ for a measure space (Ω,μ).*

Proof. This is proved in [166, Corollary to Theorem 8, p. 178]. □

Theorem 2.5.21. *A Banach space is projective if and only if it is isomorphic to a Banach space of the form $\ell^1(\Gamma)$ for a non-empty set Γ.*

Proof. For this, see [175, p. 108]; the result is due to Köthe [162]. □

2.6 The Krein–Milman and Radon–Nikodým properties

We shall be concerned with the extreme points of the closed unit ball and other bounded subsets of a Banach space; we shall discuss, rather briefly, the seminal notions of Banach spaces having the Krein–Milman property and the Radon–Nikodým property.

The first result is the famous *Krein–Milman theorem*; see [6, Theorem 3.31] or [218, Theorem 3.23], for example.

Theorem 2.6.1. *Let L be a non-empty, compact, convex subset of a locally convex space over $\mathbb{R}$ or $\mathbb{C}$. Then $L = \overline{\mathrm{co}}(\mathrm{ex}\, L)$.* □

Corollary 2.6.2. *Let E be a normed space. Then the set $\mathrm{co}(\mathrm{ex}\, E'_{[1]})$ is weak*-dense in $E'_{[1]}$.* □

Corollary 2.6.3. *Let E be a Banach space such that* $\operatorname{ex} E_{[1]} = \emptyset$. *Then E is not isometrically a dual space.*

Proof. Assume that $E \cong F'$ for a Banach space F. Set $L = E_{[1]}$, so that L is a non-empty, compact, convex subset of the locally convex space $(E, \sigma(E,F))$, and hence, by the theorem, $\operatorname{ex} L \neq \emptyset$, a contradiction. □

We shall see in Example 6.9.1 that there are Banach spaces E such that $\operatorname{ex} E_{[1]} = \emptyset$ and E is isomorphically a dual space.

Corollary 2.6.4. *Let E be Banach space, and set $B = E'_{[1]}$. Suppose that L is a closed subset of B such that* $\operatorname{ci} L = \overline{\operatorname{ex}} B$. *Then the map*

$$J : x \mapsto \kappa_E(x) \mid L, \quad E \to C(L),$$

is an isometric embedding.

Proof. The set $\operatorname{ci} L$ is a circled subspace of B with $\overline{\operatorname{co}}(\operatorname{ci} L) = B$, and so it follows from Proposition 2.2.14(ii) that, for each $x \in E$, there exist $\lambda \in L$ and $\zeta \in \mathbb{T}$ such that $\|x\| = |\langle x, \zeta\lambda \rangle|$. But then $\|x\| = |\langle x, \lambda \rangle|$, and so J is an isometry. □

We now give a geometric property, that of 'dentability', of subsets of a Banach space. This is a notion that was introduced by Rieffel in [208].

Definition 2.6.5. Let E be a Banach space. Then a bounded subset S of E is *dentable* if, for each $\varepsilon > 0$, there exists $x \in S$ such that $x \notin \overline{\operatorname{co}}(S \setminus B_\varepsilon(x))$.

The next theorem, Theorem 2.6.7, is due to Rieffel [208, Theorem 3]; it will be used in the proof of Corollary 2.6.12.

Lemma 2.6.6. *Let E be a Banach space, and let S be a bounded subset of E. Suppose that $\overline{\operatorname{co}} S$ is dentable. Then S is dentable.*

Proof. Take $\varepsilon > 0$. Then there exists $x_0 \in (\overline{\operatorname{co}} S) \setminus Q$, where

$$Q = \overline{\operatorname{co}}((\overline{\operatorname{co}} S) \setminus B_{\varepsilon/2}(x_0)).$$

Assume that $S \subset Q$. Then $\overline{\operatorname{co}} S \subset Q$ and $x_0 \in Q$, a contradiction. So $S \not\subset Q$, and there exists an element $x_1 \in S \setminus Q$; necessarily $x_1 \in B_{\varepsilon/2}(x_0)$. Thus $B_{\varepsilon/2}(x_0) \subset B_\varepsilon(x_1)$, and so $S \setminus B_\varepsilon(x_1) \subset Q$, whence $\overline{\operatorname{co}}(S \setminus B_\varepsilon(x_1)) \subset Q$. This shows that

$$x_1 \in S \setminus \overline{\operatorname{co}}(S \setminus B_\varepsilon(x_1)),$$

and so S is dentable. □

Theorem 2.6.7. *Let Γ be any non-empty set. Then every non-empty, bounded subset of $\ell^1(\Gamma)$ is dentable.*

Proof. We shall work in the underlying real-linear space of $\ell^1(\Gamma)$.

By Lemma 2.6.6, it suffices to show that every non-empty, closed, convex, bounded set in $\ell^1(\Gamma)$ is dentable. Let S be such a set, and suppose without loss of generality that $\sup\{\|f\|_1 : f \in S\} = 1$. Take $\varepsilon > 0$.

Choose $f \in S$ with $\|f\|_1 > 1 - \varepsilon/6$. Then there is a finite subset F of Γ such that $\sum_{\gamma \in F} |f(\gamma)| > 1 - \varepsilon/6$. Let $P : \ell^1(\Gamma) \to \ell^1(F)$ be the natural projection, so that $\|Pf\|_1 > 1 - \varepsilon/6$.

The set $P(S)$ is convex and bounded in the finite-dimensional space $\ell^1(F)$, and so $\overline{P(S)}$ is convex and compact. By the Krein–Milman theorem, Theorem 2.6.1, there is an extreme point g_0 of $\overline{P(S)}$ with $\|g_0\|_1 > 1 - \varepsilon/6$, and so $g_0 \notin \overline{\mathrm{co}}\,(P(S) \setminus B_{\varepsilon/6}(g_0))$. By the Hahn–Banach theorem, Theorem 2.1.2(ii), there is a real-linear functional λ in the underlying real-linear space of $\ell^\infty(\Gamma)$ such that

$$\langle g_0, \lambda \rangle > 1 \quad \text{and} \quad \langle g, \lambda \rangle < 1 \quad (g \in P(S) \setminus B_{\varepsilon/6}(g_0)).$$

Choose $g \in S$ with $\|Pg - g_0\|_1 < \varepsilon/6$ and $\langle Pg, \lambda \rangle > 1$. We *claim* that

$$g \notin \overline{\mathrm{co}}\,(S \setminus B_\varepsilon(g)). \tag{2.11}$$

Indeed, take $h \in S$ with $\langle Ph, \lambda \rangle \geq 1$. Then $\|Ph - g_0\|_1 \leq \varepsilon/6$, and so we have $\|Ph - Pg\|_1 \leq \varepsilon/3$ and $\|Ph\|_1 > 1 - \varepsilon/3$; also, $\|Pg\|_1 > 1 - \varepsilon/3$. Since

$$\|Pg\|_1 + \|g - Pg\|_1 = \|g\|_1 \leq 1,$$

we have $\|g - Pg\|_1 < \varepsilon/3$; similarly, $\|h - Ph\|_1 < \varepsilon/3$. Thus $\|g - h\|_1 < \varepsilon$. It follows that $\langle Ph, \lambda \rangle < 1$ for each $h \in S \setminus B_\varepsilon(g)$, and hence $\langle Ph, \lambda \rangle < 1$ for each element $h \in \overline{\mathrm{co}}\,(S \setminus B_\varepsilon(g))$. Since $\langle Pg, \lambda \rangle > 1$, our claim that (2.11) holds is valid.

It follows that S is dentable. □

We remark that the following related theorem of Rieffel is proved in [190]; see also [123, Appendix 2].

Theorem 2.6.8. *Let E be a separable Banach space. Then every weakly compact, convex subset of E is dentable.* □

Definition 2.6.9. Let K be a closed, bounded, convex set in a Banach space E. Then K has the *Krein–Milman property* if $L = \overline{\mathrm{co}}(\mathrm{ex}\,L)$ for every closed, convex subset L of K. A Banach space E has the *Krein–Milman property* if $E_{[1]}$ has the Krein–Milman property.

Suppose that E has the Krein–Milman property. Then every closed, bounded, convex set in E has the Krein–Milman property. Suppose, further, that F is a Banach space with $F \sim E$. Then F has the Krein–Milman property; the Krein–Milman property is an isomorphic invariant.

The study of the Krein–Milman property is assisted by the *Bishop–Phelps theorem* from [37]; we state an extension of the theorem given by Bollobás [40]. For a proof, see [85, VII, Theorem 4] and [100, Theorem 7.41], for example.

Theorem 2.6.10. *Let E be a real Banach space. Suppose that $x \in S_E$, that $\lambda \in S_{E'}$, and that $\varepsilon > 0$. Then there exist $y \in S_E$ and $\mu \in S_{E'}$ such that $\langle y, \mu \rangle = 1$, such that $\|\mu - \lambda\| < \varepsilon$, and such that $\|y - x\| < \varepsilon + \varepsilon^2$.* □

A short, direct proof of the following theorem, using the Bishop–Phelps theorem, is given in the Handbook article of Johnson and Lindenstrauss [148, p. 35] and in [85, Theorem 5, p. 190]; see also [100, Theorem 11.3].

Theorem 2.6.11. *Let E be a Banach space for which every non-empty, bounded subset is dentable. Then E has the Krein–Milman property.* □

Corollary 2.6.12. *Let Γ be a non-empty set. Then $\ell^1(\Gamma)$ has the Krein–Milman property.*

Proof. This follows from Theorem 2.6.7 and the above theorem. □

The above results give examples of Banach spaces that do have the Krein–Milman property. We shall now show, in Theorem 2.6.15, that the spaces $C_0(K)$ never have the Krein–Milman property whenever K is infinite.

Proposition 2.6.13. *Let K be a non-empty, locally compact space, and suppose that $f \in C_0(K)$. Then $f \in \operatorname{ex} C_0(K)_{[1]}$ if and only if $|f(x)| = 1$ $(x \in K)$.*

Proof. Set $B = C_0(K)_{[1]}$. Suppose that $f \in B$ and that there exists $x_0 \in K$ such that $|f(x_0)| < 1$. Set $\varepsilon = (1 - |f(x_0)|)/2$. Then there exists $U \in \mathscr{N}_{x_0}$ with $|f(x)| < 1 - \varepsilon$ for $x \in U$. Take $g \in C_{\mathbb{R}}(K)$ such that $0 \leq g \leq \chi_U$ and $g(x_0) = 1$. Then $f \pm \varepsilon g \in B$ and

$$f = \frac{1}{2}(f + \varepsilon g) + \frac{1}{2}(f - \varepsilon g),$$

and so $f \notin \operatorname{ex} B$.

It is easy to see that each $f \in C_0(K)$ with $|f(x)| = 1$ $(x \in K)$ belongs to $\operatorname{ex} B$. □

Corollary 2.6.14. *Let K be a locally compact space that is not compact. Then $C_0(K)_{[1]}$ has no extreme points, and $C_0(K)$ is not isometrically a dual space.* □

In particular, we see again that c_0 is not isometrically a dual space.

Theorem 2.6.15. *Let K be an infinite, locally compact space. Then $C_0(K)$ does not have the Krein–Milman property.*

Proof. By Corollary 2.6.14, we may suppose that K is compact. Since K is infinite, there is a non-isolated point, say x_0, of K. Consider the set

$$\{f \in C(K)_{[1]} : f(x_0) = 0\} :$$

this set is closed, bounded, and convex in $C(K)$, but it follows from Proposition 2.6.13 that it has no extreme points. □

The results in the remainder of this section require more background than our guidelines indicate, and so we shall omit most proofs.

The main theorem relating the above properties is the following.

Theorem 2.6.16. *Let E be a Banach space. Then the following conditions on E are equivalent:*

(a) *E' has the Krein–Milman property;*

(b) *each bounded subset of E' is dentable;*

(c) *each separable subspace of E has a separable dual space.*

Proof. The implication (b) $\Rightarrow$ (a) follows from Theorem 2.6.11. For proofs of the other implications, see [85, pp. 190, 198], where histories of the theorems are also given. A key original source is a paper of Stegall [230]; see also [100, Theorem 11.14]. □

Corollary 2.6.17. *Let E be a separable Banach space. Then E' has the Krein–Milman property if and only if E' is separable.*

Proof. This follows from the equivalence (a) $\Leftrightarrow$ (c) of the above theorem. □

Corollary 2.6.18. *Let Γ be a non-empty set. Then $\ell^1(\Gamma)$ is isomorphically the dual of a separable Banach space if and only if Γ is countable.*

Proof. We have $\ell^1(\Gamma) \cong (c_0(\Gamma))'$ and $c_0(\Gamma)$ is separable whenever Γ is countable.

Now suppose that $\ell^1(\Gamma) \sim E'$ for a separable Banach space E. By Corollary 2.6.12, $\ell^1(\Gamma)$ has the Krein–Milman property, and so E' has this property. By Corollary 2.6.17, E' is separable, and so $\ell^1(\Gamma)$ is separable. Hence Γ is countable. □

Corollary 2.6.19. *Let E be a separable Banach space such that* $\operatorname{ex} E_{[1]} = \emptyset$. *Then E is not isomorphically a dual space.*

Proof. Assume that $E \sim F'$ for a Banach space F. By Proposition 2.1.6, F is separable, and so, by Corollary 2.6.17, F' has the Krein–Milman property, and hence E has this property. In particular, $\operatorname{ex} E_{[1]} \neq \emptyset$, a contradiction. □

We outline, without defining terms, a proof of one implication in Corollary 2.6.17, namely, of the fact that E' has the Krein–Milman property whenever E' is separable; this implication will be used in the proof of Theorem 4.4.17(i). The proof uses an idea of Bessaga and Pełczyński [35] concerning a re-norming theorem of Kadec and Klee for spaces with a separable dual. The full proofs are available in readily accessible texts, but this argument may not be as well known as some others.

The first step is as follows.

Proposition 2.6.20. *Let E be a Banach space such that E' is separable. Then E admits an equivalent norm which is Fréchet differentiable at every $x \in E$ with $x \neq 0$.*

Proof. An explicit formula for such an equivalent norm on E' is given in [33, Theorem 4.13, p. 89]. This norm is shown to be the dual of the desired equivalent Fréchet-differentiable norm on E. □

Proposition 2.6.21. *Let E be a Banach space whose norm is is Fréchet differentiable at every $x \in E$ with $x \neq 0$. Then E' is dentable.*

Proof. This is also a standard result; it is again a straightforward application of the Bishop–Phelps theorem, Theorem 2.6.10. See [100, Proposition 8.11, p. 391], for example. □

The stated implication in Corollary 2.6.17 now follows from Theorem 2.6.11.

There is another elegant proof of the above proposition due to Namioka [189]. This article introduced and crystallized the important concept of points of weak*-to-norm continuity of the identity map on a dual Banach space (although the concept was already implicit in Bessaga–Pełczyński [35]). Namioka's proof is reproduced in [83, p. 159]; the original article is not cited in [83].

Let E be a Banach space. The *Radon–Nikodým property* for E delineates when there is an E-valued version of the standard Radon–Nikodým theorem: see [85, III.1] and [100], for example. The fine text [85] contains many different characterizations of the Radon–Nikodým property. See Chapters III, IV, and VII of [85] for a discussion of this property and some of its variants; in particular, pages 217/218 summarize many equivalent formulations of this property, and pages 218/219 specify many spaces that do and do not have the property. Each Banach space with the Radon–Nikodým property has the Krein–Milman property; it is not known whether the converse of this statement holds.

It is shown in [85, pp. 190, 198] and [100, Theorem 11.14] that the three clauses in Theorem 2.6.16 are also equivalent to the condition that E' have the Radon–Nikodým property.

Chapter 3
Banach Algebras and C^*-Algebras

This chapter will first give the basic background that we shall require concerning Banach algebras, C^*-algebras, and von Neumann algebras. In particular, in §3.1, we shall discuss the bidual of a Banach algebra, taken with its Arens products. In §3.3, we shall exhibit the Baire classes as examples of commutative C^*-algebras. We shall conclude the chapter in §3.4 with a few remarks on the generalizations of some of our discussions concerning the commutative C^*-algebras $C_0(K)$ to general (non-commutative) C^*-algebras; as we said, these generalizations will not be used within our main text.

3.1 Banach algebras

For the theory of Banach algebras, we refer to the monograph of Dales [68]; here we recall a few very basic facts.

Let A be a (complex, associative) algebra. The algebra formed by adjoining an identity to A is denoted by $A^\#$, with $A^\# = A$ when A already has an identity, in which case A is *unital*. The identity of $A^\#$ is denoted by e_A.

Let S be a subset of an algebra A. Then the *commutant* of S is

$$S^c = \{b \in A : ab = ba \ (a \in S)\},$$

so that S^c is a subalgebra of A; we set $S^{cc} = (S^c)^c$, the *double commutant* of S. Clearly $S \subset S^{cc}$.

A linear subspace I of an algebra A is a *left* (respectively, *right*) *ideal* if $ax \in I$ (respectively, $xa \in I$) whenever $a \in A$ and $x \in I$; I is an *ideal* if it is both a left and a right ideal. A left/right ideal I in an algebra A is *modular* if there exists $u \in A$ such that $a - au \in I / a - ua \in I$ for each $a \in A$; every proper modular left/right ideal is contained in a maximal modular left/right ideal. The (Jacobson) *radical* of A, denoted by $J(A)$, is defined to be the intersection of the maximal modular left ideals

H.G. Dales et al., *Banach Spaces of Continuous Functions as Dual Spaces*,
CMS Books in Mathematics, DOI 10.1007/978-3-319-32349-7_3

of A; it is also equal to the intersection of the maximal modular right ideals of A, and so it is an ideal in A; the algebra A is *semi-simple* if $J(A) = \{0\}$ and *radical* if there are no maximal modular left ideals, so that $J(A) = A$. By [68, Theorem 1.5.32(ii)], $J(A)$ consists of the elements $a \in A$ such that $e_A - ba$ is invertible in $A^\#$ for each $b \in A^\#$. The quotient algebra $A/J(A)$ is a semi-simple algebra.

An element a in an algebra A is *nilpotent* if $a^n = 0$ for some $n \in \mathbb{N}$. The *spectrum*, $\sigma(a)$, of a is the complement in $\mathbb{C}$ of the set

$$\{z \in \mathbb{C} : ze_A - a \text{ is invertible in } A^\#\};$$

an element $a \in A$ is *quasi-nilpotent* if $\sigma(a) \subset \{0\}$, and the set of quasi-nilpotent elements of A is denoted by $\mathscr{Q}(A)$. Clearly each nilpotent element is quasi-nilpotent and $J(A) \subset \mathscr{Q}(A)$.

Let A and B be algebras. A *homomorphism* $\theta : A \to B$ is a linear map such that

$$\theta(ab) = \theta(a)\theta(b) \quad (a, b \in A);$$

a bijective homomorphism is an *isomorphism*, and the algebras A and B are *isomorphic* if there is an isomorphism from A onto B. Now suppose that A and B are unital algebras. Then a homomorphism $\theta : A \to B$ is *unital* if $\theta(e_A) = e_B$.

Suppose that A and B are unital algebras. We note the triviality that an isomorphism $\theta : A \to B$ is necessarily unital. For suppose that $\theta(e_A) = p \in B$ and $\theta^{-1}(e_B) = q \in A$. Then $e_B = \theta(q) = \theta(e_A)\theta(q) = pe_B = p$.

Definition 3.1.1. A *character* on an algebra A is a homomorphism from A onto $\mathbb{C}$. The set of all characters on A is denoted by Φ_A; this is the *character space* of A.

Definition 3.1.2. An algebra A that is also a normed space for a norm $\|\cdot\|$ is a *normed algebra* if

$$\|ab\| \leq \|a\| \, \|b\| \quad (a, b \in A);$$

it is a *Banach algebra* if $(A, \|\cdot\|)$ is a Banach space; it is a *unital* Banach algebra if, further, A is a unital algebra and $\|e_A\| = 1$.

For example, let E be a Banach space. Then $\mathscr{B}(E)$ is a unital Banach algebra with respect to the composition of operators; $\mathscr{B}(E)$ is semi-simple. Let E and F be non-zero Banach spaces. It is an interesting old theorem of Eidelheit (see [68, Theorem 2.3.7]) that $\mathscr{B}(E)$ and $\mathscr{B}(F)$ are algebraically isomorphic (by an isomorphism that is automatically continuous) if and only if $E \sim F$.

Let A be a Banach algebra. Then the spectrum $\sigma(a)$ is a non-empty, compact subset of $\mathbb{C}$ for each $a \in A$. Further, the radical $J(A)$ is always a closed ideal in A, and $A/J(A)$ is a semi-simple Banach algebra. All characters φ on A are continuous, with $\|\varphi\| \leq 1$, and Φ_A is a locally compact space with respect to the relative weak* topology, $\sigma(A', A)$.

Now suppose that A is a unital Banach algebra. Then Φ_A is compact; in this case, we shall consider the set

$$K_A = \{\lambda \in A' : \|\lambda\| = \langle e_A, \lambda\rangle = 1\};$$

clearly, K_A is a weak*-compact, convex subset of A', and $\Phi_A \subset K_A$.

Now suppose that A is a commutative Banach algebra. Then

$$\sigma(a) = \{\varphi(a) : \varphi \in \Phi_{A^\#}\};$$

the maximal modular ideals of A have codimension 1, and they are exactly the kernels of the characters on A. In the case where A is not radical, so that $\Phi_A \neq \emptyset$, define $\widehat{a}$ for $a \in A$ by

$$\widehat{a}(\varphi) = \varphi(a) \quad (\varphi \in \Phi_A).$$

Then $\widehat{a} \in C_0(\Phi_A)$, and the *Gel'fand transform*, defined by

$$\mathscr{G} : a \mapsto \widehat{a}, \quad A \to C_0(\Phi_A),$$

is a contractive homomorphism. The kernel of $\mathscr{G}$ is $J(A)$, and it consists of the quasi-nilpotent elements, so that A is semi-simple if and only if $\mathscr{G}$ is an injection if and only if $\mathscr{Q}(A) = \{0\}$.

The following result is immediate.

Proposition 3.1.3. *Let A and B be Banach algebras, and suppose that $\theta : A \to B$ is a continuous homomorphism. Then $\theta' \mid \Phi_B : \Phi_B \to \Phi_A \cup \{0\}$ is a continuous map.* □

Let K be a non-empty, locally compact space, and take $x \in K$. Then $C_0(K)$ is a commutative Banach algebra, and

$$\varepsilon_x : f \mapsto f(x), \quad C_0(K) \to \mathbb{C},$$

is a character on $C_0(K)$, called the *evaluation character* at x, and

$$M_x := \{f \in C_0(K) : f(x) = 0\} = \ker \varepsilon_x$$

is a maximal modular ideal of $C_0(K)$; further, every maximal modular ideal of $C_0(K)$ has this form, equivalently, all characters on $C_0(K)$ are evaluation characters, and so $\Phi_{C_0(K)} = K$.

We take J_x to be the set of functions in $C_{00}(K)$ that vanish on a neighbourhood of x, so that J_x is an ideal in $C_0(K)$ and J_x is dense in M_x. It is easy to see that x is a P-point of K if and only if $M_x = J_x$ and that K is an F-space if and only if J_x is a prime ideal in $C_0(K)$ for each $x \in K$ [68, Proposition 4.2.18].

The notions of a positive linear functional, a state, and a pure state on an ordered linear space were given in §1.3; see page 9. Let K be a non-empty, compact space. Then 1_K is of course an order unit for the ordered linear space $C_{\mathbb{R}}(K)$. A linear functional on $C(K)$ is *positive* or a *state* or a *pure state* if its restriction to $C_{\mathbb{R}}(K)$ has the corresponding property (with respect to the order unit 1_K).

The following result relates characters and pure states.

Proposition 3.1.4. *Let K be a non-empty, compact space. Then a linear functional on $C(K)$ is a pure state if and only if it is a character on $C(K)$.*

Proof. A character on $C(K)$ has the form ε_x for some $x \in K$. Take a positive linear functional μ on $C(K)$ with $\mu \leq \varepsilon_x$. For each $f \in M_x^+$, we have $\mu(f) = 0$. Since $M_x = \lim M_x^+$, we see that $\mu \mid M_x = 0$. Thus μ is a scalar multiple of ε_x. By Lemma 1.3.3, ε_x is a pure state.

Conversely, suppose that λ is a pure state on $C(K)$. For each $f \in C(K, \mathbb{I})$, set $\mu(g) = \lambda(fg)$ $(g \in C(K))$. Then μ is a positive linear functional on $C(K)$ such that $\mu \leq \lambda$, and so $\mu = t\lambda$ for some $t \in \mathbb{R}^+$. Suppose that $h \in \ker \lambda$. Then $\lambda(fh) = \mu(h) = t\lambda(h) = 0$, and so $\ker \lambda$ is an ideal in $C(K)$. Since it has codimension 1 in $C(K)$, it is a maximal ideal, and so λ is an evaluation character. □

Corollary 3.1.5. *Let K and L be non-empty, compact spaces. Then each Riesz isomorphism $\theta : C(L) \to C(K)$ such that $\theta(1_L) = 1_K$ has the form $\theta = \eta^\circ$ for some homeomorphism $\eta : K \to L$.*

Proof. For each $x \in K$, the character ε_x is a pure state on $C(K)$, and so $\varepsilon_x \circ \theta$ is a pure state on $C(L)$, noting that $\theta(1_L) = 1_K$. By Proposition 3.1.4, there exists a unique point in L, say $\eta(x)$, such that $\varepsilon_x \circ \theta = \varepsilon_{\eta(x)}$. The map $\eta : K \to L$ is a bijection, and $(f \circ \eta)(x) = \theta(f)(x)$ $(x \in K)$.

Sets of the form $f^{-1}(U)$, with $f \in C(L)$ and U open in $\mathbb{C}$, form a subbase of the topology of L. Since $\eta^{-1}(f^{-1}(U)) = (\theta(f))^{-1}(U)$, an open set in K, it follows that η is continuous, and so η is a homeomorphism. Clearly $\theta = \eta^\circ$. □

We recall the following standard fact about the closed ideals of $C_0(K)$ [68, Theorem 4.2.1(iii)].

Proposition 3.1.6. *Let K be a non-empty, locally compact space, and take F to be a closed subspace of K. Then*

$$I(F) := \{f \in C_0(K) : f \mid F = 0\}$$

is a closed ideal in $C_0(K)$; further, every closed ideal in $C_0(K)$ has the form $I(F)$ for some closed subspace F of K. □

Let A be a Banach algebra. Then a construction originally due to Arens [11, 12] from 1951 shows that the bidual space A'' of A is a Banach algebra with respect to two products, which we shall denote by $\Box$ and $\Diamond$, respectively, and that the natural embedding of A into its bidual identifies A as a closed subalgebra of both $(A'', \Box)$ and $(A'', \Diamond)$. These products are called the *first* and *second Arens products* on A''. We recall briefly the definitions of $\Box$ and $\Diamond$; for further details, see [68, §2.6].

Let A be a Banach algebra. First, for $\lambda \in A'$, we have

$$\langle b, a \cdot \lambda \rangle = \langle ba, \lambda \rangle, \quad \langle b, \lambda \cdot a \rangle = \langle ab, \lambda \rangle \quad (a, b \in A). \tag{3.1}$$

We see that the Banach space A' is a bimodule over A and that $\|a \cdot \lambda\| \leq \|a\| \|\lambda\|$ and $\|\lambda \cdot a\| \leq \|a\| \|\lambda\|$ for all $a \in A$ and $\lambda \in A'$. Thus A' is a *Banach A-bimodule* in the sense of [68, §2.6].

Now, for $\lambda \in A'$ and $\mathrm{M} \in A''$, define $\lambda \cdot \mathrm{M}$ and $\mathrm{M} \cdot \lambda$ in A' by

$$\langle a, \lambda \cdot \mathrm{M} \rangle = \langle \mathrm{M}, a \cdot \lambda \rangle, \quad \langle a, \mathrm{M} \cdot \lambda \rangle = \langle \mathrm{M}, \lambda \cdot a \rangle \quad (a \in A). \tag{3.2}$$

Again, $\|\lambda \cdot \mathrm{M}\| \leq \|\lambda\| \|\mathrm{M}\|$ and $\|\mathrm{M} \cdot \lambda\| \leq \|\lambda\| \|\mathrm{M}\|$ for all $\lambda \in A'$ and $\mathrm{M} \in A''$.

Finally, for $\mathrm{M}, \mathrm{N} \in A''$, define

$$\langle \mathrm{M} \Box \mathrm{N}, \lambda \rangle = \langle \mathrm{M}, \mathrm{N} \cdot \lambda \rangle, \quad \langle \mathrm{M} \Diamond \mathrm{N}, \lambda \rangle = \langle \mathrm{N}, \lambda \cdot \mathrm{M} \rangle \quad (\lambda \in A'). \tag{3.3}$$

Again, $\|\mathrm{M} \Box \mathrm{N}\| \leq \|\mathrm{M}\| \|\mathrm{N}\|$ and $\|\mathrm{M} \Diamond \mathrm{N}\| \leq \|\mathrm{M}\| \|\mathrm{N}\|$ for all $\mathrm{M}, \mathrm{N} \in A''$.

With respect to the maps $(a, \mathrm{M}) \mapsto a \Box \mathrm{M}$ and $(a, \mathrm{M}) \mapsto \mathrm{M} \Box a$ from $A \times A'' \to A''$, the space A'' is a Banach A-bimodule, and the two products $\Box$ and $\Diamond$ extend these module operations.

The following theorem of Arens from [12] is proved in [68, Theorem 2.6.15].

Theorem 3.1.7. *Let A be a Banach algebra. Then both $(A'', \Box)$ and $(A'', \Diamond)$ are Banach algebras containing A as a closed subalgebra.* □

It follows from the definition that, for $\mathrm{M}, \mathrm{N} \in A''$, we have

$$\mathrm{M} \Box \mathrm{N} = \lim_\alpha \lim_\beta a_\alpha b_\beta \quad \text{and} \quad \mathrm{M} \Diamond \mathrm{N} = \lim_\beta \lim_\alpha a_\alpha b_\beta \quad \text{in} \quad (A'', \sigma(A'', A'))$$

whenever (a_α) and (b_β) are nets in A which are weak*-convergent to M and N, respectively. In the case where A is a commutative algebra,

$$\mathrm{M} \Diamond \mathrm{N} = \mathrm{N} \Box \mathrm{M} \quad (\mathrm{M}, \mathrm{N} \in A''),$$

so that $(A'', \Diamond)$ is just the algebra $(A'', \Box)$ with the 'opposite' multiplication.

Definition 3.1.8. A Banach algebra A is *Arens regular* if the two products $\Box$ and $\Diamond$ agree on A''.

Thus a commutative Banach algebra A is Arens regular if and only if $(A'', \Box)$ is commutative.

Definition 3.1.9. Let A be an algebra, and suppose that $* : A \to A$ is a linear involution on A. Then $*$ is an *involution* on A if

$$(ab)^* = b^* a^* \quad (a, b \in A).$$

An algebra with an involution is a *$*$-algebra*, and a Banach algebra with an isometric involution is a *Banach $*$-algebra*.

For studies of Banach $*$-algebras, see [68, §3.1] and the monumental monographs [194, 195] of Palmer. For example, the group algebra $L^1(G)$ on a locally compact abelian group G (see page 132) is a Banach $*$-algebra with respect to the involution defined by $f \mapsto f^*$, where

$$f^*(s) = \overline{f(s^{-1})} \quad (s \in G).$$

Let $(A, *)$ be a $*$-algebra. Recall that $A_{\mathrm{sa}} = \{a \in A : a^* = a\}$ is the real-linear subspace of A consisting of the self-adjoint elements of A. We set

$$A^+ = \left\{ \sum_{j=1}^{n} a_j^* a_j : a_1, \dots, a_n \in A, n \in \mathbb{N} \right\} \subset A_{\mathrm{sa}};$$

the algebra A is *ordered* if $A^+ \cap (-A^+) = \{0\}$. For $a, b \in A_{\mathrm{sa}}$, we say that $a \le b$ if $b - a \in A^+$. In the case where A is ordered, $(A_{\mathrm{sa}}, \le)$ is an ordered linear space. A linear functional λ on A is *positive* if

$$\lambda(a^*a) \ge 0 \quad (a \in A).$$

Suppose that $(A, *)$ is a unital $*$-algebra. Then a positive linear functional λ on A is a *state* if $\langle e_A, \lambda \rangle = 1$.

Let A and B be $*$-algebras. A homomorphism $\theta : A \to B$ is a *$*$-homomorphism* if it is $*$-linear, so that $\theta(a^*) = \theta(a)^*$ $(a \in A)$.

Let A be a Banach $*$-algebra. Then, as we explained in §2.1, the map $*$ extends to an isometric linear involution on A''. In general, this extended linear involution is not an involution on $(A'', \Box)$.

Theorem 3.1.10. *Let A be a Banach $*$-algebra. Then the extended linear involution on $(A'', \Box)$ is an involution if and only if A is Arens regular.*

Proof. Let $\mathrm{M}, \mathrm{N} \in A''$. Then there are nets (a_α) and (b_β) in A such that $\mathrm{M} = \lim_\alpha a_\alpha$ and $\mathrm{N} = \lim_\beta b_\beta$ in $(A'', \sigma(A'', A'))$. We have

$$(\mathrm{M} \Box \mathrm{N})^* = \lim_\alpha \lim_\beta (a_\alpha b_\beta)^* = \lim_\alpha \lim_\beta b_\beta^* a_\alpha^* = \mathrm{N}^* \Diamond \mathrm{M}^*.$$

Thus $(\mathrm{M} \Box \mathrm{N})^* = \mathrm{N}^* \Box \mathrm{M}^*$ for all $\mathrm{M}, \mathrm{N} \in A''$ if and only if A is Arens regular. □

We shall see in Theorems 4.5.5 and 5.4.1 that the Banach algebras $C_0(K)$ are Arens regular. Indeed all C^*-algebras are Arens regular; see Theorem 5.6.1. The standard examples of Banach algebras which are not Arens regular are the group algebra $L^1(G)$ (see page 132) and the measure algebra (see page 112) $M(G)$ of an infinite, locally compact group G. Indeed, these Banach algebras are *strongly Arens irregular*; see [72] for a definition of 'strongly Arens irregular', a discussion, and a proof that group algebras are strongly Arens irregular; and see Example 4.2.15, below, for a specific related example. A proof that the measure algebra $M(G)$ is strongly Arens irregular for each locally compact group G, so resolving a long-standing question, is given in [177]. For examples of commutative Banach algebras that are neither Arens regular nor strongly Arens irregular, see [70].

3.2 C^*-algebras

A fundamentally important class of Banach algebras is that consisting of the C^*-algebras. Standard texts on C^*-algebras include those of Kadison and Ringrose [149, 150], Sakai [222], and Takesaki [234].

Definition 3.2.1. A Banach algebra A is a *C^*-algebra* if it has an involution, denoted by $*$, and if

$$\|a^*a\| = \|a\|^2 \quad (a \in A).$$

A *C^*-subalgebra* of A is a subalgebra of A which is $*$-closed and norm-closed.

Let A be a C^*-algebra. It follows immediately that $\|a^*\| = \|a\| \ (a \in A)$, and so a C^*-algebra is a Banach $*$-algebra.

For each non-empty, locally compact space K, the Banach spaces $C_0(K)$ and $C^b(K)$ are commutative C^*-algebras with respect to the pointwise product and involution given by the conjugation of functions, so that $f^* = \overline{f} \ (f \in C^b(K))$.

Let A be a C^*-algebra. Then A is an ordered algebra, so that $(A_{\rm sa}, \leq)$ is an ordered linear space. We have defined A^+; in fact, in this case, $a \in A_{\rm sa}$ belongs to A^+ if and only if $\sigma(a) \subset \mathbb{R}^+$, equivalently, if and only if $a = b^*b$ for some $b \in A$. In the case where A is unital, e_A is an order unit for $(A_{\rm sa}, \leq)$. Further, a linear functional λ on A is positive if and only if it is continuous and $\|\lambda\| = \langle e_A, \lambda \rangle$. See [68, Propositions 3.2.8 and 3.2.14] for these results. It follows that a linear functional λ on A is a state if and only if $\lambda \mid A_{\rm sa}$ is a state in the sense of page 9, and so the sets K_A defined on pages 9 and 95 can be identified for a unital C^*-algebra A; by [234, Exercise IV.6.2(b)], the space K_A is a simplex if and only if A is commutative.

We shall use the following result, which combines Corollary 3.2.4 and Theorem 3.2.23 of [68].

Proposition 3.2.2. *Let A and B be C^*-algebras, and let $\theta : A \to B$ be a $*$-homomorphism. Then θ is a contraction and the range $\theta(A)$ is a C^*-subalgebra of B. A $*$-monomorphism is an isometry onto its range.* □

Let A and B be C^*-algebras. A *C^*-homomorphism* from A to B is a $*$-homomorphism; by Proposition 3.2.2, such a map is necessarily contractive. The map is a *C^*-embedding* if, further, it is an injection and a *C^*-isomorphism* if, further, it is a bijection; A and B are *C^*-isomorphic* if there is a C^*-isomorphism from A onto B.

We shall also use the following famous *Gel'fand–Naimark theorem* for commutative C^*-algebras; see [68, Theorem 3.2.6].

Theorem 3.2.3. *Let A be a non-zero, commutative C^*-algebra. Then $\Phi_A \neq \emptyset$ and the Gel'fand transform*

$$\mathscr{G} : A \to C_0(\Phi_A)$$

is a C^*-isomorphism. In the case where A is unital, the space Φ_A is compact and $\mathscr{G} : A \to C(\Phi_A)$ is a unital C^*-isomorphism. □

For example, let X be a non-empty, completely regular topological space. Then $C^b(X)$ is a commutative, unital C^*-algebra, with character space βX, so that the Gel'fand transform identifies $C^b(X)$ with $C(\beta X)$. The connection between points p in βX, z-ultrafilters on X, and maximal ideals in $C^b(X)$ was described in the Gel'fand–Kolmogorov theorem, Theorem 1.5.2; now we further identify βX with the character space of the C^*-algebra $C^b(X)$. In particular, for each non-empty set S, the Banach space $\ell^\infty(S)$ is a commutative, unital C^*-algebra for the pointwise product, and its character space is homeomorphic to βS. We shall now generalize this latter remark.

Let $\{K_\alpha : \alpha \in A\}$ be a family of non-empty, compact spaces, and set

$$\mathfrak{A} = \bigoplus_\infty \{C(K_\alpha) : \alpha \in A\},$$

with coordinatewise algebraic operations; as before, the norm is given by

$$\|(f_\alpha)\| = \sup\{|f_\alpha|_{K_\alpha} : \alpha \in A\} \quad ((f_\alpha) \in \mathfrak{A}).$$

Then it is clear that $\mathfrak{A}$ is a commutative, unital C^*-algebra.

We define $U_\mathfrak{A}$ as a set to be the disjoint union of the sets K_α, and give $U_\mathfrak{A}$ the topology in which each K_α is a compact and open subspace of $U_\mathfrak{A}$. Then $U_\mathfrak{A}$ is a locally compact space, and clearly $\mathfrak{A}$ is C^*-isomorphic to $C^b(U_\mathfrak{A})$ and hence to $C(\beta U_\mathfrak{A})$.

Alternatively, in the language of commutative C^*-algebras, take $\Phi_\mathfrak{A}$ to be the character space of $\mathfrak{A}$, so that $\mathscr{G} : \mathfrak{A} \to C(\Phi_\mathfrak{A})$ is a C^*-isomorphism. Take $\alpha \in A$, and write 1_α for the element (f_β) in $\mathfrak{A}$ such that $f_\alpha = 1_{K_\alpha}$ and $f_\beta = 0$ for $\beta \neq \alpha$, so that each 1_α is an idempotent in $\mathfrak{A}$. Take $\alpha \in A$ and $x \in K_\alpha$. Then the map

$$\varphi_x : (f_\beta) \mapsto f_\alpha(x), \quad \mathfrak{A} \to \mathbb{C},$$

is a character on $\mathfrak{A}$, and the map $x \mapsto \varphi_x$, $K_\alpha \to \Phi_\mathfrak{A}$, is a homeomorphism onto a compact subspace of $\Phi_\mathfrak{A}$, which we identify with K_α. Clearly $K_\alpha \cap K_\beta = \emptyset$ when $\alpha, \beta \in A$ with $\alpha \neq \beta$. For each $\alpha \in A$, we have $K_\alpha = \{\varphi \in \Phi_\mathfrak{A} : \varphi(1_\alpha) = 1\}$, and so K_α is clopen in $\Phi_\mathfrak{A}$. Further, $U_\mathfrak{A} = \bigcup\{K_\alpha : \alpha \in A\}$ and $U_\mathfrak{A}$ is a dense, open subspace of $\Phi_\mathfrak{A}$.

Theorem 3.2.4. *Let $\{K_\alpha : \alpha \in A\}$ and $\mathfrak{A}$ be as above. Then $\Phi_\mathfrak{A}$ is homeomorphic to $\beta U_\mathfrak{A}$, and $\mathfrak{A}$ is C^*-isomorphic to $C(\beta U_\mathfrak{A})$ and to $C^b(U_\mathfrak{A})$.*

Proof. Take $F \in C^b(U_{\mathfrak{A}})$. Then $(F \mid K_\alpha : \alpha \in A) \in \mathfrak{A}$, and so F defines an element, say $\widetilde{F}$, of $C(\Phi_{\mathfrak{A}})$; clearly $\widetilde{F} \mid K_\alpha = F \mid K_\alpha \;\; (\alpha \in A)$, and so $\widetilde{F}$ is a continuous extension of F to $\Phi_{\mathfrak{A}}$. Further,

$$\left|\widetilde{F}\right|_{\Phi_{\mathfrak{A}}} = \|(F \mid K_\alpha)\| = |F|_{U_{\mathfrak{A}}} .$$

The result follows. □

Let K and L be two non-empty, compact spaces, and suppose that $\eta : K \to L$ is a continuous surjection, so that (K, η) is a cover of L. As in equation (2.9), we define the map

$$\eta^\circ : f \mapsto f \circ \eta , \quad C(L) \to C(K);$$

we remarked on page 83 that η° is an isometric embedding of $C(L)$ onto the closed subspace $\eta^\circ(C(L))$ of $C(K)$. It is clear that this map is a unital C^*-embedding and that $\eta^\circ(C(L))$ is a C^*-subalgebra of $C(L)$.

Conversely, suppose that A is a C^*-subalgebra of $C(K)$. For $x, y \in K$, set

$$x \sim y \quad \text{if} \quad f(x) = f(y) \quad (f \in A).$$

Then $\sim$ is an equivalence relation on K, and the character space Φ_A is identified with the compact space $L = K/\sim$, the quotient of K; the quotient map $\eta : K \to L$ is a continuous surjection, and $\eta^\circ(C(L))$ is equal to A.

Definition 3.2.5. Let K and L be two non-empty, compact spaces and suppose that $\eta : K \to L$ is a continuous surjection. Then a continuous operator $P \in \mathscr{B}(C(K), C(L))$ such that $P \circ \eta^\circ = I_{C(L)}$ is an *averaging operator for* η.

Let us regard $C(L)$ as a C^*-subalgebra of $C(K)$. Then an averaging operator is a projection in $\mathscr{B}(C(K))$ with range $A = C(L)$, and so such an averaging operator exists if and only if A is a complemented subspace of $C(K)$.

The above terminology was developed by Pełczyński in [198, §2], and has been used by several authors; for example, see [39, §8.5]. We shall return to averaging operators in §6.8, and we shall extend the following proposition by giving more equivalent conditions in Theorem 6.1.4.

Theorem 3.2.6. *Let K and L be two non-empty, compact spaces. Then the following are equivalent:*

(a) *the spaces K and L are homeomorphic;*

(b) *there is a Banach-lattice isometry from $C(L)$ onto $C(K)$;*

(c) *there is a Riesz isomorphism from $C(L)$ onto $C(K)$;*

(d) *$C(L)$ and $C(K)$ are C^*-isomorphic;*

(e) *there is an algebra isomorphism from $C(L)$ onto $C(K)$.*

Proof. (a) $\Rightarrow$ (b), (d) Suppose that $\eta : K \to L$ is a homeomorphism. Then η° is a unital C^*-isomorphism, giving (d), and η° is an isometry. The map

$$\eta^\circ \mid C_{\mathbb{R}}(L) : C_{\mathbb{R}}(L) \to C_{\mathbb{R}}(K)$$

is positive and hence a Riesz isomorphism. Thus $\eta^\circ : C(L) \to C(K)$ is a Banach-lattice isomorphism, giving (b).

(b) $\Rightarrow$ (c) and (d) $\Rightarrow$ (e) These are trivial.

(c) $\Rightarrow$ (a) Let $\theta : C(L) \to C(K)$ be a Riesz isomorphism, and set $f_0 = \theta(1_L)$, so that $f_0 \in C(K)^+$.

For each $g \in C_{\mathbb{R}}(L)$, there exists $\alpha > 0$ such that $-\alpha f_0 \leq \theta(g) \leq \alpha f_0$ in $C_{\mathbb{R}}(K)$. Assume that there exists $x \in K$ with $f_0(x) = 0$. Then $\theta(g)(x) = 0$ $(g \in C_{\mathbb{R}}(L))$, a contradiction of the fact that θ is a surjection. Thus $\mathbf{Z}_K(f_0) = \emptyset$, and f_0 is invertible in $C(K)$ with $1/f_0 \in C(K)^+$. By replacing θ by the Riesz isomorphism $g \mapsto \theta(g)/f_0$, we may suppose that $\theta(1_L) = 1_K$.

By Corollary 3.1.5, there is a homeomorphism $\eta : K \to L$ with $\theta = \eta^\circ$.

(e) $\Rightarrow$ (a) Suppose that $\theta : C(L) \to C(K)$ is an algebra isomorphism. For each $x \in K$, the functional $\varepsilon_x \circ \theta$ is a character on $C(L)$, and so $\varepsilon_x \circ \theta = \varepsilon_{\eta(x)}$ for a unique point $\eta(x)$ in L. As in Corollary 3.1.5, $\eta : K \to L$ is a homeomorphism. □

3.3 Borel functions and Baire classes

We now give some examples of commutative C^*-algebras involving Borel and Baire functions. We recall that $\mathfrak{B}_X$ denotes the Boolean algebra of all Borel subsets of a topological space X; see Example 1.7.17.

Definition 3.3.1. Let X be a non-empty topological space. A complex-valued function f on X is a *Borel function* if $f^{-1}(U) \in \mathfrak{B}_X$ for each open set U in $\mathbb{C}$. The space of all bounded Borel functions on X is denoted by $B^b(X)$; the space of real-valued functions in $B^b(X)$ is $B^b_{\mathbb{R}}(X)$.

Let X be a non-empty topological space. Then the space $(B^b_{\mathbb{R}}(X), |\cdot|_X)$ is a real Banach lattice for the usual definitions of $\vee$ and $\wedge$, and $B^b(X)$ is its complexification, so that $B^b(X)$ is a Banach lattice. It is clear that $B^b(X)$ is Dedekind σ-complete. However, in the case where there is a non-Borel set S in X (for example, when X is an uncountable Polish space, as in Corollary 1.4.15), the lattice $B^b(X)$ is not Dedekind complete because the family of characteristic functions of finite subsets of S, ordered by inclusion, is an increasing net in $B^b(X)$ that does not have a supremum. It is also clear that $(B^b(X), |\cdot|_X)$ is a commutative, unital C^*-algebra with respect to the pointwise product; $C^b(X)$ is a C^*-subalgebra of $(B^b(X), |\cdot|_X)$, and $B^b(X)$ is a C^*-subalgebra of $(\ell^\infty(X), |\cdot|_X)$.

Definition 3.3.2. Let X be a non-empty topological space. The character space of $B^b(X)$ is denoted by $\Phi_b(X)$.

Thus $B^b(X)$ is identified with $C(\Phi_b(X))$ as a unital C^*-algebra and as a Banach lattice, so that $C(\Phi_b(X))$ is a Dedekind σ-complete Banach lattice.

Proposition 3.3.3. (i) *Let X be a non-empty topological space. Then the space $\Phi_b(X)$ is basically disconnected and homeomorphic to $St(\mathfrak{B}_X)$.*

(ii) *Let X be an infinite, Hausdorff topological space. Then $|\Phi_b(X)| \geq 2^{\mathfrak{c}}$; further, $|\Phi_b(X)| = 2^{\mathfrak{c}}$ whenever $|\mathfrak{B}_X| = \mathfrak{c}$.*

Proof. (i) That $\Phi_b(X)$ is basically disconnected follows from Theorem 2.3.3.

For $B \in \mathfrak{B}_X$, we recall that $S_B = \{p \in St(\mathfrak{B}_X) : B \in p\}$ is a clopen subspace of $St(\mathfrak{B}_X)$; we define $T(\chi_B) = \chi_{S_B}$. The map T clearly extends to an algebra $*$-isomorphism from $\mathrm{lin}\{\chi_B : B \in \mathfrak{B}_X\}$ onto $\mathrm{lin}\{\chi_{S_B} : B \in \mathfrak{B}_X\}$, and this extension is an isometry. Since $\mathrm{lin}\{\chi_B : B \in \mathfrak{B}_X\}$ and $\mathrm{lin}\{\chi_{S_B} : B \in \mathfrak{B}_X\}$ are dense in $B^b(X)$ and $C(St(\mathfrak{B}_X))$, respectively, we can further extend T to obtain a C^*-isomorphism from $B^b(X)$ onto $C(St(\mathfrak{B}_X))$. Thus, by Theorem 3.2.6, (d) $\Rightarrow$ (a), $\Phi_b(X)$ is homeomorphic to $St(\mathfrak{B}_X)$.

(ii) By Theorem 1.7.2(v), $|St(B)| \leq 2^{|B|}$ for every Boolean ring B. As we remarked in Example 1.7.17, $|St(\mathfrak{B}_X)| \geq 2^{\mathfrak{c}}$, and so $|\Phi_b(X)| = 2^{\mathfrak{c}}$ when $|\mathfrak{B}_X| = \mathfrak{c}$. □

Let K be a non-empty, compact space, and define M_K to be the set of functions $f \in B^b(K)$ such that $\{x \in K : f(x) \neq 0\}$ is meagre. Then M_K is a closed ideal in the C^*-algebra $B^b(K)$.

Definition 3.3.4. Let K be a non-empty, compact space. Then

$$D(K) = B^b(K)/M_K\,;$$

$D(K)$ is the *Dixmier algebra* of K.

Thus $D(K)$ is a commutative, unital C^*-algebra and its character space is exactly G_K, the Gleason cover of K and the Stone space of the complete Boolean algebra $\mathfrak{B}_K/\mathfrak{M}_K$, as in Example 1.7.17.

Theorem 3.3.5. *Let K be a non-empty, Stonean space. Then:*

(i) *for each $f \in B^b(K)$, there is a unique $g \in C(K)$ such that $\{x \in K : f(x) \neq g(x)\}$ is meagre;*

(ii) *there is a C^*-homomorphism that is a bounded projection from $B^b(K)$ onto $C(K)$;*

(iii) *$D(K)$ and $C(K)$ are C^*-isomorphic.*

Proof. (i) First consider a simple, bounded Borel function f, so that f has the form $f = \sum_{i=1}^n \alpha_i \chi_{B_i}$, where $\alpha_1, \dots, \alpha_n \in \mathbb{C}$ and $B_1, \dots, B_n \in \mathfrak{B}_K$ are pairwise disjoint. By Proposition 1.4.4, there exist $U_1, \dots, U_n \in \mathfrak{C}_K$ such that $B_i \equiv U_i$ $(i \in \mathbb{N}_n)$. Clearly the sets $U_1, \dots, U_n$ are pairwise disjoint. We define $g = \sum_{i=1}^n \alpha_i \chi_{U_i}$, so that indeed the set $\{x \in K : f(x) \neq g(x)\}$ is meagre.

Now consider a general function $f \in B^b(K)$. There is a sequence (f_n) of simple, bounded Borel functions that converges uniformly to f on K. For each $n \in \mathbb{N}$, choose $g_n \in C(K)$ such that $S_n := \{x \in K : f_n(x) \neq g_n(x)\}$ is a meagre subset of K. The set $S := \bigcup\{S_n : n \in \mathbb{N}\}$ is also meagre in K, and $g_n(x) = f_n(x)$ for all $n \in \mathbb{N}$ and $x \in K \setminus S$, and so (g_n) is a Cauchy sequence in $(C(K \setminus S), |\cdot|_{K\setminus S})$; the sequence converges uniformly to a function, say g, in $C(K \setminus S)$. By Theorem 1.4.11, $K \setminus S$ is dense in K and, by Lemma 1.5.7, g has an extension, also called g, in $C(K)$. The function g has the required properties; clearly g is uniquely specified.

(ii) For each $f \in B^b(K)$, take Pf to be the unique $g \in C(K)$ specified in (i), and consider the map $P : B^b(K) \to C(K)$. Clearly the restriction of P to the simple functions is a $*$-homomorphism; since the simple functions are dense in $B^b(K)$ and $Pf = f$ $(f \in C(K))$, the map P is a C^*-homomorphism that is a bounded projection from $B^b(K)$ onto $C(K)$.

(iii) Let P be as in (ii). Clearly $\ker P = M_K$, and so the map

$$\overline{P} : D(K) = B^b(K)/M_K \to C(K)$$

is a C^*-isomorphism. □

We shall now define spaces of Baire functions on a topological space. The study of the Baire classes as Banach spaces was initiated by Bade in 1973 [25].

Definition 3.3.6. Let X be a non-empty topological space. The *Baire functions of order* 0 are the functions in $C^b(X)$; we now write $B_0(X)$ for the space consisting of these functions. Let $\alpha > 0$ be an ordinal. Given a definition of the Baire class of order β for each ordinal $\beta < \alpha$, we define $B_\alpha(X)$, the *Baire class of order* α on X, to be the space of bounded functions on X which are pointwise limits of sequences of functions in the union of the earlier classes. The recursive construction terminates at $\alpha = \omega_1$; the *Baire functions* on X are the members of the final class, $B_{\omega_1}(X)$.

It is easy to see that $B_{\omega_1}(X)$ is precisely the family of all bounded, complex-valued functions such that $f^{-1}(U) \in \mathfrak{Ba}_X$ for every open subset U of $\mathbb{C}$, where $\mathfrak{Ba}_X$ denotes the σ-algebra of Baire sets specified in Definition 1.4.29. As we remarked on page 24, $\mathfrak{Ba}_X = \mathfrak{B}_X$ whenever each closed subset of K is a zero set, and so the algebra of Baire functions on X is equal to $B^b(X)$ in these cases; for example, this holds whenever X is metrizable.

In fact, it is shown in [6, §7.1] that, in the case where K is locally compact, the space of real-valued functions in $B_{\omega_1}(K)$ is the 'bounded-monotone class' generated by $C_{0,\mathbb{R}}(K)$. This space is relevant to the construction of the Daniell integral and the Borel functional calculus for normal operators on a Hilbert space; see [6, Chapter 7], [39, §7.8], and [216, Chapter 16].

Let X be a non-empty topological space, and take an ordinal α such that $0 \leq \alpha \leq \omega_1$. Then the Baire class $B_\alpha(X)$ is clearly a $*$-subalgebra of $B^b(X)$, and it is complete with respect to uniform convergence, and so it is a C^*-subalgebra of $B^b(X)$, with character space, $\Phi_\alpha(X)$, say. Thus, $(B_\alpha(X), |\cdot|_X)$ is C^*-isomorphic to $(C(\Phi_\alpha(X)), |\cdot|_{\Phi_\alpha(X)})$. For $1 \leq \alpha \leq \omega_1$, the compact space $\Phi_\alpha(X)$ is totally disconnected: to see this, note that the linear span of the set of idempotents in $B_\alpha(X)$ is a dense subspace of $(B_\alpha(X), |\cdot|_X)$, as is easily seen [164, §31, VIII, Theorem 3].

We shall discuss some of the central ideas associated with Banach spaces of Baire functions and the proof of some parts of the following theorems in §6.7, when we shall have more terminology at our disposal. Clause (iii) of Theorem 3.3.7, below, was first proved in [25, Theorem 3.4]; clauses (i), (ii), (iv), and Theorem 3.3.8 are taken from [75, 77]; the results are contained in the thesis of Dashiell, written under the direction of William Bade. Theorem 3.3.9 is proved in [76].

Theorem 3.3.7. *Let X be an uncountable Polish space.*

(i) *For each ordinal α with $1 \leq \alpha < \omega_1$, the space $B_\alpha(X)$ is not isomorphic to any complemented subspace of $B_{\omega_1}(X)$.*

(ii) *For each ordinal α with $2 \leq \alpha < \omega_1$, there is no injective, bounded operator from $B_\alpha(X)$ into $B_1(X)$.*

(iii) *For each ordinals α, β with $1 \leq \alpha < \beta \leq \omega_1$, the space $B_\alpha(X)$ is not complemented in $B_\beta(X)$.*

(iv) *For each ordinals α, β with $1 \leq \alpha < \beta \leq \omega_1$, the space $B_\alpha(X)$ is not linearly isometric to $B_\beta(X)$.* □

Theorem 3.3.8. *Let K be a non-empty, locally compact space. Then the following are equivalent:*

(a) *K is basically disconnected;*

(b) *$C^b(K)$ is 1-complemented in $B_1(K)$;*

(c) *$C^b(K)$ is 1-complemented in $B_{\omega_1}(K)$;*

(d) *there is a C^*-homomorphism that projects $B_{\omega_1}(K)$ onto $C^b(K)$.* □

Theorem 3.3.9. *Let X be a non-empty topological space. Then the Baire class $B_\alpha(X)$ is a Grothendieck space for each α with $1 \leq \alpha \leq \omega_1$.* □

It will be seen that Theorem 3.3.7 leaves open the following question, specifically raised by Dashiell in [75] (and later mentioned in [186, p. 181] and in [209, Problem 56, p. 488]).

Question 1: *Are any or all of the Banach spaces $B_\alpha(\mathbb{I})$ and $B_\beta(\mathbb{I})$ pairwise isomorphic in the cases where $2 \leq \alpha < \beta < \omega_1$?*

3.4 General C^*-algebras

We now recall some standard definitions and facts about arbitrary C^*-algebras that generalize some of our above remarks about the commutative C^*-algebras $C_0(K)$. However, these remarks (save for the definition of a 'von Neumann algebra') will not be used for any of our subsequent results, and could be omitted.

Let H be a Hilbert space, with the inner product $[\cdot,\cdot]$. For $T \in \mathscr{B}(H)$, define T^* by

$$[T^*x,y] = [x,Ty] \quad (x,y \in H).$$

Then $T^* \in \mathscr{B}(H)$, the map $T \mapsto T^*$ is an involution on $\mathscr{B}(H)$, and $(\mathscr{B}(H),*)$ is a unital C^*-algebra.

Let H be a Hilbert space. The *weak operator topology* and the *strong operator topology* on the C^*-algebra $\mathscr{B}(H)$ are defined by the semi-norms

$$p_{x,y} : T \mapsto |[Tx,y]| \quad (x,y \in H)$$

and

$$p_x : T \mapsto \|Tx\| \quad (x \in H),$$

respectively; they are denoted by wo- and so-, respectively. Thus $(\mathscr{B}(H),\ \mathrm{wo})$ and $(\mathscr{B}(H),\mathrm{so})$ are locally convex spaces, in the terminology of §2.1. We note that $\mathscr{B}(H)_{[1]}$ is wo-compact. For each convex subset K of $\mathscr{B}(H)$, we have $\overline{K}^{\mathrm{so}} = \overline{K}^{\mathrm{wo}}$, and so K is wo-closed if and only if it is so-closed. The Banach space $\mathscr{B}(H)$ is isometrically the dual Banach space of $\mathscr{N}(H)$, the space of nuclear, or trace class, operators on H (see [234, p. 63]), and the weak* topology $\sigma(\mathscr{B}(H),\mathscr{N}(H))$ specified by this duality is called the *σ-weak topology* on $\mathscr{B}(H)$; it is stronger than the weak operator topology, but these two topologies agree on the closed unit ball $\mathscr{B}(H)_{[1]}$ of $\mathscr{B}(H)$. For a discussion of these and several other topologies on $\mathscr{B}(H)$, see [150, Chapter 7] and [234, Chapter II, §2].

A *$*$-representation* of a C^*-algebra A on a Hilbert space H is a $*$-homomorphism $\pi : A \to \mathscr{B}(H)$. By Proposition 3.2.2, π is a contraction and $\pi(A)$ is closed in $\mathscr{B}(H)$. A $*$-representation π is *faithful* if it is a monomorphism, and, in the case where A is unital, π is *universal* if π is unital and isometric and if each state on A has the form $a \mapsto [\pi(a)x,x]$ for some $x \in H$ with $\|x\| = 1$. See [150, Chapter 10], for example.

There is a definition of 'universal $*$-representation' that applies to C*-algebras that may not have an identity; it is equivalent to the earlier one when the C*-algebra is unital. See [234, Definition III.2.3]. Indeed, a $*$-representation π of a C^*-algebra A on a Hilbert space H is *universal* if, for each $*$-representation ρ of A on a Hilbert space, there is a σ-weakly-continuous $*$-homomorphism $\widetilde{\rho}$ from $\pi(A)''$ onto $\rho(A)''$ such that $\rho(a) = (\widetilde{\rho} \circ \pi)(a)$ $(a \in A)$.

The very famous *non-commutative Gel'fand–Naimark theorem* is the following; see [68, Theorem 3.2.29], [149, Theorem 4.5.6], [234, Theorem III.2.4], and all texts on C^*-algebras. There is also a somewhat different proof in [6, §6.6].

Theorem 3.4.1. *Let A be a C^*-algebra. Then there is a Hilbert space H and a universal $*$-representation of A on H.* □

The following result is the famous *double commutant theorem* of von Neumann; see [68, Theorem 3.2.32], [149, Theorem 5.3.1], and [234, Theorem II.3.9].

Theorem 3.4.2. *Let H be a Hilbert space, and let A be a C^*-subalgebra of $\mathscr{B}(H)$. Then $\overline{A}^{\mathrm{so}} = \overline{A}^{\mathrm{wo}} = A^{cc}$.* □

We shall also refer in passing to the following *Kaplansky's density theorem*; see [68, Theorem 3.2.34], [149, Theorem 5.3.5], and [234, Theorem II.4.8].

Theorem 3.4.3. *Let H be a Hilbert space, and let A be a C^*-subalgebra of $\mathscr{B}(H)$. Then $A_{[1]}$ is so-dense in $(\overline{A}^{\mathrm{so}})_{[1]}$.* □

We shall now define von Neumann algebras and W^*-algebras; there are many equivalent definitions of these algebras. In Theorem 6.4.1, we shall determine when commutative C^*-algebras are von Neumann algebras.

Definition 3.4.4. A C^*-algebra A is a *von Neumann algebra* if there is a Hilbert space H such that A is a C^*-subalgebra of $\mathscr{B}(H)$ with A closed in the weak operator topology. A C^*-algebra is a *W^*-algebra* if it is isometrically a dual space.

Thus a C^*-subalgebra of $\mathscr{B}(H)$ is a von Neumann algebra if and only if it is equal to its double commutant; this is the definition of a von Neumann algebra given by Takesaki in [234, Definition II.3.2], for example.

We see that W^*-algebras are defined abstractly, but von Neumann algebras are defined concretely. However, it is a seminal theorem of Sakai [221] that every abstractly defined W^*-algebra can be represented as a von Neumann subalgebra of $\mathscr{B}(H)$ for a suitable Hilbert space H; see the accounts in [150, Exercise 10.5.87], [195, §9.3], [222], and [234, Chapter III, §3]. In the future, using standard terminology, we shall use the term 'von Neumann algebra' for a W^*-algebra, as defined in Definition 3.4.4. In particular, we shall say that $C_0(K)$ is a von Neumann algebra if and only if it is isometrically a dual Banach space.

Suppose that A is a commutative W^*-algebra. Then we shall see in Theorem 6.4.1 that there is a locally compact space Γ and a decomposable measure ν on Γ such that A is C^*-isomorphic to $L^\infty(\Gamma,\nu)$. In this case, take H to be the Hilbert space $H = L^2(\Gamma,\nu)$, and, for $f \in L^\infty(\Gamma,\nu)$, define

$$M_f(g) = fg \quad (g \in H).$$

Then $M_f \in \mathscr{B}(H)$ and $B := \{M_f : f \in L^\infty(\Gamma,\nu)\}$ is a C^*-subalgebra of $\mathscr{B}(H)$. Further, the map

$$f \mapsto M_f, \quad L^\infty(\Gamma,\nu) \to B,$$

is a C^*-isomorphism and a homeomorphism when $L^\infty(\Gamma,\nu)$ has the weak* topology and B has the weak operator topology, and so the C^*-algebra B is wo-closed in $\mathscr{B}(H)$. Thus A is a von Neumann algebra in the sense of Definition 3.4.4.

This proves Sakai's theorem for commutative W^*-algebras.

Let A be a C^*-algebra. Then A is *monotone complete* if every upward-directed, norm-bounded net in A^+ has a supremum. For example, let K be a non-empty, locally compact space. Then the C^*-algebra $C_0(K)$ is monotone complete if and only if it is Dedekind complete, in our earlier terminology; we proved in Theorem 2.3.3 that, for a compact space K, the Banach lattice $C(K)$ is Dedekind complete if and only if K is a Stonean space. Hence $C(K)$ is monotone complete if and only if K is a Stonean space.

Every von Neumann algebra is monotone complete, but the converse is false. For example, we shall prove in Theorem 6.4.2 that, for a non-empty, compact space K, the C^*-algebra $C(K)$ is a von Neumann algebra if and only if the space K is 'hyper-Stonean' (see Definition 5.1.1), and we shall note in Example 5.1.4(ii) that the Gleason cover $G_{\mathbb{I}}$ of $\mathbb{I}$ is Stonean, but not hyper-Stonean.

Another class of C^*-algebras is that of the AW^*-algebras; these are characterized as the unital C^*-algebra in which every maximal abelian C^*-subalgebra is monotone complete. Every unital C^*-algebra that is monotone complete is an AW^*-algebra, and the classes coincide for unital, commutative C^*-algebra. It seems to be an open question whether every AW^*-algebra is monotone complete.

For an account of the above matters, see the recent text [220].

Chapter 4
Measures

In this chapter, we shall study the (complex) Banach lattice $M(K)$ consisting of all complex-valued, regular Borel measures on a locally compact space K and, in particular, the positive measures in $M(K)$, which form the cone $M(K)^+$. The Banach space $M(K)$ is isometrically isomorphic to the dual of $C_0(K)$. In §4.2, we shall discuss the linear spaces of discrete measures and of continuous measures on K.

In §4.3, we shall show that a specific quotient of the lattice $M(K)^+$ is a Dedekind complete Boolean ring B such that the Banach space of bounded, continuous functions on the Stone space of B is isometrically isomorphic to the dual space of $M(K)$, and hence to the bidual of $C_0(K)$; this Boolean ring will reappear in §5.4.

We shall also describe, in §4.4, the Banach lattices $L^p(K,\mu)$ and the Boolean algebra $\mathfrak{B}_\mu$ for $\mu \in M(K)^+$ and $1 \leq p \leq \infty$. Important features to be discussed will include consideration of when spaces of the form $C(K)$ are Grothendieck spaces (in §4.5); maximal singular families of measures in $M(K)^+$ (in §4.6), to be used in a later explicit construction of $C_0(K)''$; and the closed subspace $N(K)$ of $M(K)$ consisting of the normal measures (in §4.7). We shall give several examples of spaces with $N(K) = \{0\}$; for example, we shall show in Theorem 4.7.23 that $N(K) = \{0\}$ whenever K is a locally connected, compact space without isolated points. However, we shall show in Theorem 4.7.26 that there is a non-empty, connected, compact space K with $N(K) \neq \{0\}$.

4.1 Measures

Let K be a non-empty, locally compact space. We recall that a *Borel measure* μ on K is a function $\mu : \mathfrak{B}_K \to \mathbb{C}$ such that $\mu(\emptyset) = 0$ and μ is σ-additive, in the sense that

$$\mu(B) = \sum\{\mu(B_n) : n \in \mathbb{N}\}$$

H.G. Dales et al., *Banach Spaces of Continuous Functions as Dual Spaces*,
CMS Books in Mathematics, DOI 10.1007/978-3-319-32349-7_4

whenever (B_n) is a sequence of pairwise-disjoint sets in $\mathfrak{B}_K$ with $\bigcup\{B_n : n \in \mathbb{N}\} = B$. Thus a Borel measure on K is just the same as a σ-normal measure on the Boolean algebra $\mathfrak{B}_K$ in the sense of Definition 1.7.12. Further, in the case where $\mu(B) \geq 0$ $(B \in \mathfrak{B}_K)$, the triple $(K, \mathfrak{B}_K, \mu)$ is a measure space.

Definition 4.1.1. Let K be a non-empty, locally compact space, and take a Borel measure μ defined on $\mathfrak{B}_K$. Then

$$|\mu|(B) = \sup \sum_{i=1}^{\infty} |\mu(B_i)| \quad (B \in \mathfrak{B}_K),$$

where the supremum is taken over all partitions of a Borel set B by a countable family $\{B_i : i \in \mathbb{N}\}$ in $\mathfrak{B}_K$. Then $|\mu|$ is the *total variation measure* of μ. The measure μ is *regular* if, for each $B \in \mathfrak{B}_K$ and each $\varepsilon > 0$, there is a compact subset $L \subset B$ and an open set $U \supset B$ with $|\mu|(U \setminus L) < \varepsilon$.

The total variation measure of μ is indeed a Borel measure on K that is regular when μ is regular. On a locally compact space with a countable basis, every Borel measure is regular, but there are compact spaces on which there are Borel measures which are not regular; see [39, §7.1].

Definition 4.1.2. Let K be a non-empty, locally compact space. Then we denote by $M(K)$ the space of complex-valued, regular Borel measures on K, and we set

$$\|\mu\| = |\mu|(K) \quad (\mu \in M(K)).$$

Henceforth, we shall just write 'measure on K' for 'complex-valued, regular Borel measure on K'. The pair $(M(K), \|\cdot\|)$ is a Banach space.

Let L be a closed subspace of K, and take $\mu \in M(L)$. Then we regard μ as an element of $M(K)$ by setting $\mu(B) = \mu(B \cap L)$ $(B \in \mathfrak{B}_K)$. Thus $M(L)$ is a closed subspace of $M(K)$.

The following *Riesz representation theorem* (of F. Riesz) identifies $M(K)$ as the dual space of $C_0(K)$.

Theorem 4.1.3. *Let K be a non-empty, locally compact space. Then the dual space to $C_0(K)$ is identified isometrically with $M(K)$ via the duality specified by*

$$\langle f, \mu \rangle = \int_K f \, \mathrm{d}\mu \quad (f \in C_0(K), \mu \in M(K)).$$

□

In particular, we have the identifications

$$(c_0)''' = (\ell^1)'' = (\ell^\infty)' = C(\beta\mathbb{N})' = M(\beta\mathbb{N}).$$

For details of the Riesz representation theorem, see the recent text of Bogachev [39, §§7.10, 7.11] and the classic texts of Halmos [132] and Rudin [217, Theorem 6.19], for example. The latter two texts were the congenial companions of the authors' distant youths.

Let K be a non-empty, locally compact space. The space of real-valued measures in $M(K)$ is $M_{\mathbb{R}}(K)$. For $\mu, \nu \in M_{\mathbb{R}}(K)$, set

$$\begin{cases} (\mu \vee \nu)(B) = \sup\{\mu(C) + \nu(B \setminus C) : C \in \mathfrak{B}_K, C \subset B\}, \\ (\mu \wedge \nu)(B) = \inf\{\mu(C) + \nu(B \setminus C) : C \in \mathfrak{B}_K, C \subset B\}, \end{cases} \quad (B \in \mathfrak{B}_K). \tag{4.1}$$

Then $M_{\mathbb{R}}(K)$ is a real Banach lattice with respect to the operations $\vee$ and $\wedge$. The definitions in (4.1) agree with those in equation (2.8) when we regard $M_{\mathbb{R}}(K)$ as the dual lattice to $C_{0,\mathbb{R}}(K)$, and so $M_{\mathbb{R}}(K)$ and $M(K)$ are Dedekind complete lattices.

As before, for $\mu \in M_{\mathbb{R}}(K)$, we set $\mu^+ = \mu \vee 0$, $\mu^- = (-\mu) \vee 0$, and

$$|\mu| = \mu^+ + \mu^- = \mu \vee (-\mu),$$

so that $\mu = \mu^+ - \mu^-$, and $|\mu|$ coincides with the total variation measure of Definition 4.1.1; the two measures μ^+ and μ^- are uniquely characterized by the facts that $\mu = \mu^+ - \mu^-$ and $\|\mu\| = \|\mu^+\| + \|\mu^-\|$.

Now take $\mu \in M(K)$. Then we shall write $\Re\mu$ and $\Im\mu$ for the real and imaginary parts of μ, respectively, so that $\mu = \Re\mu + \mathrm{i}\Im\mu$; the *conjugate* of μ is defined to be $\overline{\mu} = \Re\mu - \mathrm{i}\Im\mu$. The measure $|\mu|$ defined in equation (2.5) is indeed the total variation measure of μ defined in Definition 4.1.1. Further, the space $M(K)$, the complexification of $M_{\mathbb{R}}(K)$, is a Banach lattice, and the norm defined by equation (2.7) agrees with that defined in Definition 4.1.2. Clearly the Banach lattice $M(K)$ is an *AL*-space.

The set of positive measures in $M(K)$ is denoted by $M(K)^+$; this set $M(K)^+$ is weak*-closed in $M(K)$. We note that positive measures correspond to positive linear functionals on $C_0(K)$, in the sense that, for $\mu \in M(K)$, we have $\mu \in M(K)^+$ if and only if $\langle f, \mu \rangle \geq 0$ $(f \in C_0(K)^+)$. We also note that, in the case where K is compact and $\mu \in M(K)$, we have

$$\mu \in M(K)^+ \quad \text{if and only if} \quad \langle 1_K, \mu \rangle = \|\mu\|. \tag{4.2}$$

A measure $\mu \in M(K)^+$ with $\|\mu\| = 1$ is a *probability measure*; the set of these measures is denoted by $P(K)$. In the case where K is compact, $P(K)$ can be identified with the state space $K_{C(K)}$ of the unital C^*-algebra $C(K)$, and $P(K)$ is then clearly a Choquet simplex in the ambient space $(M(K), \sigma(M(K), C(K))$, and so, as in Example 1.7.15, $\mathrm{Comp}_{P(K)}$ is a complete Boolean algebra.

Let K and L be two non-empty, locally compact spaces, and take $\mu \in M(K)$ and $\nu \in M(L)$. Then there is a unique measure $\mu \otimes \nu \in M(K \times L)$ such that

$$(\mu \otimes \nu)(B \times C) = \mu(B)\nu(C) \quad (B \in \mathfrak{B}_K, C \in \mathfrak{B}_L);$$

$\mu \otimes \nu$ is the *product* of μ and ν. In the case where $\mu \in P(K)$ and $\nu \in P(L)$, we have $\mu \otimes \nu \subset P(K \times L)$.

There is one special measure $m \in P(\mathbb{I})$ that we shall use.

Definition 4.1.4. Denote by m the Lebesgue measure on the interval $\mathbb{I} = [0, 1]$.

As well as integrating continuous functions, we can integrate Borel functions against a measure. Recall from Definition 3.3.1 that $B^b(K)$ denotes the space of bounded Borel functions on a locally compact space K.

Definition 4.1.5. Let K be a non-empty, locally compact space. For $f \in B^b(K)$, define $\kappa(f)$ on $M(K)$ by

$$\langle \kappa(f), \mu \rangle = \int_K f \,\mathrm{d}\mu \quad (\mu \in M(K)). \tag{4.3}$$

We see immediately that $\kappa(f) \in M(K)' = C_0(K)''$ and that

$$\mu(B) = \langle \kappa(\chi_B), \mu \rangle \quad (B \in \mathfrak{B}_K, \mu \in M(K)).$$

Indeed, we are regarding each $\mu \in M(K)$ as a continuous linear functional on $B^b(K)$ which extends μ defined on $C_0(K)$; we note that this extension of $\mu \in C_0(K)'$ to $B^b(K)$ is usually not unique.

Let G be a group. Then the identity of G is denoted by e_G. For an element $t \in G$ and subsets S and T of G, we set

$$tS = \{ts : s \in S\}, \quad S^{-1} = \{s^{-1} : s \in S\}, \quad ST = \{st : s \in S, t \in T\}.$$

A *locally compact group* is a group that is also a locally compact topological space such that the group operations are continuous. For example, the Cantor cube $\{0,1\}^\kappa = \mathbb{Z}_2^\kappa$ of weight κ, where κ is an infinite cardinal, is a compact group.

Let G be a locally compact group. Then the Banach space $M(G)$ of all measures on G is a Banach algebra with respect to the *convolution product* $\star$: given measures $\mu, \nu \in M(G)$, we must define $\mu \star \nu$, and we do this by specifying the action of $\mu \star \nu$ on an element $f \in C_0(G)$ and using the Riesz representation theorem. Indeed,

$$\langle f, \mu \star \nu \rangle = \int_G \int_G f(st) \,\mathrm{d}\mu(s) \,\mathrm{d}\nu(t) \quad (f \in C_0(G)).$$

It is standard that $M(G) = (M(G), \star, \|\cdot\|)$ is a unital Banach algebra; the identity is δ_{e_G}. This Banach algebra is called the *measure algebra* of G. For a study of this algebra, see the books [68, 137, 194, 195], and the memoir [72], for example.

Let G be a locally compact group. Then there is a positive measure m_G defined on $\mathfrak{B}_G$ such that $m_G(U) > 0$ for each non-empty, open subset U of G and such that m_G is left-translation invariant, in the sense that $m_G(sB) = m_G(B)$ for each $s \in G$ and $B \in \mathfrak{B}_G$. Such a measure is a *left Haar measure* on G; it is unique up to multiplication by a positive constant. For constructions of this measure, see the classic texts of Hewitt and Ross [137] and Rudin [218].

For example, Haar measure on $(\mathbb{R}, +)$ is the usual Lebesgue measure. Also, set $L = \mathbb{Z}_2^{\mathfrak{c}}$, and let m_L be the product measure on L from the measure on $\{0,1\}$ that gives the value $1/2$ to each of the two points. Then m_L is the Haar measure on L, with $m_L(L) = 1$.

We now return to the spaces $M(K)$. Let K be a non-empty, locally compact space. A measure $\mu \in M(K)$ is *supported* on a Borel subset B of K if $|\mu|(K \setminus B) = 0$. The *support* of a measure $\mu \in M(K)$ is denoted by $\operatorname{supp} \mu$: it is the complement of the union of the open sets U in K such that $|\mu|(U) = 0$, and so is a closed subset of K.

Proposition 4.1.6. *Let K be a non-empty, locally compact space, and suppose that μ is a non-zero measure in $M(K)^+$. Then* $\operatorname{supp} \mu$ *satisfies* CCC. *In the case where K is a compact F-space,* $\operatorname{supp} \mu$ *is Stonean.*

Proof. It follows quickly from the definition of $\operatorname{supp} \mu$ that $\mu(U) > 0$ for each non-empty, open subset U of $\operatorname{supp} \mu$. Thus $\operatorname{supp} \mu$ satisfies CCC. In the case where K is a compact F-space, $\operatorname{supp} \mu$ is Stonean by Proposition 1.5.14. □

A measure $\mu \in M(K)^+$ is *strictly positive* on K if $\mu(U) > 0$ for each non-empty, open subset U of K, equivalently, if $\operatorname{supp} \mu = K$.

We shall use *Hahn's decomposition theorem* and *Lusin's theorem* in the following forms; see [217, Theorems 2.24 and 6.14], for example.

Theorem 4.1.7. *Let K be a non-empty, locally compact space, and take $\mu \in M_{\mathbb{R}}(K)$.*

(i) *There exist Borel subsets P and N of K such that $\{P,N\}$ is a partition of K, such that $\mu(B) \geq 0$ for each Borel subset B of P, and such that $\mu(B) \leq 0$ for each Borel subset B of N.*

(ii) *For each Borel function f on K and each $\varepsilon > 0$, there is a compact subset L of K such that $|\mu|(K \setminus L) < \varepsilon$ and $f \mid L$ is continuous.* □

The partition $\{P,N\}$ in clause (i) of Theorem 4.1.7 is called a *Hahn decomposition of K with respect to μ;* it is unique up to sets of measure zero.

Proposition 4.1.8. *Let K be a non-empty, compact space, and let E be a real-linear subspace of $M_{\mathbb{R}}(K)$ such that*

$$|f|_K = \sup\{|\langle f, \mu\rangle| : \mu \in E_{[1]}\} \quad (f \in C_{\mathbb{R}}(K)).$$

For each non-empty, open subset U of K and each $\varepsilon > 0$, there exists $\mu \in S_E$ with $\mu(U \cap P) > 1 - \varepsilon$, where $\{P,N\}$ is a Hahn decomposition of K with respect to μ.

Proof. Let U be a non-empty, open subset of K, and take $\varepsilon > 0$. Choose $f \in C(K)^+$ with $|f|_K = 1$ and $\operatorname{supp} f \subset U$, and then take $\mu \in S_E$ with $\langle f, \mu \rangle > 1 - \varepsilon$. We see that

$$1 - \varepsilon < \int_K f \, d\mu = \int_U f \, d\mu \leq \int_{U \cap P} f \, d\mu \leq \mu(U \cap P),$$

which gives the result. □

We shall also require the following version of *Choquet's theorem*; we state a general form, which is the *Choquet–Bishop–de Leeuw theorem*; see, for example, [4, §1.4], [104, Theorem 2.10], or [201, §4]. In the case where the specified space

K is metrizable, $\operatorname{ex} K$ is a G_δ-set (by Proposition 2.1.9), and hence a Borel set. As explained in [178, Remark 2.32(c), p. 16], the case of complex scalars is a simple extension of the real case.

Theorem 4.1.9. *Let K be a non-empty, compact, convex subset of a locally convex space E over $\mathbb{R}$ or $\mathbb{C}$, and let $x_0 \in K$. Then there exists $\mu \in P(K)$ such that*

$$\langle x_0, \lambda \rangle = \int_K \lambda \, \mathrm{d}\mu = \int_K \langle x, \lambda \rangle \, \mathrm{d}\mu(x) \quad (\lambda \in E') \tag{4.4}$$

and such that μ vanishes on every Baire subset and on every G_δ-subset of K which is disjoint from $\operatorname{ex} K$. In the case where K is metrizable, $\mu(\operatorname{ex} K) = 1$. □

In the above setting, x_0 is termed the *resultant* or *barycentre* of the measure μ.

We shall use the following known application of the Choquet–Bishop–de Leeuw theorem. It is given in [104, Theorem 2.18]; the proof here is somewhat shorter.

Theorem 4.1.10. *Let E be a normed space, and let K be a weak*-compact, convex subset of E'. Suppose that D is a countable, $\|\cdot\|$-dense subset of $\operatorname{ex} K$. Then K is the $\|\cdot\|$-closure of the convex hull of D, and so K is $\|\cdot\|$-separable.*

Proof. The result is trivial when D is finite, and so we may suppose that D is infinite, say $D = \{\lambda_i : i \in \mathbb{N}\}$. Fix $\varepsilon > 0$, and, for each $i \in \mathbb{N}$, set

$$K_i = \{\lambda \in K : \|\lambda - \lambda_i\| \leq \varepsilon\},$$

so that K_i is a weak*-compact subspace of E' and $\operatorname{ex} K \subset \bigcup\{K_i : i \in \mathbb{N}\} \subset K$. Take $\lambda_0 \in K$. By Theorem 4.1.9, there exists $\mu_0 \in P(K)$ such that

$$\langle x, \lambda_0 \rangle = \int_K \langle x, \lambda \rangle \, \mathrm{d}\mu_0(\lambda) \quad (x \in E)$$

and such that μ_0 vanishes on each G_δ-subset of K that is disjoint from $\operatorname{ex} K$. Clearly $\bigcap\{K \setminus K_i : i \in \mathbb{N}\}$ is such a G_δ-set, and so $\mu_0(\bigcup\{K_i : i \in \mathbb{N}\}) = 1$.

Choose pairwise-disjoint Borel sets B_i for $i \in \mathbb{N}$ such that $B_i \subset K_i$ $(i \in \mathbb{N})$ and $\bigcup_{i\in\mathbb{N}} B_i = \bigcup_{i\in\mathbb{N}} K_i$, and set $\alpha_i = \mu_0(B_i) \in \mathbb{I}$ $(i \in \mathbb{N})$, so that $\sum_{i=1}^\infty \alpha_i = 1$. Next set

$$\Lambda = \sum_{i=1}^\infty \alpha_i \lambda_i \in \overline{\mathrm{co}}\, D.$$

Take $x \in E_{[1]}$. For each $i \in \mathbb{N}$ and $\lambda \in B_i$, we have $|\langle x, \lambda_i \rangle - \langle x, \lambda \rangle| < \varepsilon$, and so

$$\left| \langle x, \alpha_i \lambda_i \rangle - \int_{B_i} \langle x, \lambda \rangle \, \mathrm{d}\mu_0(\lambda) \right| \leq \alpha_i \varepsilon.$$

It follows that $|\langle x, \Lambda \rangle - \langle x, \lambda_0 \rangle| \leq \varepsilon$, and hence $\|\Lambda - \lambda_0\| \leq \varepsilon$. Thus $K = \overline{\mathrm{co}}\, D$. □

Definition 4.1.11. Let K be a non-empty, compact, convex subset of a locally convex space, and suppose that $\mu, \nu \in M(K)^+$. Then

$$\mu \approx \nu \quad \text{if} \quad \langle h, \mu\rangle = \langle h, \nu\rangle \tag{4.5}$$

for each affine function $h \in C_{\mathbb{R}}(K)$, and

$$\mu \prec \nu \quad \text{if} \quad \langle h, \mu\rangle \leq \langle h, \nu\rangle \tag{4.6}$$

for each convex function $h \in C_{\mathbb{R}}(K)$.

Let K be a non-empty, compact, convex subset of a locally convex space. The relation $\prec$ is a partial order on $M(K)^+$; a measure $\mu \in M(K)^+$ is *maximal* if it is maximal in the partially ordered set $(M(K)^+, \prec)$. It is shown in [201, Lemma 4.1] that, for each $\nu \in M(K)^+$, there is a maximal measure $\mu \in M(K)^+$ with $\nu \prec \mu$.

The following result combines Propositions 3.1 and 10.3 of [201] and the *Choquet–Meyer theorem* from [201, p. 56]. Recall that a Choquet simplex was defined within Example 1.7.15.

Theorem 4.1.12. *Let K be a non-empty, compact, convex subset of a locally convex space. Suppose that $\mu \in P(K)$ is such that* $\operatorname{supp}\mu \subset \operatorname{ex} K$. *Then μ is a maximal measure on K. Suppose further that K is a Choquet simplex. Then, for each $x \in K$, there is a unique maximal measure μ such that $\mu \approx \varepsilon_x$.* □

Proposition 4.1.13. *Let K be a non-empty, locally compact space. Suppose that (μ_α) is a net in $M(K)$ which converges to $\mu \in M(K)$ in the weak* topology $\sigma(M(K), C_0(K))$. Then*

$$|\mu|(U) \leq \liminf_\alpha |\mu_\alpha|(U)$$

for each open set U in K. In particular, $\|\mu\| \leq \liminf_\alpha \|\mu_\alpha\|$.

Further, the following maps from $(M(K), \sigma(M(K), C_0(K)))$ to $\mathbb{R}$ are lower semi-continuous: $\mu \mapsto |\mu|(U)$, for each fixed open subset U of K; $\mu \mapsto \int_K g \,\mathrm{d}|\mu|$, for each fixed $g \in C_0(K)^+$; $\mu \mapsto \|\mu\|$.

Proof. Let U be a non-empty, open set in K, and choose $\varepsilon > 0$. Then there exists $f \in C_{00}(K)_{[1]}$ such that $|f| \leq \chi_U$ and $|\int_K f \,\mathrm{d}\mu| > |\mu|(U) - \varepsilon$. For each α, we have

$$|\mu_\alpha|(U) = \int_K \chi_U \,\mathrm{d}|\mu_\alpha| \geq \int_K |f| \,\mathrm{d}|\mu_\alpha| \geq \left|\int_K f \,\mathrm{d}\mu_\alpha\right|,$$

and so

$$\liminf_\alpha |\mu_\alpha|(U) \geq \lim_\alpha \left|\int_K f \,\mathrm{d}\mu_\alpha\right| = \left|\int_K f \,\mathrm{d}\mu\right| > |\mu|(U) - \varepsilon,$$

giving the main result. The remainder is clear. □

Note that the map $\mu \mapsto |\mu|$ on $M(K)$ is not always weak*-weak*-continuous. For example, for $n \in \mathbb{N}$, set

$$s_n(t) = \sin(nt) \quad (t \in \mathbb{I}),$$

and regard (s_n) as a sequence in $L^1(\mathbb{I}) \subset M(\mathbb{I})$. Then (s_n) converges weakly to 0 in $L^1(\mathbb{I})$. To see this, let J be a subinterval of $\mathbb{I}$. Then $\int_J s_n(t)\,\mathrm{d}t \to 0$ as $n \to \infty$, and so $\int_{\mathbb{I}} f(t)s_n(t)\,\mathrm{d}t \to 0$ as $n \to \infty$ whenever f is a finite linear combination of characteristic functions of intervals. Since each $f \in L^\infty(\mathbb{I})$ is the limit in $\|\cdot\|_1$ of such functions, $\int_{\mathbb{I}} f(t)s_n(t)\,\mathrm{d}t \to 0$ as $n \to \infty$ for each $f \in L^\infty(\mathbb{I})$. In particular, (s_n) converges weak* to 0 in $M(\mathbb{I})$. But of course $(|s_n|)$ does not converge weak* to 0.

Let K and L be two non-empty, compact spaces, and again suppose that $\eta : K \to L$ is a continuous surjection. For $\mu \in M(K)$, there is a measure $\nu = (\eta^\circ)'(\mu) \in M(L)$, called the *image* of μ, such that

$$\int_K \eta^\circ(f)(x)\,\mathrm{d}\mu(x) = \int_K (f \circ \eta)(x)\,\mathrm{d}\mu(x) = \int_L f(y)\,\mathrm{d}\nu(y) \quad (f \in C_{00}(L)).$$

It is proved in [132, Theorem 39 (C)] and [138, Theorem (12.46(i))] that

$$\nu(B) = \mu(\eta^{-1}(B)) = \int_K (\chi_B \circ \eta)(x)\,\mathrm{d}\mu(x) \quad (B \in \mathfrak{B}_L). \tag{4.7}$$

We write $\eta[\mu]$ for the image measure ν, so that $\eta[\mu] \in M(L)$; in the case where $\mu \in P(K)$, we have $\eta[\mu] \in P(L)$. The following three results are taken from [206]; see Theorem 4.7.26 for our application of the results.

Proposition 4.1.14. *Let L be a non-empty, connected, compact space. Suppose that $\nu \in P(L)$ is a strictly positive measure and that F is a closed subset of L such that $\nu(F) > 0$. Then there are a non-empty, connected, compact space K containing L as a closed subspace, a strictly positive measure $\mu \in P(K)$, and a continuous surjection $\eta : K \to L$ such that $\eta[\mu] = \nu$ and $\mathrm{int}_K \eta^{-1}(F) \neq \emptyset$.*

Proof. Let $F_0 = \mathrm{supp}(\nu \mid F)$, so that

$$F_0 = F \setminus \bigcup\{U : U \text{ open in } L,\ \nu(F \cap U) = 0\}.$$

Set $K = (F_0 \times \mathbb{I}) \cup (L \times \{0\})$, so that K is a non-empty, connected, compact subspace of $F \times \mathbb{I}$. The map η is defined by $\eta(x,t) = x$ $((x,t) \in K)$, so that $\eta : K \to L$ is a continuous surjection. The set $\eta^{-1}(F)$ contains $F_0 \times (0,1]$, and the latter is a non-empty, open subset of K, and so $\mathrm{int}_K \eta^{-1}(F) \neq \emptyset$.

Let $C \in \mathfrak{B}_K$, and define $\mu(C)$ by setting

$$\mu(C) = \nu(C \cap (L \setminus F_0)) + (\nu \otimes m)((F_0 \times \mathbb{I}) \cap C),$$

where we recall that m denotes Lebesgue measure on $\mathbb{I}$. Then it is clear that $\mu \in P(K)$ and that μ is strictly positive. Further, $\mu(\eta^{-1}(B)) = \nu(B)$ $(B \in \mathfrak{B}_L)$, and so $\eta[\mu] = \nu$. □

The notion of an inverse limit of an inverse system of compact spaces arose in Definition 1.4.31.

Let κ be an ordinal. An *inverse system with measures* is an inverse system of compact spaces $(K_\alpha, \pi_\alpha^\beta : 0 \le \alpha \le \beta < \kappa)$, together with measures $\mu_\alpha \in P(K_\alpha)$ for each α with $0 \le \alpha < \kappa$ such that $\pi_\alpha^\beta[\mu_\beta] = \mu_\alpha$ for $0 \le \alpha \le \beta < \kappa$; such a system is denoted by

$$(K_\alpha, \mu_\alpha, \pi_\alpha^\beta : 0 \le \alpha \le \beta < \kappa).$$

Proposition 4.1.15. *Let κ be an ordinal, let $(K_\alpha, \mu_\alpha, \pi_\alpha^\beta : 0 \le \alpha \le \beta < \kappa)$ be an inverse system of compact spaces with measures, and take (K, π_α) to be the inverse limit of $(K_\alpha, \pi_\alpha^\beta : 0 \le \alpha \le \beta < \kappa)$. Then there is a unique measure $\mu \in P(K)$ such that $\pi_\alpha[\mu] = \mu_\alpha$ for $0 \le \alpha < \kappa$. In the case where each μ_α is strictly positive, the measure μ is strictly positive.*

Proof. For each ordinal α with $0 \le \alpha < \kappa$, the map π_α° identifies $C(K_\alpha)$ with a unital, self-adjoint, closed subalgebra, say A_α, of $C(K)$. Set $A = \bigcup\{A_\alpha : 0 \le \alpha < \kappa\}$. Then A separates the points of K, and so, by the Stone–Weierstrass theorem, Theorem 1.4.26(ii), A is dense in $(C(K), |\cdot|_K)$. Set

$$\lambda(f) = \int_{K_\alpha} f \, \mathrm{d}\mu_\alpha \quad (f \in A_\alpha).$$

Since $\pi_\alpha^\beta[\mu_\beta] = \mu_\alpha$ for $0 \le \alpha \le \beta < \kappa$, the value of $\lambda(f)$ is independent of the choice of α. It is clear that λ is a positive, continuous linear functional on $(A, |\cdot|_K)$ with $\|\lambda\| = 1$, and so λ extends to a positive, continuous linear functional on $(C(K), |\cdot|_K)$ with $\|\lambda\| = 1$. By the Riesz representation theorem, there exists $\mu \in P(K)$ such that $\lambda(f) = \langle f, \mu \rangle$ $(f \in C(K))$. The measure μ has the required properties. □

Theorem 4.1.16. *Let L be a non-empty, connected, compact space, and suppose that $\nu \in P(L)$ is a strictly positive measure. Then there are a non-empty, connected, compact space $L^\#$, a strictly positive measure $\mu^\# \in P(L^\#)$, and a continuous surjection $\eta^\# : L^\# \to L$ such that $\eta^\#[\mu^\#] = \nu$ and $\mathrm{int}_{L^\#}(\eta^\#)^{-1}(Z) \ne \emptyset$ for each $Z \in \mathbf{Z}(L)$ with $\nu(Z) > 0$.*

Proof. Let $\{Z_\alpha : 0 \le \alpha < \kappa\}$ be an enumeration of the sets $Z \in \mathbf{Z}(L)$ with $\nu(Z) > 0$, where κ is a cardinal. We shall define inductively an inverse system with strictly positive measures

$$(K_\alpha, \mu_\alpha, \pi_\alpha^\beta : 0 \le \alpha \le \beta < \kappa)$$

such that $K_0 = L$ and $\mu_0 = \nu$.

In the case where $0 \le \gamma < \kappa$ is such that $(K_\alpha, \mu_\alpha, \pi_\alpha^\beta : 0 \le \alpha \le \beta \le \gamma)$ is an inverse system with non-empty, connected, compact spaces K_α and strictly positive measures μ_α (for $0 \le \alpha \le \gamma$), we define $K_{\gamma+1}$ and $\mu_{\gamma+1}$ by applying Proposition 4.1.14 with $L = K_\gamma$, with $\nu = \mu_\gamma$, and with $F = (\pi_0^\gamma)^{-1}(Z_\gamma)$ (and also defining the maps

$\pi_\alpha^{\gamma+1}$ to be $\eta \circ \pi_\alpha^\gamma$ for $0 \le \alpha \le \gamma$, where η arises in Proposition 4.1.14, and $\pi_{\gamma+1}^{\gamma+1}$ to be the identity on $K_{\gamma+1}$).

In the case where $0 \le \gamma \le \kappa$, γ is a limit ordinal, and $(K_\alpha, \mu_\alpha, \pi_\alpha^\beta : 0 \le \alpha \le \beta < \gamma)$ is an inverse system with non-empty, connected, compact spaces K_α and strictly positive measures μ_α, we define $(K_\gamma, \pi_\alpha^\gamma : 0 \le \alpha < \gamma)$ to be the inverse limit of $(K_\alpha, \pi_\alpha^\beta : 0 \le \alpha \le \beta < \gamma)$ (and take π_α^γ to be the continuous surjections that arise in Theorem 1.4.32), so that K_γ is compact and connected by Theorem 1.4.32; we take $\mu_\gamma \in P(K_\gamma)$ to be the measure specified in Proposition 4.1.15. In the special case in which $\gamma = \kappa$, we set $L^\# = K_\gamma$, $\mu^\# = \mu_\gamma \in P(L^\#)$, and $\eta^\# = \pi_0^\kappa : L^\# \to L$, so that $\eta^\#[\mu^\#] = \nu$. Then $L^\#$, $\mu^\#$ and $\eta^\#$ have the required properties.

Now suppose that $Z \in \mathbf{Z}(L)$ with $\nu(Z) > 0$. Then $Z = Z_\alpha$ for some $\alpha < \kappa$. The interior of the set

$$(\pi_0^{\alpha+1})^{-1}(Z_\alpha) = \left((\pi_\alpha^{\alpha+1})^{-1} \circ (\pi_0^\alpha)^{-1}\right)(Z_\alpha)$$

is non-empty by the basic construction of Proposition 4.1.14, and so we see that $\mathrm{int}_{L^\#}(\eta^\#)^{-1}(Z) = \mathrm{int}_{L^\#}(\eta^\#)^{-1}(Z_\alpha) \neq \emptyset$, as required. □

In the case where $L = \mathbb{I}$ and $\nu = m$, we see that $|\{Z \in \mathbf{Z}(L) : \nu(Z) > 0\}| = \mathfrak{c}$, and so $\kappa = \mathfrak{c}$ in the above proof. It follows by an easy induction that $w(L^\#) = \mathfrak{c}$.

4.2 Discrete and continuous measures

We now introduce discrete, continuous, singular, and absolutely continuous measures.

Definition 4.2.1. Let K be a non-empty, locally compact space. The measures μ for which every set A with $|\mu|(A) > 0$ contains a point x with $|\mu|(\{x\}) > 0$ are the *discrete* measures, and the measures μ such that $\mu(\{x\}) = 0$ for each $x \in K$ are the *continuous* measures.

Let K be a non-empty, locally compact space. The sets of discrete and continuous measures on K are denoted by $M_d(K)$ and $M_c(K)$, respectively; they are closed linear subspaces of $M(K)$ and

$$M(K) = M_d(K) \oplus_1 M_c(K). \tag{4.8}$$

Further, both $M_d(K)$ and $M_c(K)$ are closed $C_0(K)$-submodules of $M(K)$, both are lattice ideals in $M(K)$, and it is standard that $M_d(K)$ is $\sigma(M(K), C_0(K))$-dense in $M(K)$; see Corollary 4.4.16. The point mass at $x \in K$ is denoted by δ_x, so that $\delta_x \in M_d(K)$. Indeed, $M_d(K) = \ell^1(K)$ when we identify the measure δ_x with the function $\chi_{\{x\}}$ for $x \in K$. The measure m on $\mathbb{I}$ is continuous. We set

$$P_d(K) = P(K) \cap M_d(K) \quad \text{and} \quad P_c(K) = P(K) \cap M_c(K).$$

Proposition 4.2.2. *Let K be a non-empty, locally compact space that contains a countable, dense subset Q, and suppose that $\mu \in M_c(K)^+$. Then K contains a dense G_δ-subset D such that $Q \subset D$ and $\mu(D) = 0$.*

Proof. Set $Q = \{x_n : n \in \mathbb{N}\}$. Since the measure μ is continuous, it follows that, for each $k,n \in \mathbb{N}$, there is an open neighbourhood $U_{k,n}$ of x_n such that $\mu(U_{k,n}) < 1/2^n k$. Set

$$U_k = \bigcup\{U_{k,n} : n \in \mathbb{N}\} \quad (k \in \mathbb{N}).$$

Then each U_k is an open subset of K with $\mu(U_k) < 1/k$. The set $D := \bigcap\{U_k : k \in \mathbb{N}\}$ is a G_δ-subset of K; it is dense in K because it contains $\{x_n : n \in \mathbb{N}\}$, and clearly $\mu(D) = 0$. □

Proposition 4.2.3. *Let K be an uncountable, compact, metrizable space. Then we have $|M(K)| = \mathfrak{c}$.*

Proof. By Proposition 1.4.14, $|K| = \mathfrak{c}$, and so $|M(K)| \geq |M_d(K)| \geq \mathfrak{c}$.

The topological space K has a countable base; we may suppose that this base is closed under finite unions. Each open set in K is a countable, increasing union of members of the base, and so each $\mu \in M(K)$ is determined by its values on the sets of this base. Hence $|M(K)| \leq \mathfrak{c}$. □

Definition 4.2.4. Let K be a non-empty, locally compact space, and suppose that $\mu, \nu \in M(K)$. Then $\mu \perp \nu$ if μ and ν are *mutually singular*, in the sense that there exists $B \in \mathfrak{B}_K$ with $|\mu|(B) = 0$ and $|\nu|(K \setminus B) = 0$, and $\mu \ll \nu$ if $|\mu|$ is *absolutely continuous* with respect to $|\nu|$, in the sense that $|\mu|(B) = 0$ whenever $B \in \mathfrak{B}_K$ and $|\nu|(B) = 0$.

For $\mu, \nu \in M(K)$, set

$$\mu \sim \nu \quad \text{if} \quad \mu \ll \nu \quad \text{and} \quad \nu \ll \mu .$$

We recall that $\mu \ll \nu$ if and only if, for each $\varepsilon > 0$, there exists $\delta = \delta(\varepsilon) > 0$ such that $|\mu(B)| < \varepsilon$ whenever $B \in \mathfrak{B}_K$ and $|\nu|(B) < \delta$. Suppose that $\mu, \nu \in M(K)$ with $\mu \ll \nu$. Then $\operatorname{supp} \mu \subset \operatorname{supp} \nu$.

It is easy to check that $\sim$ is an equivalence relation on the space $M(K)$. Clearly $\mu \sim |\mu|$ for each $\mu \in M(K)$.

It follows from the Hahn decomposition theorem that each $\mu \in M(K)$ has a *Jordan decomposition*:

$$\mu = \mu_1 - \mu_2 + \mathrm{i}(\mu_3 - \mu_4), \tag{4.9}$$

where $\mu_1 = (\Re\mu)^+$, $\mu_2 = (\Re\mu)^-$, $\mu_3 = (\Im\mu)^+$, and $\mu_4 = (\Im\mu)^-$. Note that $\mu_1, \mu_2, \mu_3, \mu_4 \in M(K)^+$ and $\mu_j \ll \mu$ for $j = 1,2,3,4$.

The following inequality, which follows easily, will be useful. Let K be a non-empty, locally compact space, and take $\mu \in M(K)$. Then, for each $B \subset \mathfrak{B}_K$, we have

$$|\mu|(B) \leq 4 \sup\{|\mu(C)| : C \in \mathfrak{B}_K, C \subset B\} . \tag{4.10}$$

The following two results are clear.

Proposition 4.2.5. *Let K be a non-empty, locally compact space, and suppose that $\mu, \nu \in M(K)$. Then $\mu \perp \nu$ if and only if*

$$\|\mu+\nu\| = \|\mu-\nu\| = \|\mu\| + \|\nu\| .$$

□

Corollary 4.2.6. *Let K and L be non-empty, locally compact spaces. Suppose that E is a linear subspace of $M(K)$ and that $T : E \to M(L)$ is a linear isometry. Take measures $\mu, \nu \in E$. Then $T\mu \perp T\nu$ if and only if $\mu \perp \nu$.* □

Proposition 4.2.7. *Let K be a non-empty, locally compact space, and suppose that $\mu \in M(K)$. Then μ is continuous if and only if, for each $\varepsilon > 0$, there exist $n \in \mathbb{N}$ and $\mu_1, \dots, \mu_n \in M(K)$ with*

$$\mu = \mu_1 + \cdots + \mu_n ,$$

with $\mu_i \perp \mu_j$ $(i, j \in \mathbb{N}_n, i \neq j)$, and with $\|\mu_i\| < \varepsilon$ $(i \in \mathbb{N}_n)$.

Proof. Suppose that $\mu \in M_c(K)$, and take $\varepsilon > 0$. Then there is a compact subset L of K such that $|\mu|(K \setminus L) < \varepsilon$. Each point $x \in L$ has an open neighbourhood U_x with $|\mu|(U_x) < \varepsilon$, and the union, say $\bigcup\{U_j : j \in \mathbb{N}_n\}$, of finitely many of these neighbourhoods contains L. Set $V_1 = U_1$ and $V_j = U_j \setminus (U_1 \cup \cdots \cup U_{j-1})$ for $j = 2, \dots, n$. Then set $\mu_0 = \mu \mid (K \setminus L)$ and $\mu_j = \mu \mid (V_j \cap L)$ $(j \in \mathbb{N}_n)$. We see that $\mu_0, \mu_1, \dots, \mu_n \in M(K)$, and they have the required properties (after re-labelling).

The converse is immediate. □

Corollary 4.2.8. *Let K and L be non-empty, locally compact spaces, and suppose that $T : M(K) \to M(L)$ is a linear isometry. Then $T\mu \in M_c(L)$ whenever $\mu \in M_c(K)$.*

Proof. Take $\mu \in M_c(K)$ and $\varepsilon > 0$. Then there exist $n \in \mathbb{N}$ and $\mu_1, \dots, \mu_n \in M(K)$ with $\mu = \mu_1 + \cdots + \mu_n$, with $\mu_i \perp \mu_j$ $(i, j \in \mathbb{N}_n, i \neq j)$, and with $\|\mu_i\| < \varepsilon$ $(i \in \mathbb{N}_n)$. Then $T\mu = T\mu_1 + \cdots + T\mu_n$, with $T\mu_i \perp T\mu_j$ $(i, j \in \mathbb{N}_n, i \neq j)$ by Corollary 4.2.6 and with $\|T\mu_i\| < \varepsilon$ $(i \in \mathbb{N}_n)$. Thus $T\mu \in M_c(L)$. □

The following theorem is the *Lebesgue decomposition theorem*; see [59, Theorem 4.3.2] and [217, Theorem 6.10(a)], for example.

Theorem 4.2.9. *Let K be a non-empty, locally compact space, and suppose that $\mu \in M(K)^+$ and $\nu \in M(K)$. Then there is a unique pair $\{\nu_a, \nu_s\}$ of measures in $M(K)$ with $\nu = \nu_a + \nu_s$, with $\nu_a \ll \mu$, and with $\nu_s \perp \mu$.* □

It is clear that, in the above setting, the maps $\nu \mapsto \nu_a$ and $\nu \mapsto \nu_s$ are Banach-lattice homomorphisms on $M(K)$.

Proposition 4.2.10. *Let K be a non-empty, locally compact space, and suppose that $\mu, \nu \in M(K)^+$ with $\nu \ll \mu$. Then there exists $B \in \mathfrak{B}_K$ with $\nu \sim \mu \mid B$.*

Proof. Take $\mu = \mu_a + \mu_s$ with $\mu_a \ll \nu$ and $\mu_s \perp \nu$, and partition K into two disjoint Borel subsets B and C such that $\mu_s(B) = \nu(C) = 0$. Then

$$(\mu \mid B)(E) = \mu_a(E \cap B) + \mu_s(E \cap B) = \mu_a(E) \quad (E \in \mathfrak{B}_K),$$

and so $\mu \mid B = \mu_a$. Now $\mu_a \sim \nu$ because, for each $A \in \mathfrak{B}_K$ with $\mu_a(A) = 0$, we have $\nu(A) = \nu(A \cap B) = \mu_a(A \cap B) = 0$. Hence $\nu \sim \mu \mid B$, as desired. □

Definition 4.2.11. Let K be a non-empty, locally compact space. For each measure $\mu \in M(K)$, the *disjoint complement* of μ is

$$\mu^{\perp} = \{\nu \in M(K) : \nu \perp \mu\}.$$

It is clear that $\mu^{\perp}$ is a linear subspace of $M(K)$. Further, $\mu \ll \nu$ if and only if $\nu^{\perp} \subset \mu^{\perp}$. The following proposition is easily verified by using elementary vector-lattice exercises.

Proposition 4.2.12. *Let K be a non-empty, locally compact space, and suppose that $\mu, \nu \in M(K)^+$. Then:*

(i) $\mu \perp \nu$ *if and only if* $\mu \wedge \nu = 0$;
(ii) $(\mu \vee \nu)^{\perp} = \mu^{\perp} \cap \nu^{\perp} = (\mu + \nu)^{\perp}$;
(iii) $\mu^{\perp} \cup \nu^{\perp} \subset (\mu \wedge \nu)^{\perp}$;
(iv) $\mu \sim \nu$ *if and only if* $\mu^{\perp} = \nu^{\perp}$. □

Proposition 4.2.13. *Let K be a non-empty, locally compact space, and suppose that F is a complemented face of $P(K)$. Take $\mu \in F$ and $\nu \in F^{\perp}$. Then $\mu \wedge \nu = 0$.*

Proof. Set $\lambda = \mu \wedge \nu$. Clearly $\lambda \leq \nu$, and $\lambda \neq \mu$ because $\mu \not\leq \nu$. Assume towards a contradiction that $\lambda \neq 0$. Then

$$\mu = \|\lambda\| \left(\frac{\lambda}{\|\lambda\|} \right) + \|\mu - \lambda\| \left(\frac{\mu - \lambda}{\|\mu - \lambda\|} \right),$$

and $\|\lambda\| + \|\mu - \lambda\| = 1$ because $\mu - \lambda \geq 0$ and $\|\cdot\|$ is additive on $M(K)^+$. Thus $\lambda / \|\lambda\| \in F$. Similarly, $\lambda / \|\lambda\| \in F^{\perp}$, a contradiction because $F \cap F^{\perp} = \emptyset$. Thus $\lambda = 0$. □

Proposition 4.2.14. *Let K be an infinite, locally compact space. Then*

$$M(K) \cong M(K_\infty).$$

Proof. By equation (4.8), it suffices to show that the subspaces of discrete measures and of continuous measures on K and on K_∞, respectively, are isometrically isomorphic to each other. However, $M_d(K) \cong M_d(K_\infty)$ because $|K| = |K_\infty|$, and, since $\mathfrak{B}_K \subset \mathfrak{B}_{K_\infty}$, the map $\mu \mapsto \mu | \mathfrak{B}_K$ determines a linear isometry from $M_c(K_\infty)$ onto $M_c(K)$. □

Example 4.2.15. Let S be a semigroup. In §1.5, we noted that the space βS becomes a right or left topological semigroup with respect to the operations $\Box$ and $\Diamond$, respectively. Thus the products of u and v in βS are $u \Box v$ and $u \Diamond v$.

The Banach space $(\ell^1(S), \|\cdot\|_1)$ is a Banach algebra with respect to the convolution product $\star$ defined by

$$(f \star g)(t) = \sum\{f(r)g(s) : r,s \in S,\, rs = t\} \quad (t \in S)$$

for $f,g \in \ell^1(S)$, where we take the sum to be 0 when there are no elements $r,s \in S$ with $rs = t$. It is easily checked that $(\ell^1(S), \|\cdot\|_1, \star)$ is a Banach algebra; it is called the *semigroup algebra* on S.

The bidual of the space $(\ell^1(S), \|\cdot\|_1)$ is identified with the space $M(\beta S)$ of measures on βS, and so the Arens products described in §3.1 give the products $\mu \Box \nu$ and $\mu \Diamond \nu$ for $\mu, \nu \in M(\beta S)$. In particular, we can define the products $\delta_u \Box \delta_v$ and $\delta_u \Diamond \delta_v$ of point masses for $u, v \in \beta S$. These products are easily seen to be consistent with those in βS, in the sense that

$$\delta_u \Box \delta_v = \delta_{u \Box v}, \quad \delta_u \Diamond \delta_v = \delta_{u \Diamond v} \quad (u, v \in \beta S).$$

The Banach algebras $(M(\beta S), \Box)$ and $(M(\beta S), \Diamond)$ are studied in the memoir [71]. In particular, it is shown that $\ell^1(S)$ is usually (but not always) strongly Arens irregular. The interplay between properties of the Banach algebras and the combinatorial properties of the semigroup βS is rather subtle. For further results, see [47]. $\Box$

4.3 A Boolean ring

An introduction to the general theory of Boolean rings and algebras was given in §1.7. We shall now discuss a specific Boolean ring B defined for each non-empty, locally compact space K, with the property that $C^b(St(B)) \cong M(K)'$; this Boolean ring will be used to give a new representation of $C_0(K)''$ in §5.4.

Definition 4.3.1. Let (Ω, Σ, μ) be a measure space. The family of subsets S of Ω such that $\mu(S) = 0$ is denoted by $\mathfrak{N}_\mu$. Then $\Sigma_\mu = \Sigma / \mathfrak{N}_\mu$ and $\pi_\mu : \Sigma \to \Sigma_\mu$ is the quotient map.

Clearly $\mathfrak{N}_\mu$ is a σ-complete ideal in the Boolean algebra Σ, and so Σ_μ is a σ-complete Boolean algebra. We regard μ as a measure on Σ_μ, so that

$$\mu(\pi_\mu(A)) = \mu(A) \quad (A \in \Sigma).$$

In particular, let K be a non-empty, locally compact space, and suppose that $\mu \in M(K)^+$. Then $\mathfrak{B}_\mu = \mathfrak{B}_K / \mathfrak{N}_\mu$. For example, with $K = \mathbb{I}$ and $\mu = m$, we obtain the basic example, $\mathfrak{B}_m$. Note that, when regarded as a function on the Boolean algebra $\mathfrak{B}_\mu$, the measure μ is a σ-normal measure in the sense of Definition 1.7.12.

Proposition 4.3.2. *Let (Ω, Σ, μ) be a finite measure space.*

(i) *Each increasing net $\mathscr{C}$ in Σ_μ has a supremum $B \in \Sigma_\mu$, and*

$$\mu(B) = \sup\{\mu(C) : C \in \mathscr{C}\}.$$

(ii) *The Boolean algebra Σ_μ is complete, and so $St(\Sigma_\mu)$ is a Stonean space.*

(iii) *Suppose that Σ_μ is atomless, and take $B \in \Sigma_\mu$ and $\alpha \in [0, \mu(B)]$. Then there exists $C_0 \in \Sigma_\mu$ with $C_0 \leq B$ and $\mu(C_0) = \alpha$.*

Proof. (i) Choose an increasing sequence (B_n) in $\mathscr{C}$ such that

$$\lim_{n\to\infty} \mu(B_n) = \sup\{\mu(B) : B \in \mathscr{C}\} < \infty,$$

and define $B = \bigvee\{B_n : n \in \mathbb{N}\}$, so that $B \in \Sigma_\mu$ and $\lim_{n\to\infty} \mu(B_n) = \mu(B)$.

We first *claim* that $\mu(C - B) = 0$ $(C \in \mathscr{C})$. Indeed, take $C \in \mathscr{C}$, and assume towards a contradiction that there exists $\delta > 0$ such that $\mu(C - B) > \delta$. Then $\mu(C \vee B_n) > \mu(B_n) + \delta$ $(n \in \mathbb{N})$. Choose $m \in \mathbb{N}$ with $\mu(B_m) > \mu(B) - \delta/2$. Since $C \vee B_m \subset D$ for some $D \in \mathscr{C}$, there exists $n \in \mathbb{N}$ such that $\mu(B_n) > \mu(C \vee B_m) - \delta/2$. Thus $\mu(B_n) > \mu(B_m) + \delta/2 > \mu(B)$, the required contradiction. The claim holds.

We next *claim* that $B = \bigvee\{C : C \in \mathscr{C}\}$. By the above paragraph, $C \leq B$ $(C \in \mathscr{C})$. Now suppose that $D \in \Sigma_\mu$ is such that $C \leq D$ $(C \in \mathscr{C})$. Then $B = \bigvee\{B_n : n \in \mathbb{N}\} \leq D$. It follows that $B = \bigvee\{C : C \in \mathscr{C}\}$, as claimed, and so $\mu(B) = \sup\{\mu(C) : C \in \mathscr{C}\}$.

(ii) It is immediate from (i) that Σ_μ is complete. By Corollary 1.7.5, $St(\Sigma_\mu)$ is a Stonean space.

(iii) Let $\mathscr{C}$ be a chain in Σ_μ such that $\mathscr{C}$ is maximal with respect to the properties that $C \leq B$ and that $\mu(C) \leq \alpha$ whenever $C \in \mathscr{C}$. By (i), there exists $C_0 \in \Sigma_\mu$ with

$$\mu(C_0) = \sup\{\mu(C) : C \in \mathscr{C}\}.$$

Clearly $C_0 \leq B$ and $\mu(C_0) \leq \alpha$. Assume that $\mu(C_0) < \alpha$. Since Σ_μ is atomless, it follows from a remark on page 43 that there is an element $D \in \Sigma_\mu$ with $D \leq B \setminus C_0$ such that $0 < \mu(D) < \alpha - \mu(C_0)$. But now $\mathscr{C} \cup \{C_0 \vee D\}$ is a chain with the property that $\mu(C) \leq \alpha$ $(C \in \mathscr{C} \vee \{C_0 \vee D\})$, a contradiction of the maximality of $\mathscr{C}$. Hence $\mu(C_0) = \alpha$. □

Corollary 4.3.3. *Let K be a non-empty, locally compact space.*

(i) *Suppose that $\mu \in P(K)$. Then $\mathfrak{B}_\mu$ is atomless if and only if μ is continuous.*

(ii) *Suppose that $\mu \in M_c(K)^+$ and $\mu \neq 0$. Then $\mathfrak{B}_\mu$ is not a separable Boolean algebra.*

Proof. (i) Suppose that μ is not continuous. Then there exists $x \in K$ such that $\mu(\{x\}) > 0$, and then $\pi_\mu(\delta_x)$ is an atom in $\mathfrak{B}_\mu$.

Suppose that μ is continuous. Then it follows easily from Proposition 4.2.7 that $\mathfrak{B}_\mu$ is atomless.

(ii) Since μ is a non-zero, σ-normal measure on $\mathfrak{B}_\mu$, this follows from Proposition 1.7.13. □

Definition 4.3.4. Let (Ω, Σ, μ) be a probability measure space. We set

$$\rho_\mu(B,C) = \mu(B \Delta C) \quad (B,C \in \Sigma_\mu).$$

It is easy to see that ρ_μ is a metric on the Boolean algebra Σ_μ.

Proposition 4.3.5. *Let (Ω, Σ, μ) be a probability measure space. Then the metric space (Σ_μ, ρ_μ) is complete.*

Proof. As in any metric space, it suffices to show that there exists $B \in \Sigma_\mu$ such that $\rho_\mu(B_k, B) \leq 1/2^k$ $(k \in \mathbb{N})$ whenever $(B_n : n \in \mathbb{N})$ is a sequence in Σ_μ with $\rho_\mu(B_n, B_{n+1}) < 1/2^{n+1}$ $(n \in \mathbb{N})$.

Given such a sequence (B_n), note that $\rho_\mu(B_k, B_n) < 1/2^k$ $(n \geq k)$. For each $n \in \mathbb{N}$, set $D_n = B_n \cup \bigcup_{k\in\mathbb{N}}(B_{n+k-1} \Delta B_{n+k})$. Then $D_n = B_n \cup D_{n+1} \supset D_{n+1}$ and also $B_n \Delta D_n \subset \bigcup_{k\in\mathbb{N}}(B_{n+k-1} \Delta B_{n+k})$, so that $\rho_\mu(B_n, D_n) \to 0$ as $n \to \infty$. Set $B = \bigcap_{n\in\mathbb{N}} D_n$. Then $\mu(D_n) \to \mu(B)$ by the countable additivity of the measure μ, and hence $\rho_\mu(D_n, B) \to 0$ as $n \to \infty$. We have

$$\rho_\mu(B_k, B) \leq \rho_\mu(B_k, B_n) + \rho_\mu(B_n, D_n) + \rho_\mu(D_n, B) \quad (k,n \in \mathbb{N})\,; \tag{4.11}$$

we fix $k \in \mathbb{N}$, and then take limits in (4.11) as $n \to \infty$ to see that $\rho_\mu(B_k, B) \leq 1/2^k$, giving the result. □

Theorem 4.3.6. *Let $(\Omega_1, \Sigma_1, \mu_1)$ and $(\Omega_2, \Sigma_2, \mu_2)$ be two probability measure spaces such that Σ_{μ_1} and Σ_{μ_2} are atomless Boolean algebras and $(\Sigma_{\mu_1}, \rho_{\mu_1})$ and $(\Sigma_{\mu_2}, \rho_{\mu_2})$ are separable metric spaces. Then there is an isomorphism $\theta : \Sigma_{\mu_1} \to \Sigma_{\mu_2}$ such that*

$$\mu_2(\theta(B)) = \mu_1(B) \quad (B \in \Sigma_1)\,.$$

Proof. Let $\{U_n : n \in \mathbb{N}\}$ and $\{V_n : n \in \mathbb{N}\}$ be countable, dense families in $(\Sigma_{\mu_1}, \rho_{\mu_1})$ and $(\Sigma_{\mu_2}, \rho_{\mu_2})$, respectively, where we write Σ_{μ_i} for $\Sigma_i/\mathfrak{N}_{\mu_i}$ for $i = 1,2$.

We shall first define increasing sequences $(F_n : n \in \mathbb{N})$ and $(G_n : n \in \mathbb{N})$ of finite Boolean subalgebras of Σ_{μ_1} and Σ_{μ_2}, respectively, and an isomorphism

$$\theta : \bigcup_{n=1}^{\infty} F_n \to \bigcup_{n=1}^{\infty} G_n$$

such that $\mu_2(\theta(B)) = \mu_1(B)$ $(B \in \bigcup_{n=1}^{\infty} F_n)$.

We start by setting $F_1 = \{\emptyset, \Omega_1\}$, $G_1 = \{\emptyset, \Omega_2\}$, $\theta(\emptyset) = \emptyset$, and $\theta(\Omega_1) = \Omega_2$.

Now take $n \in \mathbb{N}$, and assume inductively that $F_1, \dots, F_n$ and $G_1, \dots, G_n$ have been defined in Σ_{μ_1} and Σ_{μ_2}, respectively, and that θ has been defined on F_n.

Suppose that n is even, and choose $r \in \mathbb{N}$ to be the smallest number such that $U_r \notin F_n$. By Proposition 4.3.2(iii), for each atom $A \in F_n$, there exists $E_A \in \Sigma_{\mu_2}$ such that $E_A \leq \theta(A)$ and $\mu_2(E_A) = \mu_1(A \wedge U_r)$. We set

$$\theta(A \wedge U_r) = E_A\,, \quad \theta(A - U_r) = \theta(A) - E_A\,,$$

for each such atom A, and we define F_{n+1} to be the (finite) Boolean subalgebra of Σ_{μ_1} generated by $F_n \cup \{U_r\}$; we then extend θ to F_{n+1} in the obvious way, and finally set $G_{n+1} = \theta(F_{n+1})$.

Suppose that n is odd, and choose $r \in \mathbb{N}$ to be the smallest number such that $V_r \notin G_n$. In a similar manner, we extend θ^{-1} to the Boolean subalgebra of Σ_{μ_2} generated by $G_n \cup \{V_r\}$. This completes the inductive construction.

We observe that

$$\theta : \left(\bigcup_{n=1}^{\infty} F_n, \rho_\mu\right) \to \left(\bigcup_{n=1}^{\infty} G_n, \rho_\nu\right)$$

is an isometry and that $\bigcup_{n=1}^{\infty} F_n$ and $\bigcup_{n=1}^{\infty} G_n$ are dense in the metric spaces $(\Sigma_{\mu_1}, \rho_{\mu_1})$ and $(\Sigma_{\mu_2}, \rho_{\mu_2})$, respectively. By Proposition 4.3.5, these two metric spaces are complete, and so the map θ can be extended to an isometry, also called θ, from $(\Sigma_{\mu_1}, \rho_{\mu_1})$ onto $(\Sigma_{\mu_2}, \rho_{\mu_2})$. Clearly θ is an isomorphism between Σ_{μ_1} and Σ_{μ_1}. □

The following consequence of the above theorem, which refers to the measure space $(\mathbb{I}, \Sigma_m, m)$, is sometimes called *von Neumann's isomorphism theorem.* However, the result was essentially known in the 1930s (see Kolmogorov [158, §20] and Szpilrajn [233, Theorem I; note the reference to Jaskowski (1932)]), but apparently the first complete, published proof was by Caratheodory [52, Satz 7 (Hauptsatz)]. Several books now have a proof of this result; a short proof is in Birkhoff [36, p. 262, Corollary]; see also Bogachev [39, Theorem 9.3.4], Halmos [132, §41, Theorem C], and Royden [216, Theorem 15.4].

Corollary 4.3.7. *Let (Ω, Σ, μ) be a probability measure space such that Σ_μ is an atomless Boolean algebra and (Σ_μ, ρ_μ) is a separable metric space. Then there is an isomorphism $\theta : \Sigma_\mu \to \Sigma_m$ such that $m(\theta(B)) = \mu(B)$ $(B \in \Sigma_\mu)$.*

Proof. Since m is a continuous measure, it follows from Corollary 4.3.3(i) that the Boolean algebra Σ_m is atomless, and (Σ_m, ρ_m) is a separable metric space. Now the result follows from Theorem 4.3.6. □

Let K be a non-empty, locally compact space, and take $\mu, \nu \in M(K)$. In Definition 4.2.4, we said that $\mu \sim \nu$ if $\mu \ll \nu$ and $\nu \ll \mu$, so that $\sim$ is an equivalence relation on $M(K)$. The equivalence class containing μ is denoted by $[\mu]$. It is now trivial to check that the relation $\leq$ defined on $M(K)/\sim$ by

$$[\mu] \leq [\nu] \quad \text{if and only if} \quad \mu \ll \nu$$

is a well-defined partial order on $M(K)/\sim$.

We wish to show that the partially ordered space $(M(K)/\sim,\leq)$ is a Boolean ring with certain nice properties. In virtue of the fact that $[\mu]=[|\mu|]$, the space $(M(K)/\sim,\leq)$ is isomorphic to $(M(K)^+/\sim,\leq)$, and so we shall simplify notation and restrict the discussion to positive measures; in particular, for $\mu\in M(K)^+$, we restrict $\mu^\perp$ to $M(K)^+$.

Definition 4.3.8. Let K be a non-empty, locally compact space. We define operations $\vee$ and $\wedge$ on $M(K)^+/\sim$ by:

$$[\mu]\vee[\nu]=[\mu\vee\nu],\quad [\mu]\wedge[\nu]=[\mu\wedge\nu]\quad (\mu,\nu\in M(K)^+).$$

We have to show that the above operations are well defined.

Proposition 4.3.9. *Let K be a non-empty, locally compact space. Then*

$$(M(K)^+/\sim,\leq)$$

is a distributive lattice with a minimum element in which $\vee$ and $\wedge$ are the supremum and infimum in the partial order $\leq$. In particular, $\vee$ and $\wedge$ are well defined.

Proof. Let $\mu,\nu\in M(K)^+$, and set $L=M(K)^+/\sim$ and $S=\{[\mu],[\nu]\}$ in L.

We *claim* that $[\mu\vee\nu]$ is the supremum of S. Indeed, $\mu\ll\mu\vee\nu$ and $\nu\ll\mu\vee\nu$, and so $[\mu\vee\nu]$ is an upper bound for S. Now suppose that $\eta\in M(K)^+$ is such that $[\eta]$ is an upper bound for S. Then $\mu\ll\eta$ and $\nu\ll\eta$, and so $\mu\vee\nu\ll\eta$, whence $[\mu\vee\nu]\leq[\eta]$. The claim follows, and hence $[\mu\vee\nu]=[\mu]\vee[\nu]$.

We also *claim* that $[\mu\wedge\nu]$ is the infimum of S. Indeed, $\mu\wedge\nu\ll\mu$ and $\mu\wedge\nu\ll\nu$, and so $[\mu\wedge\nu]$ is a lower bound for S. Now suppose that $\eta\in M(K)^+$ is such that $[\eta]$ is a lower bound for S, so that $\mu^\perp\subset\eta^\perp$ and $\nu^\perp\subset\eta^\perp$. To show that $[\eta]\leq[\mu\wedge\nu]$, we must show that $(\mu\wedge\nu)^\perp\subset\eta^\perp$. For this, take $\gamma\in(\mu\wedge\nu)^\perp$. Then $\gamma\wedge\mu\wedge\nu=0$, whence $\gamma\wedge\mu\in\nu^\perp\subset\eta^\perp$, and so $\gamma\wedge\mu\wedge\eta=0$, i.e., $\gamma\wedge\eta\in\mu^\perp\subset\eta^\perp$. It follows that $\gamma\wedge\eta\wedge\eta=0=\gamma\wedge\eta$, and $\gamma\in\eta^\perp$ as desired. The claim follows, and hence $[\mu\wedge\nu]=[\mu]\wedge[\nu]$.

We have shown that L is a lattice. Clearly $[0]$ is the minimum element of L. That L is a distributive lattice follows immediately from the distributivity of the lattice $(M(K)^+,\vee,\wedge)$. □

We remark that an examination of the proof of the preceding proposition shows that an analogous result is valid for any distributive lattice with a minimum element, provided that the relation $a\ll b$ is *defined* by the formula $b^\perp\subset a^\perp$.

Theorem 4.3.10. *Let K be a non-empty, locally compact space, and suppose that $\mu\in M(K)^+$. Then*

$$\{[\nu]:\nu\in M(K)^+,\,\nu\ll\mu\}$$

is a Boolean algebra in the order $\leq$ inherited from $(M(K)^+/\sim,\leq)$, and it is isomorphic as a Boolean algebra to $\mathfrak{B}_\mu=\mathfrak{B}_K/\mathfrak{N}_\mu$.

Proof. Take $\nu \in M(K)^+$ with $\nu \ll \mu$, and, using the Lebesgue decomposition theorem, Theorem 4.2.9, write $\mu = \mu_a + \mu_s$, where $\mu_a, \mu_s \in M(K)^+$ are such that $\mu_a \ll \nu$ and $\mu_s \perp \nu$. Thus:

$$[\mu_a] \leq [\nu] \leq [\mu]; \quad [\mu_s] \leq [\mu]; \quad [\mu_s] \wedge [\nu] = [0].$$

We *claim* that $[\nu] \vee [\mu_s] = [\mu]$. Indeed, $[\nu \vee \mu_s] = [\nu + \mu_s]$ by Proposition 4.2.12(ii), and so

$$[\mu] = [\mu_a + \mu_s] \leq [\nu + \mu_s] = [\nu \vee \mu_s] = [\nu] \vee [\mu_s] \leq [\mu],$$

proving the claim.

We have shown that $[\mu_s]$ is the relative complement of $[\nu]$ with respect to $[\mu]$ and that the order interval $[[0], [\mu]]$ is a Boolean algebra. Moreover, we observe that $\mu_a \sim \nu$, i.e., $[\mu_a] = [\nu]$, because each is the (unique) relative complement of $[\mu_s]$ with respect to $[\mu]$.

The required Boolean isomorphism is as follows. Take $\nu \in M(K)^+$ with $\nu \ll \mu$. By Proposition 4.2.10, $\nu \sim \mu \mid B$ for some $B \in \mathfrak{B}_K$; the image of ν in $\mathfrak{B}_K/\mathfrak{N}_\mu$ is the equivalence class of B. Note that for $B, C \in \mathfrak{B}_K$, we have $\mu \mid B \sim \mu \mid C$ if and only if $\mu(B \Delta C) = 0$, i.e., if and only if B and C define the same equivalence class in $\mathfrak{B}_\mu$. It is now a simple matter to verify that the map so defined is a bijection which preserves the Boolean operations. □

Theorem 4.3.11. *Let K be a non-empty, locally compact space. Then*

$$(M(K)^+/\sim, \leq)$$

is a Dedekind complete Boolean ring such that, for each $\mu \in M(K)^+$, the order interval $[[0], [\mu]]$ is a complete Boolean algebra. Further, the Stone space

$$S_K := St(M(K)^+/\sim, \leq)$$

is an extremely disconnected, locally compact space. For each $\mu \in M(K)^+$, the space $St(\mathfrak{B}_\mu)$ is compact and open in S_K. Further, each compact–open subspace of S_K has the form $St(\mathfrak{B}_\mu)$ for some $\mu \in M(K)^+$, and

$$S_K = \bigcup \{St(\mathfrak{B}_\mu) : \mu \in M(K)^+\}.$$

Proof. By Proposition 4.3.9 and Theorem 4.3.10, $(M(K)^+/\sim, \leq)$ is a distributive lattice with a minimum element such that each order interval $[0, \mu]$ is a Boolean algebra, and so it is a Boolean ring.

For each $\mu \in M(K)^+$, the order interval $[[0], [\mu]]$ is isomorphic to $\mathfrak{B}_\mu$, which, by Proposition 4.3.2(ii), is a complete Boolean algebra, and so $St(\mathfrak{B}_\mu)$ is a Stonean space. The form of S_K follows from Theorem 1.7.2. Thus $(M(K)^+/\sim, \leq)$ is Dedekind complete and S_K is extremely disconnected. □

4.4 The spaces $L^p(K,\mu)$

We now define the standard spaces $L^\infty(K,\mu)$ and $L^p(K,\mu)$ for $\mu \in M(K)^+$ and p with $1 \leq p < \infty$. In fact, we have already mentioned these spaces when they are defined on a general measure space (Ω,Σ,μ); here we give more details in our special setting.

Let K be a non-empty, locally compact space, and take $\mu \in M(K)^+$. Then two bounded, Borel functions f and g are said to be *equivalent* (with respect to μ) if $\mu(\{x \in K : f(x) \neq g(x)\}) = 0$, or, equivalently, if

$$\int_K |f-g| \, \mathrm{d}\mu = 0;$$

the family of these equivalence classes is the standard Banach space

$$L^\infty(\mu) = L^\infty(K,\mu),$$

with the *essential supremum norm*, $\|\cdot\|_\infty$, so that

$$\|f\|_\infty = \inf\{\alpha > 0 : \mu(\{x \in K : |f(x)| > \alpha\}) = 0\}.$$

The equivalence class containing an element f of $B^b(K)$ is sometimes denoted by $[f]$. The collection of (equivalence classes of) real-valued functions in $L^\infty(\mu)$ is denoted by $L^\infty_{\mathbb{R}}(\mu)$, and the positive functions form the space $L^\infty(\mu)^+$.

We note that $\mathrm{lin}\,\{[\chi_B] : B \in \mathfrak{B}_K\}$ is a dense linear subspace of $L^\infty(\mu)$.

We remark that every equivalence class in $L^\infty(K,\mu)$ contains a representative in the second Baire class, $B_2(K)$, that was defined in §3.3. This is a classical fact for real functions on an interval in $\mathbb{R}$; see [39, Example 2.12.15] or [116, Theorem 4b, p. 194], for example. The argument in the case of a general locally compact space K and $\mu \in M(K)^+$ follows a parallel route based on Lusin's theorem, Theorem 4.1.7(ii).

Proposition 4.4.1. *Let K be an infinite, locally compact space, and suppose that $\mu \in M(K)^+$ with $\mathrm{supp}\,\mu = K$. Then ℓ^∞ is isometrically isomorphic to a 1-complemented subspace of $L^\infty(K,\mu)$.*

Proof. Let (U_n) be a sequence of pairwise-disjoint, non-empty, open subsets of K, so that $\mu(U_n) > 0$ $(n \in \mathbb{N})$. The map

$$(\alpha_n) \mapsto \sum_{n=1}^\infty \alpha_n \chi_{U_n}, \quad \ell^\infty \to L^\infty(K,\mu),$$

is an isometric embedding, with range E, say. The map

$$P : f \mapsto \sum_{n=1}^\infty \frac{1}{\mu(U_n)} \left(\int_{U_n} f \, \mathrm{d}\mu \right) \chi_{U_n}, \quad L^\infty(K,\mu) \to \ell^\infty \cong E,$$

is a bounded projection onto E with $\|P\| = 1$, and so E is a 1-complemented subspace of $L^\infty(K,\mu)$. □

In the following, we shall write $L^\infty(G)$ for $L^\infty(G,m_G)$ when G is a locally compact group G.

Theorem 4.4.2. *Let G be a non-discrete, locally compact group. Then $C^b(G)$ is not complemented in $L^\infty(G)$, and so $C^b(G)$ is not injective.*

Proof. Assume towards a contradiction that there is a bounded projection Q of $L^\infty(G)$ onto the closed subspace $C^b(G)$.

It is standard that there is a compact, symmetric neighbourhood U of e_G such that $G_0 := \bigcup\{U^n : n \in \mathbb{N}\}$ is an infinite, clopen subgroup of G. By replacing G by G_0 and Q by $R \circ (Q \mid L^\infty(G_0))$, where R denotes the restriction map from $C^b(G)$ onto $C^b(G_0)$, we may suppose that G is σ-compact.

By [137, Theorem (8.7)], for each countable family $\{U_n : n \in \mathbb{N}\}$ in $\mathscr{N}_{e_G}$, there is a compact, normal subgroup N of G such that $N \subset \bigcap\{U_n : n \in \mathbb{N}\}$ and the quotient group $H := G/N$ is metrizable; take $\eta : G \to H$ to be the quotient map. Since G is not discrete, we have $m_G(\{e_G\}) = 0$, and so we may suppose that $m_G(N) = 0$; this implies that N is not open in G, and so H is not discrete. Hence there is a sequence (x_n) of distinct points in H with $\lim_{n\to\infty} x_n = e_H$.

For $f \in C^b(G)$, define

$$(Pf)(x) = \int_N f(x\zeta)\,\mathrm{d}m_N(\zeta) \quad (x \in H),$$

so that $Pf \in C^b(H)$ and the map $P : C^b(G) \to C^b(H)$ is a continuous linear surjection. The map

$$R : f \mapsto (f(x_n) - f(e_H)), \quad C^b(H) \to c_0,$$

is also a continuous linear surjection. As before, there exists a sequence (f_n) in $C(H,\mathbb{I})$ with $f_n(x_n) = 1$ $(n \in \mathbb{N})$ and such that $\operatorname{supp} f_m \cap \operatorname{supp} f_n = \emptyset$ when $m, n \in \mathbb{N}$ with $m \neq n$. The map

$$T : \alpha = (\alpha_n) \mapsto \sum_{n=1}^{\infty} \alpha_n (f_n \circ \eta), \quad \ell^\infty \to L^\infty(G),$$

is an isometric embedding, and $T(c_0) \subset C^b(G)$. Thus $S := R \circ P \circ Q \circ T : \ell^\infty \to c_0$ is a bounded operator with $S \mid c_0 = I_{c_0}$. But Phillips' theorem, Theorem 2.4.11, shows that there is no such projection S.

Thus we have a contradiction, and so $C^b(G)$ is not complemented in $L^\infty(G)$. □

For a result related to the above, see [167, Theorem 4].

In fact, it is proved in [198, Theorem 8.9] that, for each infinite, compact group G, the space $C(G)$ is isomorphic to $C(\mathbb{Z}_2^\kappa)$, where $\kappa = w(G)$, so this gives another route to the fact that $C(G)$ is not injective for each infinite, compact group G: as we remarked on page 79, $C(\mathbb{Z}_2^\kappa)$ is not injective. In contrast, there are many compact, non-metrizable spaces K such that $C(K)$ is not isomorphic to a space of the form $C(\mathbb{Z}_2^\kappa)$; such a K can be any infinite Stonean space, or any non-metrizable scattered space, or any space not satisfying CCC [198, Theorem 8.13].

Corollary 4.4.3. *Let G be an infinite, locally compact group. Then $C_0(G)$ is not injective.*

Proof. This follows from Theorem 4.4.2 when G is compact and from Theorem 2.4.12 when G is not pseudo-compact. However a locally compact group that is pseudo-compact as a topological space is already compact. Indeed, take G to be a locally compact, non-compact group, and let K be a compact, symmetric neighbourhood of e_G. Then $K^2 \neq G$: take $x \in G \setminus K^2$. Then $xK \cap K = \emptyset$. Continuing, we find infinitely many, pairwise-disjoint sets x_nK, where $x_n \in G$ $(n \in \mathbb{N})$. For each $n \in \mathbb{N}$, there exists a function $f_n \in C(G,\mathbb{I})$ with $f_n(x_n) = 1$ and $\operatorname{supp} f_n \subset x_nK$, and then $\sum_{n=1}^{\infty} nf_n$ is an unbounded, continuous function on G, and so G is not pseudo-compact. □

Corollary 4.4.4. *Let G be a locally compact group that is extremely disconnected as a topological space. Then G is discrete.*

Proof. By Proposition 1.5.9(ii), βG is Stonean, and so, by Theorem 2.5.11, the space $C^b(G) = C(\beta G)$ is 1-injective. By Theorem 4.4.2, G is discrete. □

In fact, every locally compact group that is an F-space is discrete; for this, see [60, §2.12].

It is clear that each space $L^\infty(K,\mu)$, for a non-empty, locally compact space K and $\mu \in P(K)$, is a commutative, unital C^*-algebra with respect to the pointwise product and conjugation as involution.

Definition 4.4.5. Let K be a non-empty, locally compact space, and suppose that $\mu \in P(K)$. Then the character space of the C^*-algebra $L^\infty(K,\mu)$ is denoted by Φ_μ, and the Gel'fand transform is $\mathscr{G}_\mu : L^\infty(K,\mu) \to C(\Phi_\mu)$.

Thus Φ_μ is a non-empty, compact space and $\mathscr{G}_\mu$ is a unital C^*-isomorphism and a Banach-lattice isometry. It follows that $(C(\Phi_\mu), \leq)$ is a Dedekind complete Banach lattice, and so, by Theorem 2.3.3, Φ_μ is a Stonean space.

Theorem 4.4.6. *Let K be a non-empty, locally compact space, and suppose that $\mu \in P(K)$. Then $L^\infty(K,\mu)$ is a 1-injective space.*

Proof. We know that $L^\infty(K,\mu) \cong C(\Phi_\mu)$ and that Φ_μ is a Stonean space. By Theorem 2.5.11, $C(\Phi_\mu)$ is 1-injective. □

The following is a famous isomorphism theorem of Pełczyński [196].

Theorem 4.4.7. *The spaces ℓ^∞ and $L^\infty(\mathbb{I})$ are isomorphic, so that $\ell^\infty \sim L^\infty(\mathbb{I})$.*

Proof. Set $E = L^\infty(\mathbb{I})$ and $F = \ell^\infty$. By Proposition 2.2.6, $E \sim E \times E$ and $F \sim F \times F$. By Theorem 4.4.6, both E and F are injective spaces. Since E is the dual of $L^1(\mathbb{I})$, it follows from Proposition 2.2.17(iii), there is a linear isometry from E onto a closed subspace of F; by Proposition 4.4.1, there is a linear isometry from F onto a closed subspace of E. It now follows from Proposition 2.5.4 that $E \sim F$. □

The exact Banach–Mazur distance between ℓ^∞ and $L^\infty(\mathbb{I})$ seems to be unknown.

Again let K be a non-empty, locally compact space, and take $\mu \in M(K)^+$. For each p with $1 \le p < \infty$, we define

$$L^p(K,\mu) = L^p(\mu) = \left\{ f \in \mathbb{C}^K : f \text{ measurable}, \int_K |f|^p \,\mathrm{d}\mu < \infty \right\}$$

and

$$\|f\|_p = \left(\int_K |f|^p \,\mathrm{d}\mu \right)^{1/p} \quad (f \in L^p(\mu)).$$

As usual, we identify equivalent functions f and g, that is, those with $\|f - g\|_p = 0$. Then $(L^p(\mu), \|\cdot\|_p)$ is a Banach space. In particular, with $K = \mathbb{I}$ and $\mu = m$, we obtain the standard Banach spaces $L^p(\mathbb{I})$ of page 5, where we recall that every Lebesgue measurable function on $\mathbb{I}$ is equivalent to a Borel measurable function.

The real-valued and positive functions in $L^p(\mu)$ are denoted by $L^p_{\mathbb{R}}(\mu)$ and $L^p(\mu)^+$, respectively. Again $L^p(\mu)$ is a Dedekind complete Banach lattice: for an explicit proof, see [39, Corollary 4.7.2] or [180, Example 23.3(iv), p. 126], where these spaces are, in fact, shown to be *super-Dedekind complete*, which means that each subset D of these spaces that is bounded above has a supremum which is, moreover, the supremum of some countable subset of D.

We note that $C_0(K)$ and $\mathrm{lin}\,\{[\chi_B] : B \in \mathfrak{B}_K\}$ are dense linear subspaces of $L^p(\mu)$ for each p with $1 \le p < \infty$.

Proposition 4.4.8. *Let K be a non-empty, compact, metrizable space, and suppose that $\mu \in M(K)^+$ and $1 \le p < \infty$. Then $(L^p(K,\mu), \|\cdot\|_p)$ is separable.*

Proof. By Theorem 2.1.7(i), $(C(K), |\cdot|_K)$ is separable, and so this follows because $C(K)$ is dense in $L^p(K,\mu)$. □

The following theorem is the *Radon–Nikodým theorem*; see [39, Theorem 3.2.2], [59, Theorem 4.2.4] and [217, Theorem 6.10(b)], for example.

Theorem 4.4.9. *Let K be a non-empty, locally compact space, and suppose that $\mu \in M(K)^+$ and $\nu \in M(K)$ with $\nu \ll \mu$. Then there is a unique function $h \in L^1(\mu)$ such that*

$$\nu(B) = \int_B h \,\mathrm{d}\mu\,, \quad |\nu|(B) = \int_B |h| \,\mathrm{d}\mu \quad (B \in \mathfrak{B}_K).$$

Further, $\|h\|_1 = \|\nu\|$. In particular, there is a measurable function h on K with $|h(x)| = 1$ $(x \in K)$ and such that $\mathrm{d}\mu = h\,\mathrm{d}|\mu|$. □

Thus, when $\mu, \nu \in M(K)^+$ with $\nu \ll \mu$, we may regard $L^1(\nu)$ as a closed linear subspace of $L^1(\mu)$. Further, we may identify $L^1(\mu)$ with the closed subspace

$$\{\nu \in M(K) : \nu \ll \mu\}$$

of measures in $M(K)$ that are absolutely continuous with respect to μ, so that $L^1(\mu)$ is a lattice ideal in $M(K)$; we have $M(K) = L^1(\mu) \oplus_1 \mu^{\perp}$, so that $L^1(\mu)$ is 1-complemented in $M(K)$.

The measures on a locally compact group G that are absolutely continuous with respect to left Haar measure m_G are identified with the Banach space

$$L^1(G, m_G),$$

which is regarded as a closed subspace of $M(G)$. This subspace is a closed ideal in the measure algebra $(M(G), \star)$ of G, and it is called the *group algebra* of G; the formula for the product of f and g in $L^1(G, m_G)$ is:

$$(f \star g)(s) = \int_G f(t)g(t^{-1}s)\,\mathrm{d}m_G(t) \quad (s \in G).$$

There is an enormous literature on the group algebra of a locally compact group; it is the central object in the subject 'harmonic analysis'. Again, for example, see the books [68, 137, 194, 195] and the memoir [72].

The following duality theorem is given in [39, §4.4], [59, Proposition 3.5.2], [137, Theorem (12.18)], and [217, Theorem 6.16], for example. For clause (ii), see [138, Theorem (20.20)].

Theorem 4.4.10. (i) *Let (Ω, Σ, μ) be a measure space, and take p with $1 < p < \infty$. Then $(L^p(\Omega, \mu), \|\cdot\|_p)'$ is isometrically isomorphic to $(L^q(\Omega, \mu), \|\cdot\|_q)$, where q is the conjugate index to p. The duality is given by*

$$\langle f, \lambda \rangle = \int_K f\lambda\,\mathrm{d}\mu \quad (f \in L^p(\Omega, \mu),\ \lambda \in L^p(\Omega, \mu)').$$

(ii) *Let (Ω, Σ, μ) be a decomposable measure space. Then $(L^1(\Omega, \mu), \|\cdot\|_1)'$ is isometrically isomorphic to $(L^\infty(\Omega, \mu), \|\cdot\|_\infty)$.* □

Corollary 4.4.11. *Let K be a non-empty, locally compact space, and take $\mu \in P(K)$. Then $L^1(K, \mu)$ is 1-complemented in its bidual*

Proof. We may suppose that $K = \operatorname{supp} \mu$, and so $C_0(K)$ is a closed subspace of $L^\infty(K, \mu)$.

Take $\Lambda \in L^1(K, \mu)''$. Then Λ acts on $L^1(K, \mu)' = L^\infty(K, \mu)$ and hence on $C_0(K)$; we set $R(\Lambda) = \Lambda \mid C_0(K)$, so that R is a bounded projection of $L^1(K, \mu)''$ onto $C_0(K)' = M(K)$ with $\|R\| = 1$. Since $L^1(K, \mu)$ is 1-complemented in $M(K)$, the result follows. □

We now come to a certain uniqueness result for the Banach lattice $L^1(\mathbb{I},m)$. A generalization to the lattices $L^p(\mathbb{I},m)$ for $1 \leq p < \infty$ is given in the book [184, Theorem 2.7.3].

Theorem 4.4.12. *Let $(\Omega_1,\Sigma_1,\mu_1)$ and $(\Omega_2,\Sigma_2,\mu_2)$ be probability measure spaces such that Σ_{μ_1} and Σ_{μ_2} are atomless Boolean algebras and the Banach spaces $L^1(\Omega_1,\mu_1)$ and $L^1(\Omega_2,\mu_2)$ are separable. Then there is a Banach-lattice isometry from $L^1(\Omega_1,\mu_1)$ onto $L^1(\Omega_2,\mu_2)$.*

Proof. Since $L^1(\Omega_1,\mu_1)$ and $L^1(\Omega_2,\mu_2)$ are separable Banach spaces, $(\Sigma_{\mu_1},\rho_{\mu_1})$ and $(\Sigma_{\mu_2},\rho_{\mu_2})$ are separable metric spaces. By Theorem 4.3.6. there is an isomorphism $\theta : \Sigma_{\mu_1} \to \Sigma_{\mu_2}$ such that $\mu_2(\theta(B)) = \mu_1(B)$ $(B \in \Sigma_1)$. There is an extension of θ to a linear bijection from $\operatorname{lin}\{\chi_B : B \in \Sigma_1\}$ onto $\operatorname{lin}\{\chi_C : C \in \Sigma_2\}$ with $\theta(\chi_B) = \chi_{\theta(B)}$ $(B \in \Sigma_1)$, and this map is an isometry with respect to the respective norms $\|\cdot\|_1$. Finally, the map θ extends to an isometry from $L^1(\Omega_1,\mu_1)$ onto $L^1(\Omega_2,\mu_2)$. Clearly the final map θ is a lattice isomorphism. □

In fact, let us suppose just that $(\Omega_1,\Sigma_1,\mu_1)$ is a σ-finite measure space. Then, using a remark on page 6, the same conclusion follows.

Corollary 4.4.13. *Let K and L be non-empty, locally compact spaces, and suppose that $\mu \in P_c(K)$ and $\nu \in P_c(L)$ are such that $(L^1(K,\mu),\|\cdot\|_1)$ and $(L^1(L,\nu),\|\cdot\|_1)$ are separable Banach spaces. Then there is a Banach-lattice isometry from $L^1(K,\mu)$ onto $L^1(L,\nu)$.*

Proof. The Boolean algebras $\mathfrak{B}_\mu$ and $\mathfrak{B}_\nu$ are atomless by Corollary 4.3.3(i), and so this is immediate from Theorem 4.4.12. □

Theorem 4.4.14. *Let K be a non-empty, locally compact space, and suppose that $\mu \in P_c(K)$. Then there is an isometric lattice embedding of $L^1(\mathbb{I})$ into $L^1(K,\mu)$. In the case where $(L^1(K,\mu),\|\cdot\|_1)$ is separable, $L^1(K,\mu)$ is Banach-lattice isometric to $L^1(\mathbb{I},m)$.*

Proof. Since the measure μ is continuous, it follows easily from Proposition 4.2.7 that there is a separable, complete, atomless Boolean algebra B contained in $\mathfrak{B}_\mu$. The isomorphism from $\mathfrak{B}_m$ onto B extends to the required isometric lattice embedding. □

Proposition 4.4.15. *Let K be a non-empty, locally compact space.*

(i) *The extreme points of $M(K)_{[1]}$ have the form $\zeta\delta_x$, where $\zeta \in \mathbb{T}$ and $x \in K$, and the extreme points of $P(K)$ have the form δ_x, where $x \in K$.*

(ii) *Take $\mu \in M_c(K)^+$ with $\mu \neq 0$. Then $\operatorname{ex} L^1(\mu)_{[1]} = \emptyset$.*

(iii) *Take $\mu \in M(K)^+$. Then each extreme point of $L^1(\mu)_{[1]}$ has the form $\zeta\delta_x$, where $\zeta \in \mathbb{C}$, $x \in K$, and $|\zeta|\,\mu(\{x\}) = 1$. Further, $\overline{\mathrm{co}}(\operatorname{ex} L^1(\mu)_{[1]}) = L^1(\mu_d)_{[1]}$.*

Proof. (i) Take $\mu \in \mathrm{ex}\, M(K)_{[1]}$, so that $\mu \neq 0$, and assume towards a contradiction that $\operatorname{supp}\mu$ is not a singleton. Then there exists $B_0 \in \mathfrak{B}_K$ with $\alpha := |\mu|(B_0) > 0$ and $|\mu|(B_0^c) > 0$, so that $\alpha \in (0,1)$. Define

$$\mu_1(B) = \frac{1}{\alpha}\mu(B \cap B_0), \quad \mu_2(B) = \frac{1}{1-\alpha}\mu(B \cap B_0^c) \quad (B \in \mathfrak{B}_K).$$

Then $\mu_1, \mu_2 \in M(K)_{[1]}$ and $\mu = \alpha\mu_1 + (1-\alpha)\mu_2$, but $\mu_1 \neq \mu$ and $\mu_2 \neq \mu$, a contradiction of the fact that μ is an extreme point of $M(K)_{[1]}$. The result follows.

(ii) Suppose that $f \in L^1(\mu)_{[1]}$ with $\|f\|_1 = 1$. Then there exists $B \in \mathfrak{B}_K$ with

$$0 < \int_B |f| \, \mathrm{d}\mu < 1,$$

and now essentially the same argument as above shows that f is a convex combination of two distinct elements of $L^1(\mu)_{[1]}$. Thus $\mathrm{ex}\, L^1(\mu)_{[1]} = \emptyset$.

(iii) Trivially, the extreme points of $L^1(\mu_d)_{[1]}$ have the form $\zeta\delta_x$, where $\zeta \in \mathbb{C}$, $x \in K$ and $|\zeta|\mu(\{x\}) = 1$. By (ii) and Proposition 2.1.10, $\mathrm{ex}\, L^1(\mu)_{[1]} = \mathrm{ex}\, L^1(\mu_d)_{[1]}$, and so the result follows. □

Corollary 4.4.16. *Let K be a non-empty, locally compact space. Then $M_d(K)_{[1]}$ is weak*-dense in $M(K)_{[1]}$.*

Proof. By the Krein–Milman theorem, Theorem 2.6.1, each element of $M(K)_{[1]}$ belongs to the weak*-closure of the convex hull of the set of extreme points of $M(K)_{[1]}$. By the proposition, the extreme points of $M(K)_{[1]}$ belong to $M_d(K)_{[1]}$. □

We saw in Theorem 2.4.15 that c_0 is not isomorphically a dual space: this followed because c_0 is not complemented in its bidual. We now consider the analogous question for the spaces $L^1(K,\mu) = (L^1(K,\mu), \|\cdot\|_1)$, especially in the case where $L^1(K,\mu)$ is separable; by Proposition 4.4.8, the latter case includes that in which K is compact and metrizable. However, we cannot follow the same argument as in the case of c_0 because, by Corollary 4.4.11, $L^1(K,\mu)$ is complemented in its bidual. The fact that the Banach space $L^1(\mathbb{I})$ is not isomorphic to a subspace of a separable dual space was first proved by Gel'fand himself in 1938 [110, p. 265]. The situation for more general spaces $L^1(K,\mu)$ is given below.

Theorem 4.4.17. *Let K be a non-empty, locally compact space, and suppose that $\mu \in P(K)$.*

(i) *The following are equivalent:*

(a) *$L^1(K,\mu)$ is isomorphic to a subspace of a separable dual space;*

(b) *$L^1(K,\mu)$ is isometrically isomorphic to a subspace of a separable dual space;*

(c) *μ is a discrete measure.*

(ii) *The space $L^1(K,\mu)$ is isometrically a dual space if and only if μ is discrete.*

Proof. We may suppose that $L^1(K,\mu)$ is an infinite-dimensional space.

First, suppose that μ is discrete. Then $L^1(K,\mu)$ is isometrically isomorphic to a Banach space of the form

$$\left\{\alpha = (\alpha_n) : \|\alpha\| = \sum_{n=1}^{\infty} |\alpha_n|\,\omega_n < \infty\right\}$$

for a sequence (ω_n) in $\mathbb{R}^+ \setminus \{0\}$ such that $\sum_{n=1}^{\infty} \omega_n = 1$. This space is the dual of the Banach space

$$\{(\beta_n) : |\beta_n|/\omega_n \to 0 \text{ as } n \to \infty\},$$

taken with the norm $\|(\beta_n)\| = \sup\{|\beta_n|/\omega_n : n \in \mathbb{N}\}$, and so $L^1(K,\mu)$ is isometrically a dual space.

(i) It is sufficient to show that (a) $\Rightarrow$ (c).

Take a Banach space F with $L^1(K,\mu) \sim F$, where F is a closed subspace of a separable dual space E'. Since E' is separable, E is separable by Proposition 2.1.6. By Corollary 2.6.17, E' has the Krein–Milman property, and so F and $L^1(K,\mu)$ have the Krein–Milman property. Take $\mu_c \in M_c(K)$ and $\mu_d \in M_d(K)$ with $\mu = \mu_c + \mu_d$. Then $L^1(\mu_c)_{[1]}$ is closed, bounded, and convex in $L^1(K,\mu)$, and so, by Proposition 4.4.15(ii), $\mu_c = 0$. Hence, $\mu = \mu_d$ is discrete.

(ii) Since $\mu(K) = 1$, the set $S := \{x \in K : \mu(\{x\}) > 0\}$ is countable. Let T be a countable, dense subset of $\mathbb{T}$. Then, with the identification of Proposition 4.4.15(iii), $\{\zeta\delta_x/\mu(\{x\}) : \zeta \in T, x \in S\}$ is a countable, dense subset of $\operatorname{ex} L^1(K,\mu)_{[1]}$, and so $\operatorname{ex} L^1(K,\mu)_{[1]}$ is separable.

Now suppose that $L^1(K,\mu)$ is isometrically a dual space. By Theorem 4.1.10, the space $L^1(K,\mu)$ is separable, and so μ is discrete by (i), (b) $\Rightarrow$ (c). □

Corollary 4.4.18. *Let K be a non-empty, locally compact space, and suppose that $\mu \in M_c(K)^+$ and $L^1(K,\mu)$ is separable. Then there is no embedding of $L^1(K,\mu)$ into a space $\ell^1(D)$ for an index set D.*

Proof. Assume to the contrary that there is an embedding of $L^1(K,\mu)$ into a space $\ell^1(D)$. Since $L^1(K,\mu)$ is separable, there is a countable subset D_0 of D such that $L^1(K,\mu)$ embeds into $\ell^1(D_0)$, a separable dual space. This is a contradiction of Theorem 4.4.17(i), (a) $\Rightarrow$ (c). □

The above theorem gives *Gel'fand's theorem*, which we state explicitly.

Theorem 4.4.19. *The Banach space $L^1(\mathbb{I})$ is not isomorphic to a subspace of a separable dual space. In particular, $L^1(\mathbb{I})$ is not isomorphically a dual space.* □

There is a different, self-contained proof of the above theorem, along with some informative remarks, in [3, Theorem 6.3.7].

An alternative proof that the space $L^1(K,\mu)$ of Corollary 4.4.18 does not embed in ℓ^1 is mentioned after Corollary 4.5.8, below.

Let K be a non-empty, locally compact space. Using more sophisticated techniques than the above, Pełczyński showed in [197] that, for a σ-finite positive measure μ, the space $L^1(K,\mu)$ is isomorphically a dual space if and only if μ is discrete.

See also [168] and [211]. A different proof, for the case of *finite* measures, is given in [85, p. 83]. For positive measures μ on K that are not σ-finite, it seems to be unknown which $L^1(K,\mu)$ spaces are isomorphically dual spaces. In the isometric theory, an early result of this type is given in [94, Exercise 4, p. 458]. Let (Ω,μ) be a measure space, where μ is a σ-finite positive measure. Then $L^1(\Omega,\mu)$ is isometrically a dual space if and only if Ω is a countable union $\Omega=\bigcup\Omega_i$, where each Ω_i is a measurable subset of Ω with $\mu(\Omega_i)<\infty$ and such that, for each measurable subset A of each Ω_i, we have either $\mu(A)=0$ or $\mu(A)=\mu(\Omega_i)$. Suppose that, in fact, $\mu(\{x\})=1$ for each $x\in\Omega$. Then it follows that $L^1(\Omega,\mu)\cong\ell^1$.

We conclude this section with two well-known results on weak compactness in L^1-spaces that we shall use. The first proposition is a result on equi-continuity.

Proposition 4.4.20. *Let K be a non-empty, compact space, and take $\nu\in M(K)^+$. Suppose that (μ_n) is a sequence in $L^1(K,\nu)$ that converges weakly. Then, for each $\varepsilon>0$, there exists $\delta>0$ such that $|\mu_n|(B)\le\varepsilon$ $(n\in\mathbb{N})$ whenever $B\in\mathfrak{B}_K$ with $\nu(B)\le\delta$.*

Proof. We may suppose that $\nu\in P(K)$. By Proposition 4.3.5, the metric space $(\mathfrak{B}_\nu,\rho_\nu)$ is complete.

First, suppose that (μ_n) converges weakly to 0. Fix $\varepsilon>0$, and, for $n\in\mathbb{N}$, set

$$G_n=\{B\in\mathfrak{B}_\nu:|\mu_m(B)|\le\varepsilon\quad(m\ge n)\}\,.$$

Then each set G_n is closed in the space $(\mathfrak{B}_\nu,\rho_\nu)$, and $\bigcup\{G_n:n\in\mathbb{N}\}=\mathfrak{B}_\nu$ because $\lim_{n\to\infty}\mu_n(B)=0$ for each $B\in\mathfrak{B}_K$. By Baire's theorem, Theorem 1.4.11, there exist $n_0\in\mathbb{N}$, $B_0\in\mathfrak{B}_K$, and $\delta_0>0$ such that $|\mu_n(B)|<\varepsilon$ whenever $n\ge n_0$ and $B\in\mathfrak{B}_K$ with $\rho_\nu(B,B_0)<\delta_0$.

Suppose that $B\in\mathfrak{B}_K$ with $\nu(B)<\delta_0$. Then $\rho_\nu(B_0\cup B,B_0)=\nu(B\setminus B_0)<\delta_0$ and $\rho_\nu(B_0\setminus B,B_0)=\nu(B_0\cap B)<\delta_0$, and so

$$|\mu_n(B)|\le|\mu_n(B_0\cup B)|+|\mu_n(B_0\setminus B)|<2\varepsilon\quad(n\ge n_0)\,.$$

By inequality (4.10), $|\mu_n|(B)\le 8\varepsilon$ $(n\ge n_0)$. By reducing δ_0, if necessary, we may suppose that the same inequality holds for each $n\in\mathbb{N}_{n_0}$, and hence for all $n\in\mathbb{N}$. The result now follows in this special case.

Now suppose that (μ_n) converges weakly to some limit in $M(K)$. We *claim* that, for each $\varepsilon>0$, there exist $\delta_0>0$ and $n_0\in\mathbb{N}$ such that $|\mu_m-\mu_n|(B)\le\varepsilon/2$ whenever $m,n\ge n_0$ and $B\in\mathfrak{B}_K$ with $\nu(B)\le\delta_0$. Assume that this is not the case. Then there exist $\varepsilon>0$, strictly increasing sequences (m_k) and (n_k) in $\mathbb{N}$, and sets B_k in $\mathfrak{B}_K$ such that $\nu(B_k)\le 1/k$ and $\left|\mu_{m_k}-\mu_{n_k}\right|(B_k)\ge\varepsilon$ for each $k\in\mathbb{N}$. Since

$$\lim_{k\to\infty}(\mu_{m_k}-\mu_{n_k})(B)=0\quad(B\in\mathfrak{B}_K)\,,$$

this contradicts the result in the special case. Thus the claim holds.

Finally, choose $\delta\in(0,\delta_0)$ such that $|\mu_n|(B)<\varepsilon/2$ whenever $n\in\mathbb{N}_{n_0}$ and $B\in\mathfrak{B}_K$ with $\nu(B)\le\delta$. Then the required conclusion follows. □

Theorem 4.4.21. *Let K be a non-empty, compact space, and take $\nu \in M(K)^+$. Suppose that S is a subset of $L^1(K,\nu)$. Then S is relatively weakly compact if and only if:*

(i) *S is norm-bounded;*

(ii) *for each $\varepsilon > 0$, there exists $\delta > 0$ such that $|\mu(B)| < \varepsilon$ $(\mu \in S)$ whenever $B \in \mathfrak{B}_K$ with $\nu(B) \leq \delta$.*

Proof. Suppose that S is relatively weakly compact. Then S is weakly bounded, and hence norm-bounded by Corollary 2.2.2, so that (i) holds. Assume towards a contradiction that (ii) fails. Then there exist $\varepsilon > 0$, a sequence (μ_n) in S, and a sequence (B_n) in $\mathfrak{B}_K$ with $\nu(B_n) \leq 1/n$ and $|\mu_n(B_n)| > \varepsilon$ for all $n \in \mathbb{N}$. By the Eberlein–Šmulian theorem, Theorem 2.1.4(vii), (μ_n) has a weakly convergent subsequence, say (μ_{n_k}). By Proposition 4.4.20, there exists $\delta > 0$ with $|\mu_{n_k}|(B) < \varepsilon/2$ $(k \in \mathbb{N})$ whenever $B \in \mathfrak{B}_K$ with $\nu(B) \leq \delta$. Take $k \in \mathbb{N}$ with $1/n_k < \delta$. Then

$$\varepsilon \leq |\mu_{n_k}(B_{n_k})| \leq |\mu_{n_k}|(B_{n_k}) \leq \frac{\varepsilon}{2},$$

a contradiction. Thus (ii) holds.

Conversely, suppose that S satisfies clauses (i) and (ii). We regard $E := L^1(K,\nu)$ and S as subsets of E''. Then S is norm-bounded in E'', and so has a weak*-limit point, say M, in E''. Define

$$\lambda(B) = \langle \chi_B, \mathrm{M} \rangle \quad (B \in \mathfrak{B}_K).$$

Take $\varepsilon > 0$, and choose $\delta = \delta(\varepsilon) > 0$ as specified in (ii). Now take $\eta > 0$. For each $B \in \mathfrak{B}_K$ with $\nu(B) \leq \delta$, we have $\chi_B \in E'$, and so there exists $\mu \in S$ with

$$|\langle \chi_B, \mathrm{M} \rangle - \mu(B)| < \eta,$$

and then $|\lambda(B)| \leq \varepsilon + \eta$. This holds for each $\eta > 0$, and so $|\lambda(B)| \leq \varepsilon$.

Suppose that (B_n) is a sequence in $\mathfrak{B}_K$ with $\nu(B_n) \searrow 0$. Then $|\lambda(B_n)| \searrow 0$, and so λ is countably additive on $\mathfrak{B}_K$, and hence $\lambda \in M(K)$. Also $\lambda \ll \nu$, and so, by the Radon–Nikodým theorem, Theorem 4.4.9, $\lambda \in E$. It follows that M is a weak-limit point of S in E, and hence that S is relatively weakly compact. □

4.5 The space $C(K)$ as a Grothendieck space

We now consider when a space $C(K)$ for K compact is a Grothendieck space. Of course we have characterized such spaces in the (unproved) Proposition 2.4.7. We shall show in Corollary 4.5.10 that $C(K)$ is a Grothendieck space whenever it is an injective space.

First note that $C(K)$ is certainly not a Grothendieck space whenever K contains a convergent sequence (x_n) of distinct points, say with limit $x \in K$. Indeed, the sequence $(\delta_{x_n} - \delta_x)$ in $M(K)$ converges weak* to 0, but it does not converge weakly to 0, as can be seen by considering the linear functional $\mu \mapsto \sum_{n=1}^{\infty} \mu(\{x_n\})$ on $M(K)$.

We shall also use the following result of Grothendieck from [124] about relative weak compactness in the Banach space $M(K)$.

Theorem 4.5.1. *Let K be a non-empty, compact space, and take S to be a norm-bounded subset of $M(K)$. Then the following conditions are equivalent:*

(a) *S is relatively weakly compact;*

(b) *for each sequence (μ_n) in S, necessarily $\lim_{n\to\infty} \mu_n(U_n) = 0$ for each sequence (U_n) of pairwise-disjoint, open sets in K.*

An early proof of this theorem is contained in Bade's notes [24, §9]; see also [3, §5.3], [94, Theorem IV.9.1], and [184, Theorem 2.5.5], for example.

We shall first prove two lemmas, in which we suppose that the set S is a norm-bounded subset of $M(K)$ that satisfies clause (b) of Theorem 4.5.1.

Lemma 4.5.2. *Let (μ_n) be a sequence in S. Then $\lim_{n\to\infty} |\mu_n|(U_n) = 0$ for each sequence (U_n) of pairwise-disjoint, open sets in K.*

Proof. For $n \in \mathbb{N}$, take ν_n to be either $\Re\mu_n$ or $\Im\mu_n$. Then $\lim_{n\to\infty} \nu_n(U_n) = 0$ for each sequence (U_n) of pairwise-disjoint, open sets in K.

Assume to the contrary that there is a sequence (U_n) of pairwise-disjoint, open sets in K such that $(|\nu_n|(U_n))$ does not converge to 0. Set $\nu_n = \nu_n^+ - \nu_n^-$ $(n \in \mathbb{N})$; we may suppose that $(\nu_n^+(U_n))$ does not converge to 0, and, by passing to a subsequence, we may suppose that there exists $\delta > 0$ with $\nu_n^+(U_n) > \delta$ $(n \in \mathbb{N})$. By Hahn's decomposition theorem, Theorem 4.1.7(i), for each $n \in \mathbb{N}$, there is a Borel subset B_n of U_n with $\nu_n(B_n) = \nu_n^+(U_n)$, and, by the regularity of ν_n, there is an open set V_n with $B_n \subset V_n \subset U_n$ and $\nu_n(V_n) > \delta$, a contradiction.

Thus $\lim_{n\to\infty} |\nu_n|(U_n) = 0$ for each sequence (U_n) of pairwise-disjoint, open sets, and then the result follows. □

The second lemma states that the subset S of $M(K)$ is *uniformly regular*.

Lemma 4.5.3. *For each compact subset L of K and each $\varepsilon > 0$, there is an open subset U of K with $U \supset L$ such that $|\mu|(U \setminus L) \leq \varepsilon$ $(\mu \in S)$.*

Proof. Assume that the conclusion fails. Then there is a compact subset L of K and $\varepsilon > 0$ such that, for each open neighbourhood U of L, there exists $\mu \in S$ with $|\mu|(U \setminus L) > \varepsilon$.

We *claim* that there are a sequence (W_n) of open subsets of K such that the sets $\overline{W_n}$ are contained in $K \setminus L$ and are pairwise disjoint and a sequence (μ_n) in S such that $|\mu_n(W_n)| > \varepsilon/4$ $(n \in \mathbb{N})$.

Indeed, take $V_1 = K$, and choose $\mu_1 \in S$ with $|\mu_1|(V_1 \setminus L) > \varepsilon$. By the regularity of $|\mu_1|$, there is an open set W_1 in K with $\overline{W_1} \subset V_1 \setminus L$ and with $|\mu_1(W_1)| > \varepsilon/4$, where we are using inequality (4.10). Now take $k \in \mathbb{N}$, and assume that $W_1,\dots,W_k$ and $\mu_1,\dots,\mu_k$ have been determined to satisfy the claim for each $n \in \mathbb{N}_k$. Set $V_{k+1} = \bigcup_{j=1}^k (K \setminus \overline{W_j})$, and then choose $\mu_{k+1} \in S$ and an open set W_{k+1} such that $\overline{W_{k+1}} \subset V_{k+1} \setminus L$ and $|\mu_{k+1}(W_{k+1})| > \varepsilon/4$. This continues the inductive construction, and hence the claim holds.

However, the claim contradicts clause (b) of Theorem 4.5.1, and so the conclusion holds. □

Proof of Theorem 4.5.1. We first show that clause (b) of Theorem 4.5.1 holds whenever S is relatively weakly compact.

Indeed, take a sequence (μ_n) in S. By the Eberlein–Šmulian theorem, Theorem 2.1.4(vii), we may suppose, by passing to a subsequence, that (μ_n) converges weakly in $M(K)$. Define

$$\nu = \sum_{n=1}^{\infty} \frac{|\mu_n|}{2^n} \in M(K)^+ . \tag{4.12}$$

For each $n \in \mathbb{N}$, we have $\mu_n \ll \nu$, and so, by the Radon–Nikodým theorem, Theorem 4.4.9, we may suppose that $\mu_n \in L^1(K,\nu)$ $(n \in \mathbb{N})$. Clearly the sequence (μ_n) converges weakly in $L^1(K,\nu)$, and so, by Proposition 4.4.20, clause (b) holds.

We now show that clause (b) implies that S is relatively weakly compact.

By the Eberlein–Šmulian theorem, it is sufficient to show that each countable subset of S is relatively weakly compact in $M(K)$; we take such a countable set $T := \{\mu_n : n \in \mathbb{N}\}$, and define ν as in equation (4.12). Clearly, it suffices to show that the set T is relatively weakly compact in $L^1(K,\nu)$; for this, we shall show that T satisfies clauses (i) and (ii) of Theorem 4.4.21.

By hypothesis, S is norm-bounded in $M(K)$, and so T satisfies clause (i) of 4.4.21.

Assume towards a contradiction that T does not satisfy clause (ii). Then, by using the regularity of ν and passing to a subsequence of (μ_n), we may suppose that there are $\varepsilon > 0$ and a sequence (B_n) of sets in $\mathfrak{B}_K$ such that

$$\nu(B_n) \leq \frac{1}{n} \quad \text{and} \quad |\mu_n|(B_n) \geq |\mu_n(B_n)| > \varepsilon$$

for all $n \in \mathbb{N}$.

For each $m \in \mathbb{N}$, we have $\lim_{n\to\infty} |\mu_m|(B_n) = 0$, and so, by passing to a further subsequence, we may suppose that

$$|\mu_m|(B_n) < \frac{\varepsilon}{2^{n+2}} \quad (n > m,\, m,n \in \mathbb{N}) .$$

Take $m \in \mathbb{N}$, and set $C_m = B_m \setminus \bigcup\{B_n : n \geq m+1\}$. Then C_m is a Borel subset of B_m such that $|\mu_m|(C_m) > \varepsilon/2$. Further, the sets C_m are pairwise disjoint. By the regularity of the measures μ_m, we can choose compact subsets L_m of C_m such that $|\mu_m|(L_m) > \varepsilon/2$. It follows from Lemma 4.5.3 that there is an open set W_m with $W_m \supset L_m$ such that $|\mu_n|(W_m \setminus L_m) < \varepsilon/2^{m+4}$ $(n \in \mathbb{N})$. We can then choose an open set V_m such that $L_m \subset V_m \subset \overline{V_m} \subset W_m$.

Now take $m,n \in \mathbb{N}$ with $m < n$. Then

$$(\overline{V_m} \cap V_n) \subset (\overline{V_m} \setminus L_m) \cup (V_n \setminus L_n) \subset (W_m \setminus L_m) \cup (W_n \setminus L_n),$$

and so $|\mu_n|(V_n \cap \overline{V_m}) < \varepsilon/2^{m+3}$. For $n \geq 2$, set $G_n = V_n \setminus \overline{V_1 \cup \cdots \cup V_{n-1}}$. Then the sequence $(G_n : n \geq 2)$ consists of pairwise-disjoint, open subsets of K, and $|\mu_n|(G_n) > \varepsilon/2 - \varepsilon/4 = \varepsilon/4$. This is a contradiction of Lemma 4.5.2, and so T satisfies clause (ii) of Theorem 4.4.21. By Theorem 4.4.21, T is relatively weakly compact in $L^1(K,\nu)$, as required. □

Corollary 4.5.4. *Let K be a non-empty, compact space, and take $\nu \in M(K)$. Then the set $\{\mu \in M(K) : |\mu| \leq |\nu|\}$ is weakly compact.*

Proof. This result follows immediately from Theorem 4.5.1. □

We shall use Corollary 4.5.4 to give the following direct, elementary proof that each space $C_0(K)$ is Arens regular; in fact, this result will also follow from the construction of the bidual of $C_0(K)$, to be given in Theorem 5.4.1.

Theorem 4.5.5. *Let K be a non-empty, locally compact space. Then the C^*-algebra $C_0(K)$ is Arens regular, and $(C_0(K)'', \Box)$ is commutative.*

Proof. Take $\mathrm{M} \in C_0(K)'' = M(K)'$ and $\mu \in C_0(K)'_{[1]} = M(K)_{[1]}$, and consider the continuous linear functional

$$\theta : \mathrm{N} \mapsto \langle \mathrm{M} \Box \mathrm{N}, \mu \rangle = \langle \mathrm{M}, \mathrm{N} \cdot \mu \rangle, \quad M(K)' \to \mathbb{C}.$$

We *claim* that θ is weak*-continuous on $M(K)'_{[1]}$. For suppose that $\mathrm{N}_\alpha \to \mathrm{N}_0$ in $(M(K)'_{[1]}, \sigma(M(K)', M(K)))$. Then $(\mathrm{N}_\alpha \cdot \mu)$ is a net in $\{\nu \in M(K) : |\nu| \leq |\mu|\}$; by Corollary 4.5.4, this latter set is weakly compact, and so $(\mathrm{N}_\alpha \cdot \mu)$ has a weakly convergent subnet, say $(\mathrm{N}_{\alpha_\beta} \cdot \mu)$. For each $f \in C_0(K)$, we have

$$\langle f, \mathrm{N}_0 \cdot \mu \rangle = \langle \mathrm{N}_0, \mu \cdot f \rangle = \lim_\alpha \langle \mathrm{N}_\alpha, \mu \cdot f \rangle = \lim_\beta \langle \mathrm{N}_{\alpha_\beta}, \mu \cdot f \rangle = \lim_\beta \langle f, \mathrm{N}_{\alpha_\beta} \cdot \mu \rangle,$$

and hence $\lim_\beta \mathrm{N}_{\alpha_\beta} \cdot \mu = \mathrm{N}_0 \cdot \mu$ in $(M(K), \sigma(M(K), C_0(K)))$. This implies that the net $(\mathrm{N}_\alpha \cdot \mu)$ converges weakly to $\mathrm{N}_0 \cdot \mu$, and so

$$\lim_\alpha \theta(\mathrm{N}_\alpha) = \lim_\alpha \langle \mathrm{M}, \mathrm{N}_\alpha \cdot \mu \rangle = \langle \mathrm{M}, \mathrm{N}_0 \cdot \mu \rangle = \theta(\mathrm{N}_0),$$

giving the claim.

It follows from Theorem 2.1.4(iv), (c) ⇒ (a), that there exists $\nu \in M(K)$ such that

$$\theta(\mathrm{N}) = \langle \mathrm{N}, \nu \rangle \quad (\mathrm{N} \in M(K)').$$

For each $f \in C_0(K)$, we have $\langle f, \nu \rangle = \langle \mathrm{M} \cdot f, \mu \rangle = \langle \mathrm{M}, f \cdot \mu \rangle = \langle f, \mu \cdot \mathrm{M} \rangle$, and so $\nu = \mu \cdot \mathrm{M}$. We have shown that

$$\langle \mathrm{M}\Box\mathrm{N},\mu\rangle = \theta(N) = \langle \mathrm{N},\mu\cdot\mathrm{M}\rangle = \langle \mathrm{M}\Diamond\mathrm{N},\mu\rangle \quad (\mathrm{M},\mathrm{N}\in C_0(K)'',\ \mu\in C_0(K)'),$$

and hence $\mathrm{M}\Box\mathrm{N} = \mathrm{M}\Diamond\mathrm{N}$ $(\mathrm{M},\mathrm{N}\in C_0(K)'')$. Thus $C_0(K)$ is Arens regular.

Since $C_0(K)$ is commutative, $(C_0(K)'',\Box)$ is commutative. □

The next result is a classic theorem of Grothendieck [124]. Grothendieck's proof utilized a lemma of Phillips [202] on sequential convergence in the space of finitely additive measures on $\mathscr{P}(\mathbb{N})$, as described in [24]; we give a direct and self-contained proof.

Theorem 4.5.6. *Let K be a Stonean space. Then $C(K)$ is a Grothendieck space.*

Proof. Let (μ_n) be a sequence in $C(K)' = M(K)$ that converges weak* to 0; we must show that (μ_n) converges weakly, and, for this, it suffices to show that the set $\{\mu_n : n\in\mathbb{N}\}$ is relatively weakly compact in $M(K)$.

Assume to the contrary that this fails. Then, it follows from Theorem 4.5.1 that, after passing to a subsequence and rescaling, we may suppose that there is a pairwise-disjoint sequence (U_n) of open subsets of K with $|\mu_n(U_n)| > 1$ $(n\in\mathbb{N})$. Since K is Stonean and each μ_n is regular, we may suppose that all the sets U_n are clopen.

We shall define inductively a subsequence (μ_{n_k}) of (μ_n) such that (n_k) is strictly increasing in $\mathbb{N}$ and

$$|\mu_{n_r}(U_{n_s})| < \frac{1}{2^{s+1}} \quad (r,s\in\mathbb{N},\ r\neq s). \tag{4.13}$$

First, take $n_1 = 1$. Now suppose that $k\in\mathbb{N}$, and assume that $n_1,\dots,n_k$ have been defined such that (4.13) holds whenever $r,s\in\mathbb{N}_k$ and $r\neq s$. For each $j\in\mathbb{N}_k$, the set

$$\left\{n\in\mathbb{N} : \left|\mu_{n_j}(U_n)\right| \geq \frac{1}{2^{k+2}}\right\}$$

is finite and $\lim_{n\to\infty}\mu_n(U_{n_j}) = 0$, and so we can choose $n_{k+1} > n_k$ such that $\left|\mu_{n_j}(U_{n_{k+1}})\right| < 1/2^{k+2}$ and $\left|\mu_{n_{k+1}}(U_{n_j})\right| < 1/2^{j+1}$ for $j\in\mathbb{N}_k$. This continues the inductive construction of the sequence (n_k). The sequence satisfies (4.13); set $\nu_k = \mu_{n_k}$ and $V_k = U_{n_k}$ for $k\in\mathbb{N}$.

As in Proposition 1.5.5, there are an index set A such that $|A| = \mathfrak{c}$ and a family $\{S_\alpha : \alpha\in A\}$ of infinite subsets of $\mathbb{N}$ such that $S_\alpha\cap S_\beta$ is finite whenever $\alpha,\beta\in A$ with $\alpha\neq\beta$. For each $\alpha\in A$, set

$$W_\alpha = \overline{\bigcup\{V_k : k\in S_\alpha\}},$$

a clopen subset of K, and set $V = \bigcup\{V_k : k\in\mathbb{N}\}$, an open subset of K. We note that $\{W_\alpha\setminus V : \alpha\in A\}$ is a family of pairwise-disjoint, closed subsets of K. For each $k\in\mathbb{N}$, it is the case that $\nu_k(W_\alpha\setminus V)\neq 0$ for only countably many values of $\alpha\in A$, and so there exists $\alpha \in A$ with $\nu_k(W_\alpha \setminus V) = 0$ $(k \in \mathbb{N})$. Thus, for each $k \in S_\alpha$, we have

$$|\langle \chi_{W_\alpha}, \nu_k \rangle| = |\nu_k(W_\alpha \cap V)| \geq |\nu_k(V_k)| - \sum \{ |\nu_k(V_j)| : j \in \mathbb{N}, j \neq k \} > 1/2,$$

using (4.13), a contradiction of the fact that (ν_k) converges weak* to 0.

The result follows. □

Definition 4.5.7. A Banach space E has the *Schur property* if every weakly convergent sequence in E is norm-convergent.

Corollary 4.5.8. *Let S be a non-empty set. Then $\ell^\infty(S)$ is a Grothendieck space. Further, suppose that (μ_n) is sequence in $M(\beta S)$ that is weak*-convergent to 0. Then*

$$\lim_{n \to \infty} \|\mu_n \mid S\| = 0,$$

and $\ell^1(S)$ has the Schur property.

Proof. Since $\ell^\infty(S) \cong C(\beta S)$ and βS is a Stonean space, certainly $\ell^\infty(S)$ is a Grothendieck space by Theorem 4.5.6.

Suppose that (μ_n) in $M(\beta S)$ is weak*-convergent to 0, and assume towards a contradiction that it is not true that $\lim_{n\to\infty} \|\mu_n \mid S\| = 0$. By passing to a subsequence and rescaling, we may suppose that $\|\nu_n\| > 1$ $(n \in \mathbb{N})$, where $\nu_n = \mu_n \mid S$. Essentially as in the above proof, there is a sequence (F_n) of pairwise-disjoint, finite subsets of S such that $|\mu_n(F_n)| = |\nu_n(F_n)| > 1$ $(n \in \mathbb{N})$. By Theorem 4.5.1, the sequence (μ_n) is not relatively weakly compact, and this contradicts Theorem 4.5.6.

In the case where (μ_n) is weakly convergent to 0 in $\ell^1(S)$, it follows that (μ_n), regarded as a sequence in $M(\beta S)$, is weak*-convergent to 0, and so (μ_n) is norm-convergent to 0 in $\ell^1(S)$. □

The fact that ℓ^1 has the Schur property goes back to Schur in 1921 and is included in Banach's book [30, Table (property 17), p. 245; also, p. 239]; for a modern discussion, see [2, Theorem 2.3.6 and p. 102].

It is easily seen that $L^1(\mathbb{I})$ does not have the Schur property, and hence also that the spaces $L^1(K,\mu)$ for K locally compact and $\mu \in P_c(K)$ do not have the Schur property. Indeed, consider the sequence (s_n) of page 116. This sequence is weakly convergent to 0 in $L^1(\mathbb{I})$. However, (s_n) is certainly not norm-convergent to 0 in $L^1(\mathbb{I})$. Hence $L^1(K,\mu)$ does not embed in ℓ^1.

The above results give a slightly different proof of Phillips' theorem, Theorem 2.4.11. Indeed, assume towards a contradiction that $P : \ell^\infty \to c_0$ is a bounded projection, so that $P' : c_0' \to M(\beta\mathbb{N})$ is a bounded operator. Regard δ_n as a continuous linear functional on c_0 for $n \in \mathbb{N}$. Then

$$\langle f, P'(\delta_n) \rangle = \langle Pf, \delta_n \rangle \to 0 \quad \text{as} \quad n \to \infty \quad (f \in \ell^\infty),$$

and so $P'(\delta_n) \to 0$ weak* in $M(\beta\mathbb{N})$. By Corollary 4.5.8, $|P'(\delta_n)(\{n\})| \to 0$ as $n \to \infty$. But $P'(\delta_n)(\{n\}) = 1 \ (n \in \mathbb{N})$, a contradiction.

The following corollary of Theorem 4.5.6 was noted by Seever in [224]; see also [184, Corollary 2.5.17].

Corollary 4.5.9. *Let K be a compact F-space. Then C(K) is a Grothendieck space.*

Proof. Let (μ_n) be a sequence in $M(K) = C(K)'$ that converges weak* to 0, and define $\mu = \sum_{n=1}^{\infty} \mu_n/2^n \in M(K)$. Set $L = \operatorname{supp} \mu$. By Proposition 4.1.6, L is a Stonean space. Then, by Theorem 4.5.6, $(\mu_n \mid L)$ converges weak* to 0 in $M(L)$, and so it converges weakly to 0 in $M(L)$, i.e., (μ_n) converges weakly to 0 in $M(K)$. Hence $C(K)$ is a Grothendieck space. □

Corollary 4.5.10. *Each injective space is a Grothendieck space.*

Proof. Let E be a Banach space. By Proposition 2.2.14(i), there is a set S and an isometric embedding of E onto a subspace, say F, of $\ell^\infty(S)$. In the case where E is injective, F is complemented in $\ell^\infty(S)$. Since $\ell^\infty(S)$ is a Grothendieck space and complemented subspaces of Grothendieck spaces are also Grothendieck spaces (see page 73), E is a Grothendieck space. □

We shall see in Example 6.8.17 that there are compact spaces K such that $C(K)$ is a Grothendieck space, but $C(K)$ is not injective. The Baire classes $B_\alpha(\mathbb{I})$ for ordinals α with $1 \leq \alpha \leq \omega_1$ are examples of $C(K)$-spaces that are Grothendieck spaces (see Theorem 3.3.9), but are such that K is not an F-space when $\alpha < \omega_1$ [76].

A beautiful generalization of Theorem 4.5.1 characterizing weak compactness in the dual of a C^*-algebra was given by Pfitzner in [200]. For a shorter proof, see [101]; see also [2]. It follows that each von Neumann algebra is a Grothendieck space; it is proved in [219] that each monotone σ-complete C^*-algebra is a Grothendieck space.

4.6 Singular families of measures

We now introduce singular families and maximal singular families of measures.

Definition 4.6.1. Let K be a non-empty, locally compact space. A family $\mathscr{F}$ of measures in $M(K)^+$ is *singular* if $\mu \perp \nu$ whenever $\mu, \nu \in \mathscr{F}$ and $\mu \neq \nu$.

The collection of such singular families in $M(K)^+$ is ordered by inclusion.

Let S be a non-empty subset of $M(K)^+$. It is clear from Zorn's lemma that the collection of singular families contained in S has a maximal member that contains any specific singular family in S; this is a *maximal singular family in S*. In the case where $S = P(K)$, we may suppose that such a maximal singular family contains all the measures that are point masses and that all other members are continuous measures, so that, in the case where K is discrete, the family of point masses is a maximal singular family in $P(K)$.

We shall see in Proposition 5.2.7 that any two infinite, maximal singular families of continuous measures have the same cardinality.

Proposition 4.6.2. (i) *Let K be a non-empty, locally compact space, and suppose that S is a separable subspace of $M(K)^+$. Then each singular family of measures in S is countable.*

(ii) *The space $M_c(\Delta)$ contains a singular family in $P(\Delta)$ of cardinality $\mathfrak{c}$.*

(iii) *Let K be an uncountable, compact, metrizable space. Then there is a maximal singular family of measures in $P(K)$ consisting of exactly $\mathfrak{c}$ point masses and $\mathfrak{c}$ continuous measures.*

Proof. (i) Let $\mathscr{F}$ be a singular family of measures in S. For each $\mu, \nu \in \mathscr{F}$ with $\mu \neq \nu$, we have $\|\mu - \nu\| = \|\mu\| + \|\nu\|$. For $n \in \mathbb{N}$, set $\mathscr{F}_n = \{\mu \in \mathscr{F} : \|\mu\| > 1/n\}$. For $\mu, \nu \in \mathscr{F}_n$ with $\mu \neq \nu$, we have $\|\mu - \nu\| > 2/n$, and so the open balls $B_{1/n}(\mu)$ and $B_{1/n}(\nu)$ are disjoint. Since S is separable, it follows that $\mathscr{F}_n$ is countable for each $n \in \mathbb{N}$, and so $\mathscr{F}$ is countable.

(ii) The Cantor cube $L = \mathbb{Z}_2^{\omega}$, identified with Δ, is a compact group and so has a Haar measure, say m_L, as on page 112, and $m_L \in M_c(L)$. By Proposition 1.4.5, L contains $\mathfrak{c}$ pairwise-disjoint, closed subspaces, each homeomorphic to L. We may transfer a copy of m_L to each of these subspaces; the resulting measures are mutually singular.

(iii) By Proposition 1.4.14, K contains Δ as a closed subspace. Let $\mathscr{F}$ be a maximal singular family of measures in $P(K)$ containing the family specified in (ii), so that $\mathscr{F}$ contains at least $\mathfrak{c}$ continuous measures. By Proposition 4.2.3, $|M(K)| = \mathfrak{c}$, and so $|\mathscr{F}| \leq \mathfrak{c}$. By Corollary 1.4.15, $|K| = \mathfrak{c}$, and hence $\mathscr{F}$ contains exactly $\mathfrak{c}$ point masses. □

We note that, under some mild set-theoretic axioms, such as Martin's axiom, there exists a compact space K with $|K| = \mathfrak{c}$ such that there is a maximal singular family in $P(K)$ of cardinality $2^{\mathfrak{c}}$: see [108].

Lemma 4.6.3. *Let K be a non-empty, locally compact space, and let $\mathscr{F}$ be a maximal singular family in $P(K)$. Then, for each $\nu \in M(K)$, there exist a countable subset Γ of $\mathscr{F}$ and $\nu_\mu \in M(K)$ for each $\mu \in \Gamma$ such that $\nu_\mu \ll \mu$ $(\mu \in \Gamma)$, such that $\nu = \sum\{\nu_\mu : \mu \in \Gamma\}$, and such that*

$$\|\nu\| = \sum\{\|\nu_\mu\| : \mu \in \Gamma\}.$$

The correspondence $\nu \mapsto (\nu_\mu)$, $M(K) \to M(K)^{\mathscr{F}}$, is a lattice homomorphism.

Proof. Take $\nu \in M(K)$. By the Lebesgue decomposition theorem, Theorem 4.2.9, for each $\mu \in \mathscr{F}$, there exist $\nu_\mu \ll \mu$ and $\sigma_\mu \perp \mu$ such that $\nu = \nu_\mu + \sigma_\mu$. Set $\Gamma = \{\mu \in \mathscr{F} : \nu_\mu \neq 0\}$.

For distinct elements $\mu_1,\ldots,\mu_n \in \mathscr{F}$, we have $\mu_i \perp \mu_j$ whenever $i,j \in \mathbb{N}_n$ with $i \neq j$, and so $\nu = \nu_{\mu_1} + \cdots + \nu_{\mu_n} + \sigma$ for some $\sigma \in M(K)$ with $\sigma \perp \nu_{\mu_i}$ $(i \in \mathbb{N}_n)$, and then $\sum_{i=1}^n \|\nu_{\mu_i}\| \leq \|\nu\|$. It follows that Γ is countable, that we can define $\rho = \sum\{\nu_\mu : \mu \in \Gamma\}$ in $M(K)$, and that $\sum\{\|\nu_\mu\| : \mu \in \Gamma\} \leq \|\nu\|$.

Clearly $|\nu - \rho| \perp \mu$ for each $\mu \in \mathscr{F}$, and so $\nu - \rho = 0$ by the maximality of $\mathscr{F}$. Thus $\nu = \sum\{\nu_\mu : \mu \in \Gamma\}$, and so $\|\nu\| \leq \sum\{\|\nu_\mu\| : \mu \in \Gamma\}$.

It follows that $\|\nu\| = \sum\{\|\nu_\mu\| : \mu \in \Gamma\}$.

Clearly, the correspondence $\nu \mapsto (\nu_\mu)$, $M(K) \to M(K)^{\mathscr{F}}$, is a lattice homomorphism. □

Let K be a non-empty, locally compact space, and take $\mu \in P(K)$. As in Definition 4.4.5, Φ_μ denotes the character space of the C^*-algebra $L^\infty(K,\mu)$.

Definition 4.6.4. Let K be a non-empty, locally compact space, let S be a non-empty subset of $P(K)$, and let $\mathscr{F}$ be a maximal singular family in S. Define $U_{\mathscr{F}}$ to be the space that is the disjoint union of the sets Φ_μ for $\mu \in S$, with the topology in which each Φ_μ is a compact and open subspace of $U_{\mathscr{F}}$.

We now give our first representation of the Banach space $M(K)' = C_0(K)''$.

Theorem 4.6.5. *Let K be a non-empty, locally compact space, and let $\mathscr{F}$ be a maximal singular family in $P(K)$. Then*

$$\|\Lambda\| = \sup\{|\langle \Lambda, \nu\rangle| : \nu \ll \mu,\ \|\nu\| \leq 1,\ \mu \in \mathscr{F}\} \quad (\Lambda \in M(K)'), \tag{4.14}$$

and $M(K)' \cong C^b(U_{\mathscr{F}})$.

Proof. Set $U = U_{\mathscr{F}}$.

Take $\Lambda \in M(K)'$, say with $\|\Lambda\| = 1$. For each $\mu \in \mathscr{F}$, set $\Lambda_\mu = \Lambda \mid L^1(K,\mu)$, so that $\Lambda_\mu \in L^1(K,\mu)' = C(\Phi_\mu)$ with $\|\Lambda_\mu\| \leq 1$. Hence there exists $F_\mu \in C(\Phi_\mu)$ with $|F_\mu|_{\Phi_\mu} \leq 1$ and

$$\langle \rho, F_\mu\rangle = \langle \rho, \Lambda\rangle \quad (\rho \in L^1(K,\mu)).$$

Now define $F \in C^b(U)$ by requiring that $F \mid \Phi_\mu = F_\mu$ $(\mu \in \mathscr{F})$; set $\alpha = |F|_U$, so that $\alpha \leq 1$.

Take $\nu \in M(K)_{[1]}$. By Lemma 4.6.3, there is a countable subset Γ of $\mathscr{F}$ and $\nu_\mu \in M(K)$ for each $\mu \in \Gamma$ such that $\nu_\mu \ll \mu$ $(\mu \in \Gamma)$, such that $\nu = \sum\{\nu_\mu : \mu \in \Gamma\}$, and such that $\|\nu\| = \sum\{\|\nu_\mu\| : \mu \in \Gamma\}$. We have

$$|\langle \Lambda, \nu\rangle| = \left|\sum\{\langle \Lambda, \nu_\mu\rangle : \mu \in \Gamma\}\right| \leq \sum\{|\langle F_\mu, \nu_\mu\rangle| : \mu \in \Gamma\} \leq \alpha,$$

and so $1 \leq \alpha$. Thus $|F|_U = \|\Lambda\|$. Set $T(\Lambda) = F$, so that $T : M(K)' \to C^b(U)$ is an isometric linear map.

Conversely, given $F \in C^b(U)$, set $F_\mu = F \mid \Phi_\mu$ $(\mu \in \mathscr{F})$. For each $\nu \in M(K)$, write $\nu = \sum\{\nu_\mu : \mu \in \Gamma\}$, as before, and define

$$\Lambda(\nu) = \sum\{\langle F_\mu, \nu_\mu\rangle : \mu \in \mathscr{F}\}.$$

Then $\Lambda \in M(K)'$ and $T(\Lambda) = F$. It follows that T is a surjection, and so we have shown that $M(K)' \cong C^b(U)$.

To obtain equation (4.14), take $\Lambda \in M(K)'$ and $\varepsilon > 0$. Then there exists a measure $\mu \in \mathscr{F}$ such that $\left|T(\Lambda) \mid \Phi_\mu\right|_{\Phi_\mu} > \|\Lambda\| - \varepsilon$, and also there exists $\nu \in L^1(K,\mu)_{[1]}$ with $|\langle \Lambda, \nu \rangle| > \|\Lambda\| - \varepsilon$. Since $\nu \ll \mu$, equation (4.14) follows. □

Theorem 4.6.6. *Let K be an uncountable, compact, metrizable space. Then there are an index set J with $|J| = \mathfrak{c}$, measures $\mu_j \in P_c(K)$ for each $j \in J$, and a set Γ with $|\Gamma| = \mathfrak{c}$ such that*

$$M_c(K) \cong \bigoplus_1 \{L^1(K,\mu_j) : j \in J\} \cong \bigoplus_1 \{L^1(\mathbb{I})_j : j \in J\} \tag{4.15}$$

and

$$M(K) \cong \bigoplus_1 \{L^1(\mathbb{I})_j : j \in J\} \oplus_1 \ell^1(\Gamma), \tag{4.16}$$

where $L^1(\mathbb{I})_j = L^1(\mathbb{I})$ for each $j \in J$. Further, all the above identifications are Banach-lattice isometries.

Proof. By Proposition 4.6.2(iii), there is a maximal singular family, say $\{\mu_j : j \in J\}$, where $|J| = \mathfrak{c}$, of measures in $P_c(K)$. Set

$$E = \bigoplus_1 \{L^1(K,\mu_j) : j \in J\}.$$

Clearly E is a closed subspace of $M_c(K)$. Take $\mu \in M_c(K)$. For each $j \in J$, there exist $\rho_j, \sigma_j \in M_c(K)$ with $\rho_j \ll \mu$ and $\sigma_j \perp \mu$; we can regard each ρ_j as an element of $L^1(\mu_j)$. It follows from Lemma 4.6.3 that $\mu = \sum_{j\in J} \rho_j$, with $\|\mu\| = \sum_{j\in J} \|\rho_j\|$, so that $\mu \in E$. Thus $M_c(K) \cong \bigoplus_1 \{L^1(K,\mu_j) : j \in J\}$; the identification is a Banach-lattice isometry.

For each $j \in J$, the space $L^1(K,\mu_j)$ is separable, and so, by Theorem 4.4.14, $L^1(\mu_j)$ is Banach-lattice isometric to $L^1(\mathbb{I}, m)$. Equation (4.15) follows.

Again by Proposition 4.6.2(iii), a maximal singular family in $P(K)$ is the set $\{\mu_j : j \in J\} \cup \{\delta_x : x \in K\}$, and so equation (4.16) follows, where we set $\Gamma = K$, so that $|\Gamma| = \mathfrak{c}$ by Proposition 1.4.14. □

Corollary 4.6.7. *Let K and L be two uncountable, compact, metrizable spaces. Then $M(K)$ and $M(L)$ are Banach-lattice isometric.*

Proof. This is immediate from equation (4.16). □

A generalization of Theorem 4.6.6 for an arbitrary measure space is given in *Maharam's theorem* [182], which is discussed in [166, §14] and [225, §26].

Theorem 4.6.8. *Let K be a non-empty, locally compact space, and suppose that $\{\mu_j : j \in J\}$ is a singular family in $P_c(K)$ with J uncountable. Then there is no embedding of the Banach space*

$$\bigoplus_1 \{L^1(K,\mu_j) : j \in J\}$$

into a Banach space of the form $F \oplus_1 \ell^1(D)$ for any separable Banach space F and any set D.

Proof. Let D be an index set, and take G to be the Banach space $(\ell^1(D), \|\cdot\|_1)$, and let F be a separable Banach space.

We shall apply Proposition 2.2.31. For each $j \in J$, the Banach space $L^1(K,\mu_j)$ contains an isometric copy of $L^1(\mathbb{I})$ by Theorem 4.4.14, and so, by Corollary 4.4.18, there is no embedding of $L^1(K,\mu)$ into $G = \ell^1(D)$. Thus, by Proposition 2.2.31, there is no embedding of $\bigoplus_1 \{L^1(K,\mu_j) : j \in J\}$ into $F \oplus_1 \ell^1(D)$. □

Corollary 4.6.9. *Let K be an uncountable, compact, metrizable space. Then the spaces $M_c(K)$ and $M(K)$ are not isomorphic to any closed subspace of a space of the form $F \oplus_1 \ell^1(D)$, where F is a separable Banach space and D is any set.*

Proof. Let $M_c(K)$ and $M(K)$ have the forms specified in equations (4.15) and (4.16), respectively. By Theorem 4.6.8, there is no isomorphism from the space $\bigoplus_1 \{L^1(K,\mu_j) : j \in J\}$ into $F \oplus_1 \ell^1(D)$, and so there is no such isomorphism from either $M_c(K)$ or $M(K)$. □

4.7 Normal measures

Let K be a non-empty, locally compact space. In this section, we shall introduce the (complex) Banach lattice $N(K)$ that consists of the normal measures on K, and we shall give a variety of examples of compact spaces K such that $N(K) = \{0\}$ and such that $N(K) \neq \{0\}$. A 'normal measure' was defined by Dixmier [91] to be an order-continuous measure $\mu \in M(K)$. Thus we have the following definition.

Definition 4.7.1. Let K be a non-empty, locally compact space, and let $\mu \in M(K)$. Then μ is *normal* if $\langle f_\alpha, \mu \rangle \to 0$ for each net $(f_\alpha : \alpha \in A)$ in $(C_0(K)^+, \leq)$ with $f_\alpha \searrow 0$ in the lattice, and μ is *σ-normal* if μ is σ-order-continuous, in the sense that $\langle f_n, \mu \rangle \to 0$ for each sequence $(f_n : n \in \mathbb{N})$ in $(C_0(K)^+, \leq)$ with $f_n \searrow 0$.

Definition 4.7.2. Let K be a non-empty, locally compact space. The subset of $M(K)$ consisting of the normal measures is $N(K)$; the set of real-valued measures in $N(K)$ is $N_{\mathbb{R}}(K)$, and the set of positive measures in $N(K)$ is $N(K)^+$. The sets of continuous and discrete normal measures on K are denoted by $N_c(K)$ and $N_d(K)$, respectively; further, we set $N_c(K)^+ = N_c(K) \cap M(K)^+$ and $N_d(K)^+ = N_d(K) \cap M(K)^+$.

It follows easily that $N(K)$, $N_d(K)$, and $N_c(K)$ are closed linear subspaces of $M(K)$. The point mass at an isolated point of K is a discrete normal measure.

The following proposition was proved in [91] and in detail by Bade in [24]. At certain points these sources require that the space K be Stonean; this is also the assumption in [234, Proposition III.1.11]. However, this assumption is not necessary.

Proposition 4.7.3. *Let K be a non-empty, locally compact space. Then:*

(i) *$\mu \in M(K)$ is normal if and only if $\Re\mu$ and $\Im\mu$ are normal;*

(ii) *$\mu \in M_{\mathbb{R}}(K)$ is normal if and only if $|\mu|$ is normal if and only if μ^+ and μ^- are normal;*

(iii) *$\mu \in M(K)$ is normal if and only if $|\mu|$ is normal;*

(iv) *$N(K)$ is a lattice ideal in $M(K)$, and $N(K) = N_d(K) \oplus_1 N_c(K)$.*

Proof. (i) This is immediate.

(ii) Suppose that $\mu^+, \mu^- \in N(K)$. Then certainly $\mu, |\mu| \in N(K)$.

Suppose that $|\mu| \in N(K)$ and that $\nu \in M(K)$ with $|\nu| \leq |\mu|$. Then

$$0 \leq \left| \int_K f_\alpha \, \mathrm{d}\nu \right| \leq \int_K f_\alpha \, \mathrm{d}|\mu| \to 0 \tag{4.17}$$

when $f_\alpha \searrow 0$ in $C_0(K)^+$, and so $\nu \in N(K)$. In particular, μ, μ^+, and μ^- are normal whenever $|\mu|$ is normal.

Suppose that $\mu \in M_{\mathbb{R}}(K)$ is normal and that $f_\alpha \searrow 0$ in $C_0(K)^+_{[1]}$. Let $\{P,N\}$ be a Hahn decomposition of K with respect to μ, as in Theorem 4.1.7(i), and take $\varepsilon > 0$. Since μ is regular, there exist a compact set L and an open set U in K with $L \subset P \subset U$ and $|\mu|(U \setminus L) < \varepsilon$. By Theorem 1.4.25, there exists $g \in C_{00}(K)^+$ with $\chi_L \leq g \leq \chi_U$. Then

$$\int_K f_\alpha \, \mathrm{d}\mu^+ = \int_P f_\alpha \, \mathrm{d}\mu \leq \int_L g f_\alpha \, \mathrm{d}\mu + \int_{U \setminus L} g f_\alpha \, \mathrm{d}\mu + 2\varepsilon = \int_K g f_\alpha \, \mathrm{d}\mu + 2\varepsilon \,.$$

Since $g f_\alpha \searrow 0$ and μ is normal, $\lim_\alpha \langle g f_\alpha, \mu \rangle = 0$, and so

$$\limsup_\alpha \, \langle f_\alpha, \mu^+ \rangle \leq 2\varepsilon \,.$$

This holds true for each $\varepsilon > 0$, and so $\lim_\alpha \langle f_\alpha, \mu^+ \rangle = 0$. Thus μ^+ is normal; similarly, μ^- is normal.

(iii) Suppose that $\mu \in N(K)$. Then $|\Re\mu| + |\Im\mu| \in N(K)$ from (i) and (ii). However $|\mu| \leq |\Re\mu| + |\Im\mu|$, and so $|\mu| \in N(K)$.

(iv) This is immediate from (4.17). □

Note that $\lambda\mu \in N(K)$ for each $\lambda \in L^\infty(\mu)$ and $\mu \in N(K)^+$, and so we may regard $L^\infty(K,\mu)$ as a closed subspace of $N(K)$ for each $\mu \in N(K)^+$. In particular, the restriction of a normal measure on K to a Borel subspace of K is still a normal measure in the space $N(K)$.

The spaces of σ-normal measures on K have analogous properties to those in Proposition 4.7.3.

Let K be a locally compact space. Recall from Definition 1.4.1 that $\mathscr{K}_K$ denotes the family of compact subsets L of K such that $\mathrm{int}_K L = \emptyset$. Clause (i) of the following theorem, for Stonean spaces K, is due to Dixmier [91]; see [225, p. 341]. Clause (ii) was formulated and proved in [76, p. 405].

Theorem 4.7.4. *Let K be a non-empty, locally compact space. Then:*

(i) *a measure $\mu \in M(K)$ is normal if and only if $\mu(L) = 0$ $(L \in \mathscr{K}_K)$;*

(ii) *a measure $\mu \in M(K)$ is σ-normal if and only if $\mu(L) = 0$ for each G_δ-set $L \in \mathscr{K}_K$.*

Proof. (i) Suppose that $\mu \in N(K)$. By Proposition 4.7.3(iii), we may suppose that $\mu \in N(K)^+$. Now take $L \in \mathscr{K}_K$, and consider the non-empty set

$$\mathscr{F} = \{f \in C_{\mathbb{R}}(K) : f \geq \chi_L\}.$$

Suppose that $g = \inf \mathscr{F}$ in $C_{0,\mathbb{R}}(K)$. Then $g(x) = 0$ $(x \in K \setminus L)$, and so $g = 0$ because $\mathrm{int}_K L = \emptyset$. Thus $\inf \mathscr{F} = 0$. Since $\mu(L) = \inf\{\langle f, \mu \rangle : f \in \mathscr{F}\}$, we have $\mu(L) = 0$.

Conversely, suppose that $\mu \in M(K)$ and $\mu(L) = 0$ $(L \in \mathscr{K}_K)$. Again by Proposition 4.7.3(iii), it suffices to suppose that $\mu \in M(K)^+$. Take (f_α) in $C_0(K)^+$ with $f_\alpha \searrow 0$; we may suppose that $f_\alpha \leq 1$ for each α. Set

$$g(x) = \inf_\alpha f_\alpha(x) \quad (x \in K).$$

Then g is a Borel function because $g^{-1}(V)$ is an F_σ-set in K for each open set V in $\mathbb{R}$, and $g \geq 0$. For $n \in \mathbb{N}$, set $B_n = \{x \in K : g(x) > 1/n\}$, so that $B_n \in \mathfrak{B}_K$. For each compact subset L of B_n, we have $\mathrm{int}_K L = \emptyset$, and so $\mu(L) = 0$. Thus $\mu(B_n) = 0$, and so $\mu(\{x \in K : g(x) > 0\}) = 0$, whence $\int_K g \, \mathrm{d}\mu = 0$. Hence it suffices to show that

$$\lim_\alpha \int_K f_\alpha \, \mathrm{d}\mu = \int_K g \, \mathrm{d}\mu. \tag{4.18}$$

Take $\varepsilon > 0$. By Lusin's theorem, Proposition 4.1.7(ii), there is a compact subset L of K with $\mu(K \setminus L) < \varepsilon$ and such that $g \mid L \in C(L)$. By Dini's theorem, Theorem 1.4.28, $\lim_\alpha |f_\alpha \mid L - g \mid L|_L = 0$, and so there exists α_0 with $|f_\alpha \mid L - g \mid L|_L < \varepsilon$ $(\alpha \geq \alpha_0)$. It follows that

$$\left| \int_K f_\alpha \, \mathrm{d}\mu - \int_K g \, \mathrm{d}\mu \right| < \int_L |f_\alpha - g| \, \mathrm{d}\mu + 2\varepsilon < (\|\mu\| + 2)\varepsilon \quad (\alpha \geq \alpha_0),$$

giving (4.18).

(ii) This is similar. □

Consider Lebesgue measure m on $\mathbb{I}$. There are Cantor-type closed subsets L of $\mathbb{I}$ such that $\mathrm{int}\, L = \emptyset$ and $m(L) > 0$. This shows that m is not a σ-normal measure.

Corollary 4.7.5. *Let K be a non-empty, locally compact space, and suppose that $\mu \in M(K)$. Then the following are equivalent:*

(a) $\mu \in N(K)$;

(b) $|\mu|(\overline{B} \setminus \operatorname{int} B) = 0$ *for each* $B \in \mathfrak{B}_K$;

(c) $\mu(B_1) = \mu(B_2)$ *for each* $B_1, B_2 \in \mathfrak{B}_K$ *with* $B_1 \equiv B_2$.

Proof. We may suppose that $\mu \in M(K)^+$.

(a) $\Rightarrow$ (b) Take $B \in \mathfrak{B}_K$. For each $\varepsilon > 0$, there exists an open set U in K with $B \subset U$ and $\mu(U \setminus B) < \varepsilon$. Since $\overline{U} \setminus U \in \mathscr{K}_K$, we have $\mu(\overline{U} \setminus U) = 0$. Thus

$$\mu(B) \le \mu(\overline{B}) \le \mu(\overline{U}) = \mu(U) \le \mu(B) + \varepsilon,$$

and so $\mu(\overline{B}) = \mu(B)$. By taking complements, it follows that $\mu(\operatorname{int} B) = \mu(B)$. Hence $\mu(\overline{B} \setminus \operatorname{int} B) = 0$.

(a) $\Rightarrow$ (c) We know that $\mu(B) = 0$ for each nowhere dense set B in $\mathfrak{B}_K$, and so $\mu(B) = 0$ for each meagre set B in $\mathfrak{B}_K$. Thus $\mu(B_1) = \mu(B_2)$ whenever $B_1, B_2 \in \mathfrak{B}_K$ with $B_1 \Delta B_2$ meagre.

(b), (c) $\Rightarrow$ (a) These are immediate from Theorem 4.7.4(i). □

Corollary 4.7.6. *Let K be a Stonean space, and suppose that $\mu \in N(K) \cap P(K)$ is a strictly positive measure. Then every equivalence class in $L^\infty(K,\mu)$ contains a continuous function, the C^*-algebras $(L^\infty(K,\mu), \|\cdot\|_\infty)$ and $(C(K), |\cdot|_K)$ are C^*-isomorphic, and Φ_μ is homeomorphic to K.*

Proof. By Theorem 3.3.5(iii), there is a C^*-isomorphism $\overline{P} : B^b(K)/M_K \to C(K)$. However $\mu(B) = 0$ for each meagre set $B \in \mathfrak{B}_K$ by Corollary 4.7.5, and so $\ker \overline{P}$ is exactly the kernel of the projection of $B^b(K)$ onto $L^\infty(K,\mu)$. The result follows. □

Proposition 4.7.7. *Let K be a non-empty, locally compact space satisfying* CCC. *Then every σ-normal measure on K is normal.*

Proof. Let $\mu \in M(K)$ be σ-normal. We must show that $\mu \in N(K)$; it suffices to suppose that $\mu \in M(K)^+$. Recall from page 23 that $\mathbf{Z}(K)$ denotes the family of zero sets of K. By Theorem 4.7.4(ii), $\mu(Z) = 0$ for each $Z \in \mathscr{K}_K \cap \mathbf{Z}(K)$.

Take $L \in \mathscr{K}_K$. We *claim* that there exists $Z \in \mathscr{K}_K \cap \mathbf{Z}(K)$ such that $L \subset Z$. Indeed, let $\mathscr{F}$ be a maximal disjoint family of cozero sets contained in the open set $K \setminus L$. By CCC, $\mathscr{F}$ is countable, and so the set

$$Z := \bigcap \{K \setminus V : V \in \mathscr{F}\}$$

is a zero set containing L. Hence Z has empty interior by the maximality of $\mathscr{F}$, proving the claim.

By hypothesis, $\mu(Z) = 0$. Thus $\mu(L) = 0$, and so it follows from Theorem 4.7.4(i) that $\mu \in N(K)$. □

Consider the compact space $K := [0, \omega_1]$. Then $\delta_{\omega_1} \in M(K)^+$ and $\delta_{\omega_1}(Z) = 0$ for each $Z \in \mathscr{K}_K$ that is a zero set because each zero set that contains ω_1 has non-empty interior. Thus δ_{ω_1} is a σ-normal measure on K which is not normal (because $\{\omega_1\}$ is compact with empty interior). Another such example will be given below in Example 4.7.16.

We note that, if one asks whether such an example can be found on a Stonean space K, large cardinals come into the picture. The existence of a Stonean space K with a non-zero σ-normal measure which is not normal is equivalent to the existence of a measurable cardinal; see [107, Theorem 363S] or [179].

Theorem 4.7.8. *Let K be a non-empty, locally compact space. Then:*

(i) *$N(K)$ is a Dedekind complete lattice ideal in $M(K)$;*

(ii) *there is a closed subspace $S(K)$ of $M(K)$ such that $M(K) = N(K) \oplus_1 S(K)$ and $\nu \perp \sigma$ for each $\nu \in N(K)$ and $\sigma \in S(K)$;*

(iii) *$N(K)$ is a 1-complemented subspace of $M(K)$.*

Proof. (i) By Proposition 4.7.3(iv), $N(K)$ is a lattice ideal in $M(K)$.

Let $\mathfrak{F}$ be a family that is bounded above in $N(K)^+$, and set $\mu = \bigvee \mathfrak{F}$ in $M(K)^+$, so that

$$\mu(B) = \sup\{\nu(B) : \nu \in \mathfrak{F}\} \quad (B \in \mathfrak{B}_K).$$

This implies that $\mu(L) = 0$ $(L \in \mathscr{K}_K)$, and so $\mu \in N(K)^+$; clearly, μ is the supremum of $\mathfrak{F}$ in $N(K)^+$, and so $N(K)$ is Dedekind complete.

(ii) Set

$$S(K) = \{\sigma \in M(K) : \nu \perp \sigma \ (\nu \in N(K))\}.$$

Then $S(K)$ is a closed linear subspace of $M(K)$ and $N(K) \cap S(K) = \{0\}$.

Now take $\mu \in M(K)^+$, and set

$$\mu_{\mathrm{n}} = \bigvee \{\nu \in N(K)^+ : \nu \leq \mu\},$$

so that $\mu_{\mathrm{n}} \in N(K)^+$; set $\mu_{\mathrm{s}} = \mu - \mu_{\mathrm{n}}$. For $\nu \in N(K)^+$, we have $\mu_{\mathrm{n}} + (\mu_{\mathrm{s}} \wedge \nu) \leq \mu$, and hence $\mu_{\mathrm{n}} + (\mu_{\mathrm{s}} \wedge \nu) \leq \mu_{\mathrm{n}}$. Thus $\mu_{\mathrm{s}} \wedge \nu = 0$ $(\nu \in N(K)^+)$. It follows that $\mu_{\mathrm{s}} \in S(K)^+$.

For $\mu \in M(K)$, write $\mu = \mu_1 - \mu_2 + \mathrm{i}(\mu_3 - \mu_4)$, where $\mu_1, \dots, \mu_4 \in M(K)^+$. For $i = 1, \dots, 4$, the measure μ_i can be decomposed as $\mu_{i,\mathrm{n}} + \mu_{i,\mathrm{s}}$ with $\mu_{i,\mathrm{n}} \in N(K)^+$ and $\mu_{i,\mathrm{n}} \in S(K)^+$. Set

$$\mu_{\mathrm{n}} = \mu_{1,\mathrm{n}} - \mu_{2,\mathrm{n}} + \mathrm{i}(\mu_{3,\mathrm{n}} - \mu_{4,\mathrm{n}}) \quad \text{and} \quad \mu_{\mathrm{s}} = \mu_{1,\mathrm{s}} - \mu_{2,\mathrm{s}} + \mathrm{i}(\mu_{3,\mathrm{s}} - \mu_{4,\mathrm{s}}).$$

Then $\mu_{\mathrm{n}} \in N(K)$, $\mu_{\mathrm{s}} \in S(K)$, and $\mu = \mu_{\mathrm{n}} + \mu_{\mathrm{s}}$, so that $M(K) = N(K) \oplus S(K)$. Since $\mu_{\mathrm{n}} \perp \mu_{\mathrm{s}}$, we have $\|\mu\| = \|\mu_{\mathrm{n}}\| + \|\mu_{\mathrm{s}}\|$, and so $M(K) = N(K) \oplus_1 S(K)$.

(iii) This is immediate from (ii). □

The measures in $S(K)$ are sometimes called the *singular* measures, although this is a somewhat unfortunate term.

Proposition 4.7.9. *Let K be a non-empty, locally compact space, and suppose that $\mu \in N(K)$. Then* supp μ *is a regular–closed set.*

Proof. Since $\operatorname{supp}\mu = \operatorname{supp}|\mu|$, we may suppose that $\mu \in N(K)^+$.

Set $F = \operatorname{supp}\mu$, a closed set, and set $U = \operatorname{int}F$, so that $\overline{U} \subset F$. Since $F \setminus \overline{U}$ is nowhere dense, $\mu(F \setminus \overline{U}) = 0$ by Theorem 4.7.4(i). Thus $\mu(K\setminus\overline{U}) = 0$, and so, by the definition of $\operatorname{supp}\mu$, we have $K\setminus\overline{U} \subset K\setminus F$. Hence $\overline{U} = F$, and F is regular-closed. □

The next corollary does use the fact that K is Stonean; the result is due to Dixmier [91], and is set out by Bade in [23, Lemma 8.6].

Corollary 4.7.10. *Let K be a Stonean space, and suppose that $\mu \in N(K)^+ \setminus \{0\}$.*

(i) *The space* $\operatorname{supp}\mu$ *is clopen in K, and hence Stonean.*

(ii) *For each $B \in \mathfrak{B}_K$, there is a unique set $C \in \mathfrak{C}_K$ with $C \subset \operatorname{supp}\mu$ and $\mu(B\,\Delta\,C) = 0$, and so each equivalence class in $\mathfrak{B}_\mu$ contains a unique clopen subset of* $\operatorname{supp}\mu$.

Proof. (i) In a Stonean space, every regular–closed set is clopen.

(ii) By (i), $\operatorname{supp}\mu$ is a clopen subset of K and $\mu(K \setminus \operatorname{supp}\mu) = 0$, and so we may suppose that $K = \operatorname{supp}\mu$.

Take $B \in \mathfrak{B}_K$. By Proposition 1.4.4, there is a unique $C \in \mathfrak{C}_K$ with $B \equiv C$, and then $\mu(B\,\Delta\,C) = 0$. Suppose that $C_1, C_2 \in \mathfrak{C}_K$ are such that $\mu(B\,\Delta\,C_1) = \mu(B\,\Delta\,C_2) = 0$. Then $C_1 \Delta C_2 \subset (B\,\Delta\,C_1) \cup (B\,\Delta\,C_2)$, so that $\mu(C_1 \Delta C_2) = 0$. Since $C_1 \Delta C_2$ is an open set and $K = \operatorname{supp}\mu$, it follows from Proposition 4.1.6 that $C_1 \Delta C_2 = \emptyset$, i.e., $C_1 = C_2$. This establishes the required uniqueness of C. □

Corollary 4.7.11. *Let K be a Stonean space, and suppose that $\mu, \nu \in N(K)$. Then:*

(i) $\operatorname{supp}\nu \subset \operatorname{supp}\mu$ *if and only if* $\nu \ll \mu$;

(ii) $\operatorname{supp}\nu = \operatorname{supp}\mu$ *if and only if* $\nu \sim \mu$;

(iii) $\mu \perp \nu$ *if and only if* $\operatorname{supp}\mu \cap \operatorname{supp}\nu = \emptyset$.

Proof. (i) Always $\operatorname{supp}\nu \subset \operatorname{supp}\mu$ when $\nu \ll \mu$.

For the converse, we may suppose that $\mu, \nu \in N(K)^+$. By Proposition 1.4.4, for each $B \in \mathfrak{B}_\mu$, there exists $C \in \mathfrak{C}_K$ with $C \equiv B$. Now suppose that $B \in \mathfrak{N}_\mu$. Then, by Corollary 4.7.5(ii), $C \in \mathfrak{N}_\mu$, and so $C \cap \operatorname{supp}\nu = \emptyset$, whence $\nu(B) = \nu(C) = 0$. This shows that $\nu \ll \mu$.

(ii) This is immediate from (i).

(iii) Clearly $\mu \perp \nu$ when $\operatorname{supp}\mu \cap \operatorname{supp}\nu = \emptyset$.

Now suppose that $\mu \perp \nu$, and set $U = \operatorname{supp}\mu \cap \operatorname{supp}\nu$, so that, by Corollary 4.7.10(i), U is an open set. Then $(\nu \mid U) \perp \mu$ and, by (i), $\nu \mid U \ll \mu$. Thus $\nu \mid U = 0$, and hence $U = \emptyset$. □

We now determine the set of extreme points of the closed unit ball of the normal measures. Recall that D_X denotes the set of isolated points of a topological space X.

Proposition 4.7.12. *Let K be a non-empty, locally compact space. Then*

$$\operatorname{ex} N(K)_{[1]} = \{\zeta\delta_x : \zeta \in \mathbb{T}, x \in D_K\} \quad \text{and} \quad \operatorname{ex} N(K) \cap P(K) = \{\delta_x : x \in D_K\}.$$

Proof. By Proposition 2.1.10 and Theorem 4.7.8(ii),

$$\operatorname{ex} M(K)_{[1]} = \operatorname{ex} N(K)_{[1]} \cup \operatorname{ex} S(K)_{[1]} .$$

Thus, by Proposition 4.4.15(i), each point of $\operatorname{ex} N(K)_{[1]}$ has the form $\zeta\delta_x$ for some $\zeta \in \mathbb{T}$ and $x \in K$. By Theorem 4.7.4(i), $\operatorname{int}_K\{x\} \neq \emptyset$, and so $x \in D_K$.

Conversely, $\zeta\delta_x \in \operatorname{ex} N(K)_{[1]}$ whenever $\zeta \in \mathbb{T}$ and $x \in D_K$. □

Corollary 4.7.13. *Let K be a non-empty, locally compact space. Then we can identify $N_d(K)$ with $\ell^1(D_K)$ and $N_c(K)$ with $N(K \setminus \overline{D_K})$.*

Proof. We know that $\delta_x \in N_d(K)$ for each $x \in D_K$, and so $\ell^1(D_K) \subset N_d(K)$. Conversely, it is clear that $N_d(K) \subset \ell^1(D_K)$. Thus $N_d(K) = \ell^1(D_K)$.

For each $\mu \in N(K)$, we have $|\mu|(\overline{D_K} \setminus D_K) = 0$ by Corollary 4.7.5, and so we have $\operatorname{supp} \mu \subset K \setminus \overline{D_K}$ for each $\mu \in N_c(K)$. Conversely, take $\mu \in N(K \setminus \overline{D_K})$. Then $|\mu|(\{x\}) = 0$ $(x \in K \setminus D_K)$, and so $\mu \in N_c(K)$. □

Corollary 4.7.14. *Let S be a non-empty set. Then $N(\beta S) = N_d(\beta S) = \ell^1(S)$ and $N_c(\beta S) = N(S^*) = \{0\}$.*

Proof. By Proposition 1.5.9(ii), βS is Stonean, and $\overline{D_{\beta S}} = \overline{S} = \beta S$. By Corollary 4.7.13, $N(\beta S) = N_d(\beta S) = \ell^1(S)$ and $N_c(\beta S) = \{0\}$.

We now show that $N(S^*) = \{0\}$. Assume to the contrary that $\mu \in N(S^*)$ with $\mu \neq 0$. By Theorem 4.7.4(i), $\operatorname{supp} \mu$ has non-empty interior, and so $\operatorname{supp} \mu$ contains a clopen set of the form A^*, where A is an infinite subset of S. By Proposition 1.5.5, A^* contains an uncountable family of non-empty, pairwise-disjoint, open subsets. But this contradicts the fact that, by Proposition 4.1.6, $\operatorname{supp} \mu$ satisfies CCC. Thus $\mu = 0$. □

Corollary 4.7.15. *Let X be a non-empty, compact space such that $N(X)$ is isometrically a dual space. Suppose that D_X is countable and infinite. Then $N(X) \cong \ell^1$.*

Proof. Take E to be a Banach space with $E' \cong N(X)$; we shall apply Theorem 4.1.10 with K taken to be $E'_{[1]}$. Take a countable, dense subset T of $\mathbb{T}$, and consider the countable set

$$D = \{\zeta\delta_x : \zeta \in T, x \in D_X\} .$$

Then, using Proposition 4.7.12, we see that D is $\|\cdot\|$-dense in $\operatorname{ex} K$, and so, by Theorem 4.1.10, K is the $\|\cdot\|$-closure of the absolutely convex hull of $\{\delta_x : x \in D_X\}$. It follows that $E' \cong \ell^1$, and so $N(X) \cong \ell^1$. □

The next example gives some σ-normal measures on a space K that is such that $N(K) = \{0\}$.

Example 4.7.16. Consider the compact space $K = \mathbb{N}^*$. By Proposition 1.5.3(i), there are no non-empty G_δ-sets in $\mathscr{K}_K$. Thus all measures in $M(K)$ are σ-normal. However $N(K) = \{0\}$ by Corollary 4.7.14. □

Let K and L be non-empty, compact spaces, and again suppose that $\eta : K \to L$ is a continuous surjection. Recall that we defined

$$\eta^\circ : f \mapsto f \circ \eta\,, \quad C(L) \to C(K)\,,$$

in equation (2.9) on page 83, so that η° is a unital C^*-embedding and a lattice homomorphism. The dual of η° is therefore a surjection

$$T_\eta := (\eta^\circ)' : M(K) \to M(L)$$

with $\|T_\eta\| = 1$; of course, as in equation (4.7) on page 116,

$$(T_\eta \mu)(B) = \mu(\eta^{-1}(B)) \quad (B \in \mathfrak{B}_L,\, \mu \in M(K))\,, \tag{4.19}$$

and $T_\eta \mu$ is the image measure $\eta[\mu]$. We shall use this notation in the next result.

Note that $T_\eta \mu \in M(L)^+$ when $\mu \in M(K)^+$, and so T_η is a positive operator on the Banach lattice $M(K)$, and hence is an order homomorphism. (However, it is easily seen that T_η is not necessarily a lattice homomorphism.) Now take $\nu \in M(L)^+$. Then ν defines a positive linear functional on $\eta^\circ(C(L))$, and so has a norm-preserving extension to a linear functional on $C(K)$, and hence to a measure $\mu \in M(K)$ with $\|\mu\| = \|\nu\|$; by equation (4.2), $\mu \in M(K)^+$. In particular, this shows that $T_\eta(M(K)^+) = M(L)^+$.

Proposition 4.7.17. *Let K and L be non-empty, compact spaces, and suppose that $\eta : K \to L$ is a continuous surjection that is either open or irreducible. Then*

$$T_\eta(N(K)) \subset N(L)\,.$$

Suppose, further, that $N(L) = \{0\}$. Then $N(K) = \{0\}$.

Proof. Take $\mu \in N(K)$. For $L_0 \in \mathscr{K}_L$, set $K_0 = \eta^{-1}(L_0)$. Then K_0 is certainly compact in K. We *claim* that $\operatorname{int}_K K_0 = \emptyset$. This is obvious when η is open, and follows from Proposition 1.4.21(ii) when η is irreducible. Thus $K_0 \in \mathscr{K}_K$. By Theorem 4.7.4(i), $\mu(K_0) = 0$, and so $(T_\eta \mu)(L_0) = 0$. Again by Theorem 4.7.4(i), $T_\eta \mu \in N(L)$. Thus $T_\eta(N(K)) \subset N(L)$.

Now suppose that $N(L) = \{0\}$, and take $\mu \in N(K)^+$. Then $T_\eta \mu = 0$. But this implies that $\mu(K) = (T_\eta \mu)(L) = 0$, and hence $\mu = 0$. Thus $N(K) = \{0\}$. □

Theorem 4.7.18. *Let K and L be two non-empty, compact spaces, and suppose that $\eta : K \to L$ is an irreducible surjection. Then the map*

$$T_\eta \mid N(K) : N(K) \to N(L) \tag{4.20}$$

is a Banach-lattice isometry.

Proof. By Proposition 4.7.17, $T_\eta(N(K)) \subset N(L)$. We shall now show that the map $T_\eta : N(K) \to N(L)$ is a bijection.

Set

$$\eta^{-1}(\mathfrak{B}_L) = \{\eta^{-1}(B) : B \in \mathfrak{B}_L\},$$

so that $\eta^{-1}(\mathfrak{B}_L)$ is a subset of $\mathfrak{B}_K$.

We *claim* that each $C \in \mathfrak{B}_K$ is congruent to a set in $\eta^{-1}(\mathfrak{B}_L)$. First suppose that U is a non-empty, open set in K, and define $V = \{y \in L : F_y \subset U\}$, where $F_y = \eta^{-1}(\{y\})$ $(y \in L)$. By Proposition 1.4.21(ii), V is open in L and $\eta^{-1}(V)$ is a dense, open subset of U, and so $\eta^{-1}(V) \in \eta^{-1}(\mathfrak{B}_L)$ and $U \equiv \eta^{-1}(V)$. As on page 13, each $C \in \mathfrak{B}_L$ has the Baire property, and so there is an open set U in K with $C \equiv U$. The claim follows.

Now suppose that $\mu \in N(K)$ with $T_\eta\mu = 0$. Then $\mu(\eta^{-1}(B)) = 0$ $(B \in \mathfrak{B}_L)$, and so $\mu(C) = 0$ $(C \in \mathfrak{B}_K)$ by the claim and Corollary 4.7.5, (a) $\Rightarrow$ (c). Thus the map $T_\eta : N(K) \to N(L)$ is an injection.

We next *claim* that $T_\eta : N(K) \to N(L)$ is a surjection and that the map

$$T_\eta \mid N(K)^+ : N(K)^+ \to N(L)^+$$

is an isometry. Indeed, take $\nu \in N(L)^+$. As above, there exists $\mu \in M(K)^+$ with $\|\mu\| = \|\nu\|$ and $T_\eta\mu = \nu$. Take $K_0 \in \mathscr{K}_K$, and set $L_0 = \pi(K_0)$. By Proposition 1.4.22, $L_0 \in \mathscr{K}_L$, and so $\nu(L_0) = 0$. Thus $\mu(\pi^{-1}(L_0)) = 0$. Since $\mu \in M(K)^+$, it follows that $\mu(K_0) = 0$, and hence $\mu \in N(K)^+$ by Theorem 4.7.4(i). The claim follows.

We have shown that the map $T_\eta \mid N_{\mathbb{R}}(K) \to N_{\mathbb{R}}(L)$ is a bijection and that it is an order isomorphism, and so $T_\eta \mid N(K) : N(K) \to N(L)$ is a Banach-lattice isomorphism. By Proposition 2.3.5 and the above claim, it is a Banach-lattice isometry. □

Corollary 4.7.19. *Let L be a non-empty, compact space. Then the map*

$$T_{\pi_L} \mid N(G_L) : N(G_L) \to N(L)$$

is a Banach-lattice isometry. In particular, $N(G_L) \cong N(L)$.

Proof. As in Theorem 1.6.5, the map $\pi_L : G_L \to L$ is an irreducible surjection, and so this is a special case of the theorem. □

Later, we shall be concerned with compact spaces that have many normal measures, but first we shall give various examples of compact spaces that have no non-zero normal measures.

Proposition 4.7.20. *Let K be a non-empty, separable, locally compact space without isolated points. Then there are no non-zero σ-normal measures on K, and so $N(K) = \{0\}$.*

Proof. We first *claim* that each σ-normal measure μ on EK is a continuous measure. Indeed, take $x \in K$. Since the point x is not isolated, there is a countable subset, say $S = \{x_n : n \in \mathbb{N}\}$, of $K \setminus \{x\}$ such that S is dense in K. Choose a sequence (U_n) in $\mathscr{N}_x$

such that $\overline{U_1}$ is compact and such that $\overline{U_{n+1}} \subset U_n$ and $x_n \notin U_n$ for each $n \in \mathbb{N}$, and set $L = \bigcap \overline{U_n}$. Then L is a compact G_δ-set in K with $x \in L$, and $\mathrm{int}_K L = \emptyset$ because $L \cap S = \emptyset$. By Theorem 4.7.4(ii), $\mu(L) = 0$. This implies that $\mu(\{x\}) = 0$, and hence μ is continuous, as claimed.

Again, let $\{x_n : n \in \mathbb{N}\}$ be a dense subset of K. Fix $\varepsilon > 0$ and a compact subset L of K; take $g \in C_{0,\mathbb{R}}(K)$ with $g \geq \chi_L$ and $g(K) \subset \mathbb{I}$. For each $n \in \mathbb{N}$, take $U_n \in \mathscr{N}_{x_n}$ with $|\mu|(U_n) < \varepsilon/2^n$, choose $f_n \in C_{00}(K)$ with $\chi_{\{x_n\}} \leq f_n \leq \chi_{U_n}$, and set $g_n = g \wedge \bigvee_{j=1}^n f_j$, so that $g_n \nearrow g$ in $C_0(K)^+$. We have

$$\langle g_n, |\mu| \rangle \leq |\mu| \left(\bigcup_{k=1}^n U_k \right) \leq \sum_{k=1}^n |\mu|(U_k) < \varepsilon \quad (n \in \mathbb{N}).$$

Since $|\mu|$ is σ-normal, $\langle g_n, |\mu| \rangle \nearrow \langle g, |\mu| \rangle$ in $\mathbb{R}^+$, and so $|\mu|(L) \leq \langle g, |\mu| \rangle \leq \varepsilon$. This holds true for each $\varepsilon > 0$, and hence $|\mu|(L) = 0$. Thus $\mu = 0$.

This gives the result. □

It is natural to wonder whether $N(K) = \{0\}$ when the condition 'separable' in Proposition 4.7.20 is replaced by the weaker condition that K satisfies CCC. The example of Theorem 4.7.26, to be given below, will show that this is not the case.

Corollary 4.7.21. *There are no non-zero, σ-normal measures on $G_{\mathbb{I}}$, and hence $N(G_{\mathbb{I}}) = \{0\}$.*

Proof. As remarked within Example 1.7.16, $G_{\mathbb{I}}$ is an infinite, separable Stonean space without isolated points, and so this follows from the proposition. The result also follows from Proposition 1.7.13. □

Corollary 4.7.22. *Let G be a locally compact group that is not discrete. Then $N(G) = \{0\}$.*

Proof. Take $\mu \in N(G)^+$ and a compact subspace K of G. Then there is an infinite, clopen, σ-compact subgroup G_0 of G with $G_0 \supset K$. As in Theorem 4.4.2, there is a non-discrete, metrizable group H and a quotient map $\eta : G_0 \to H$; the map η is open. The space $\eta(K)$ is separable and has no isolated points, and so, by Proposition 4.7.20, $N(\eta(K)) = \{0\}$. By Proposition 4.7.17, $N(K) = \{0\}$, and so $\mu(K) = 0$. It follows that $N(G) = \{0\}$. □

The following result is essentially contained in [103].

Theorem 4.7.23. *Let K be a non-empty, locally connected, locally compact space without isolated points. Then $N(K) = \{0\}$.*

Proof. Assume that there exists $\mu \in N(K)^+$ with $\mu \neq 0$. Again, $\mu \in N_c(K)^+$.

For each $n \in \mathbb{N}$, let $\mathscr{F}_n$ be a family of non-empty, open subsets of K such that $\mathscr{F}_n$ is maximal with respect to the following properties:

(i) $\mu(U) < 1/n$ for each $U \in \mathscr{F}_n$; (ii) distinct sets in $\mathscr{F}_n$ are disjoint.

It is clear from Zorn's lemma that such a family $\mathscr{F}_n$ exists. Set $G_n = \bigcup\{U : U \in \mathscr{F}_n,$ an open subset of K. Since μ is continuous, each open set in K contains an open set of arbitrary small μ-measure, and so $\overline{G}_n = K$. By Theorem 4.7.4(i), $\mu(K \setminus G_n) = 0$.

Now set $H = \bigcap\{G_n : n \in \mathbb{N}\}$, a G_δ-set in K. We have $\mu(K \setminus H) = 0$, and so $\mu(H) > 0$. By Theorem 4.7.4(i), $\mu(\mathrm{int}_K H) > 0$. Assume that each $x \in \mathrm{int}_K H$ has an open neighbourhood V_x in K with $\mu(V_x) = 0$. For each compact subset L of $\mathrm{int}_K H$, there are finitely many points $x_1, \dots, x_n \in \mathrm{int}_K H$ with $L \subset V_{x_1} \cup \cdots \cup V_{x_n}$, and so $\mu(L) = 0$. But

$$\mu(\mathrm{int}_K H) = \sup\{\mu(L) : L \text{ compact}, L \subset \mathrm{int}_K H\}$$

because μ is a regular measure, and so $\mu(\mathrm{int}_K H) = 0$, a contradiction. Thus there exists $x_0 \in \mathrm{int}_K H$ such that $\mu(V) > 0$ for each $V \in \mathscr{N}_{x_0}$. Let V_0 be an open neighbourhood of x_0 with $V_0 \subset \mathrm{int}_K H$. Since K is locally connected, we may suppose that V_0 is connected. We have $V_0 \subset G_n$ for each $n \in \mathbb{N}$.

Since $\mu(V_0) > 0$, there exists $n \in \mathbb{N}$ with $\mu(V_0) > 1/n$. Choose $U \in \mathscr{F}_n$ with $x_0 \in U$, and set $V = G_n \setminus U$, so that V is open in K. Since $\mu(U) < 1/n < \mu(V_0)$, we have $V_0 \cap V \neq \emptyset$, and so $\{V_0 \cap U, V_0 \cap V\}$ is a partition of V_0 into two non-empty, disjoint, open subsets, a contradiction of the fact that V_0 is connected.

Thus $N(K) = \{0\}$, as required. □

Proposition 4.7.24. *Let K be a non-empty, connected, locally compact F-space. Then $N(K) = \{0\}$.*

Proof. Assume that there exists $\mu \in N(K)^+ \setminus \{0\}$, and choose a compact subset L of K such that $\mu(L) > 0$. Since L is a compact F-space satisfying CCC (by Proposition 4.1.6), the space L is Stonean, and so there is a non-empty, open subset U of K with $U \subset L$. Choose a non-empty, open subset V of K such that $\overline{V} \subset U$. Then $\overline{V}$ is open in U, and hence in K. We have shown that K contains a non-empty, clopen subset, and so K is not connected, the required contradiction. □

Proposition 4.7.25. *Let L be a compact space without isolated points which is either separable or a locally compact group or locally connected or a connected F-space, and suppose that K is a compact space such that there is a continuous surjection that is open or irreducible from K onto L. Then $N(K) = \{0\}$. In particular, $N(G_L) = \{0\}$ and $N(L \times R) = \{0\}$ for each compact space R.*

Proof. This follows from Proposition 4.7.17, Proposition 4.7.20, Corollary 4.7.22, Theorem 4.7.23, and Proposition 4.7.24. □

In the text [220, p. 2], a monotone complete C^*-algebra is said to be *wild* if there are no non-zero normal states. Let K be a non-empty, compact space. Then, as we remarked on page 107, $C(K)$ is a monotone complete C^*-algebra if and only if K is Stonean; $C(K)$ is wild if and only if $N(K) = \{0\}$. In [220, §4.3], it is shown that there are many examples of monotone complete C^*-subalgebras of ℓ^∞ that are wild, and so we obtain many examples of Stonean spaces K such that $N(K) = \{0\}$.

In the light of Theorem 4.7.23 and Proposition 4.7.24, it is natural to wonder whether $N(K) = \{0\}$ for each connected, compact set K. This question was answered by Grzegorz Plebanek [206] with the following counter-example; we are very grateful to him for his permission to include it here. Preliminary results on inverse systems with measures were given in §4.1.

Theorem 4.7.26. *There is a non-empty, connected, compact set K satisfying* CCC, *and such that $N(K) \neq \{0\}$. Indeed, there exists a strictly positive measure in $N(K)$.*

Proof. Let $L = \mathbb{I}$, a connected, compact space, and take m to be the strictly positive measure on $\mathbb{I}$ that is Lebesgue measure.

We shall define inductively an inverse system with strictly positive measures

$$(K_\alpha, \mu_\alpha, \pi_\alpha^\beta : 0 \leq \alpha \leq \beta < \omega_1)$$

with $K_0 = L$ and $\mu_0 = m$.

When $0 \leq \gamma < \omega_1$ is such that $(K_\alpha, \mu_\alpha, \pi_\alpha^\beta : 0 \leq \alpha \leq \beta \leq \gamma)$ is an inverse system with non-empty, connected, compact spaces K_α and strictly positive measures $\mu_\alpha \in P(K_\alpha)$ (for $0 \leq \alpha \leq \gamma$), we define $K_{\gamma+1}$ and $\mu_{\gamma+1}$ by applying Theorem 4.1.16 with $L = K_\gamma$ and $\nu = \mu_\gamma$ and by setting $K_{\gamma+1} = K_\gamma^\#$ and $\mu_{\gamma+1} = \mu_\gamma^\#$ (and defining the maps $\pi_\alpha^{\gamma+1}$ to be $\eta^\# \circ \pi_\alpha^\gamma$ for $0 \leq \alpha \leq \gamma$ and $\pi_{\gamma+1}^{\gamma+1}$ to be the identity on $K_{\gamma+1}$).

As in Theorem 4.1.16, we have $\mathrm{int}_{K_{\gamma+1}}(\pi_\gamma^{\gamma+1})^{-1}(W) \neq \emptyset$ for each $W \in \mathbf{Z}(K_\gamma)$ with $\mu_\gamma(W) > 0$.

When $0 \leq \gamma \leq \omega_1$, γ is a limit ordinal, and K_α and $\mu_\alpha \in P(K_\alpha)$ are defined for $0 \leq \alpha < \gamma$, we define $(K_\gamma, \pi_\alpha^\gamma : 0 \leq \alpha < \gamma)$ to be the inverse limit of the inverse system $(K_\alpha, \pi_\alpha^\beta : 0 \leq \alpha \leq \beta < \gamma)$ (and take π_α^γ to be the continuous surjections that arise in Theorem 1.4.32), so that K_γ is compact and connected; we take $\mu_\gamma \in P(K_\gamma)$ to be the strictly positive measure specified in Proposition 4.1.15. In the special case in which $\gamma = \omega_1$, we set $K = K_\gamma$, $\mu = \mu_\gamma \in P(K)$, and $\eta = \pi_0^\gamma$.

It follows from Corollary 1.4.33 that, for each $Z \in \mathbf{Z}(K)$, there exists $\alpha < \omega_1$ and $W \in \mathbf{Z}(K_\alpha)$ such that $Z = \pi_\alpha^{-1}(W)$. Suppose that $\mu(Z) > 0$. Then $\mu_\alpha(W) > 0$, and so $(\pi_\alpha^{\alpha+1})^{-1}(W)$ has non-empty interior. Hence

$$\mathrm{int}_K Z = \mathrm{int}_K(\pi_{\alpha+1}^{-1}((\pi_\alpha^{\alpha+1})^{-1}(W))) \neq \emptyset,$$

and so $\mu(Z) = 0$ whenever $Z \in \mathbf{Z}(K)$ and $\mathrm{int}_K Z = \emptyset$, i.e., μ is σ-normal by Theorem 4.7.4(ii). Since μ is strictly positive, K satisfies CCC, as is generally the case for the support of any $\mu \in M(K)$. By Proposition 4.7.7, $\mu \in N(K)$.

This completes the proof of the theorem. □

It can be shown, using the remark after Theorem 4.1.16, that $w(K) = \mathfrak{c}$, where K is the space of the above proof.

We have earlier defined a 'normal measure' on a Boolean algebra; see Definition 1.7.12. One might guess that a normal measure on a compact space K would give a normal measure on the Boolean algebra $\mathfrak{B}_K$. However this is not correct. Indeed, suppose that there exists $\mu \in N_c(K)^+$ with $\|\mu\| = 1$, and take the net (U_α) in $\mathfrak{B}_K$ consisting of the complements of the finite subsets of K, so that $U_\alpha \searrow 0$ in $\mathfrak{B}_K$, but $\mu(U_\alpha) = 1$ for each α, and so $\lim_{\alpha \in A} \mu(U_\alpha) \neq 0$. However, we do have the following result involving the Boolean algebra of regular–open sets, as defined in Example 1.7.16.

Theorem 4.7.27. *Let K be a non-empty, compact space. Then the map*

$$R : \mu \mapsto \mu \mid \mathfrak{R}_K, \quad N(K) \to N(\mathfrak{R}_K),$$

is a Riesz isomorphism

Proof. Take $\mu \in N(K)$. Then it is clear that $R\mu$ is a measure on the Boolean algebra $\mathfrak{R}_K$ in the sense of Definition 1.7.12.

We first *claim* that $R\mu \in N(\mathfrak{R}_K)$. For this, it suffices to suppose that $\mu \in N(K)^+$. Take a net (U_α) with $U_\alpha \searrow \emptyset$ in $\mathfrak{R}_K$, and consider the set

$$\Gamma = \bigcup_\alpha \{f \in C(K) : \chi_{U_\alpha} \leq f\},$$

regarded as a downward-directed net in $C(K)^+$. Take $g \in C(K)^+$ with $g \leq f$ $(f \in \Gamma)$; we shall show that $g = 0$. Indeed, assume towards a contradiction that $g \neq 0$. Then there is a non-empty, open set V in K with $g(x) > 0$ $(x \in V)$. Assume that α is such that $V \not\subset U_\alpha$. Then $V \not\subset \overline{U_\alpha}$ because U_α is regular–open, and so there exists $x \in V$ and $f \in C(K)$ with $f(x) = 0$ and $\chi_{U_\alpha} \leq f$, using the fact that K is compact. Thus $f \in \Gamma$, and hence $g(x) = 0$, a contradiction. This shows that $V \subset \bigcap U_\alpha$, a contradiction of the fact that $U_\alpha \searrow \emptyset$. Hence $g = 0$, and so $\inf \Gamma = 0$.

Since $\mu \in N(K)^+$, we see that $\inf\{\mu(f) : f \in \Gamma\} = 0$. However, for each $f \in \Gamma$, there exists α with $\chi_{U_\alpha} \leq f$, and so $\inf_\alpha \mu(U_\alpha) = 0$. We have shown that $R\mu$ satisfies the condition given in Definition 1.7.12 for it to be a normal measure on $\mathfrak{R}_K$, and so $R\mu \in N(\mathfrak{R}_K)^+$, giving the claim.

It is clear that $R : N(K) \to N(\mathfrak{R}_K)$ is a Riesz homomorphism.

We now *claim* that R is injective. Indeed, suppose that $\mu \in N_{\mathbb{R}}(K)$ with $R\mu = 0$. Then $R(|\mu|) = |R\mu| = 0$, and so $|\mu|(K) = R(|\mu|)(K) = 0$. Thus $\mu = 0$, and so R is injective, as claimed.

We finally *claim* that R is surjective. Indeed, take $\nu \in N(\mathfrak{R}_K)^+$, and define $\widehat{\mu}(B) = \nu(V_B)$ $(B \in \mathfrak{B}_K)$, where V_B is the unique regular–open subset of K with $B \equiv V_B$.

We *claim* that $\widehat{\mu}$ is a measure on K. First, note that, for disjoint sets $B, C \in \mathfrak{B}_K$, we have $V_B \cap V_C \equiv B \cap C = \emptyset$, and so $\widehat{\mu}(B \cup C) = \widehat{\mu}(B) + \widehat{\mu}(C)$. Now suppose that (B_n) is an increasing sequence in $\mathfrak{B}_K$ with union $B \in \mathfrak{B}_K$. Then

$$B \Delta \left(\bigcup \{V_{B_n} : n \in \mathbb{N}\} \right) \subset \bigcup \{B_n \Delta V_{B_n} : n \in \mathbb{N}\}$$

is meagre. Set $U = \bigvee\{V_{B_n} : n \in \mathbb{N}\}$ in $\mathfrak{R}_K$, so that $U \Delta B$ is meagre and $U = V_B$. Then $\widehat{\mu}(B) = \nu(V_B) = \lim_{n\to\infty} \nu(V_{B_n})$ because ν is normal, and so $\widehat{\mu}(B) = \lim_{n\to\infty} \widehat{\mu}(B_n)$. This shows that $\widehat{\mu}$ is σ-additive. Thus $\widehat{\mu} \in M(K)$, and $\widehat{\mu}(B) \geq 0$ $(B \in \mathfrak{B}_K)$. (Note that it is not immediately obvious that $\widehat{\mu}$ is regular, but $\widehat{\mu}$ does define a continuous linear functional on $C(K)$.) By the Riesz representation theorem, there exists $\mu \in M(K)^+$ with

$$\int_K f \, \mathrm{d}\mu = \langle f, \widehat{\mu} \rangle \quad (f \in C(K)).$$

Let L be a non-empty, closed subspace of K. The family $\mathscr{U}$ of sets in $\mathfrak{R}_K$ that contain L is a net with infimum $\operatorname{int} L$ in $\mathfrak{R}_K$, and so $\{\nu(U) : U \in \mathscr{U}\}$ is a net in $\mathbb{R}$ with infimum $\nu(\operatorname{int} L)$. For each $U \in \mathscr{U}$, there exists $f_U \in C(K)$ with $\chi_L \leq f_U \leq \chi_U$, and then

$$\mu(L) \leq \int_K f_U \, \mathrm{d}\mu = \langle f, \widehat{\mu} \rangle \leq \widehat{\mu}(U) = \nu(U).$$

Thus $\mu(L) \leq \nu(\operatorname{int} L)$.

Take $U \in \mathfrak{R}_K$. By the previous remark, we have $\mu(U) = \mu(\operatorname{int} \overline{U}) \leq \nu(U)$, and hence $\mu(\operatorname{int}(K \setminus U)) \leq \nu(\operatorname{int}(K \setminus U))$, i.e., $\mu(U') \leq \nu(U')$, which implies that $\mu(U) \geq \nu(U)$. It follows that $\mu(U) = \nu(U)$.

For each $B \in \mathfrak{B}_K$, the set $B \Delta V_B$ is meagre, and so $\mu(B) = \mu(V_B) = \nu(V_B) = \widehat{\mu}(B)$. Thus $\mu = \widehat{\mu}$. Clearly $R\mu = \nu$ and so R is a surjection.

We conclude that $R : N(K) \to N(\mathfrak{R}_K)$ is a Riesz isomorphism. □

Corollary 4.7.28. *Let K and L be two compact spaces such that $\mathfrak{R}_K$ and $\mathfrak{R}_L$ are isomorphic as Boolean algebras. Then $N(K)$ and $N(L)$ are Banach-lattice isometric.*

Proof. Let $\rho : \mathfrak{R}_K \to \mathfrak{R}_L$ be an isomorphism, and then define

$$\widehat{\rho}(\mu)(V) = \mu(\rho^{-1}(V)) \quad (\mu \in N(\mathfrak{R}_K), V \in \mathfrak{R}_L),$$

so that $\widehat{\rho} : N(\mathfrak{R}_K) \to N(\mathfrak{R}_L)$ is the induced Riesz isomorphism. Next, let

$$R_K : N(K) \to N(\mathfrak{R}_K) \quad \text{and} \quad R_L : N(L) \to N(\mathfrak{R}_L)$$

be the Riesz isomorphisms given by the theorem. Set

$$T = R_L^{-1} \circ \widehat{\rho} \circ R_K : N(K) \to N(L).$$

Then T is a Riesz isomorphism. Further, $\|T\mu\| = |T\mu|(L) = |\mu|(K)$ $(\mu \in N(K))$ because $\rho^{-1}(L) = K$. By Proposition 2.3.5, there is a Banach-lattice isometry from $N(K)$ onto $N(L)$. □

We recall from Example 1.7.16 that $\mathfrak{R}_K$ and $\mathfrak{R}_L$ are isomorphic as Boolean algebras if and only if the Gleason covers G_K and G_L are homeomorphic. Thus Corollary 4.7.28 also follows easily from Corollary 4.7.19.

Chapter 5
Hyper-Stonean Spaces

We shall now define, in §5.1, the compact spaces of most interest to us, the 'hyper-Stonean spaces', and characterize them in terms of the existence of category measures. For a locally compact space K and $\mu \in P(K)$, we shall discuss in §5.2 the commutative C^*-algebra $L^\infty(K,\mu)$, which has been identified with the space $C(\Phi_\mu)$. In particular, we shall describe Φ_m, where m is Lebesgue measure on $\mathbb{I}$: in this case, Φ_m is called $\mathbb{H}$, the hyper-Stonean space of the unit interval. We shall give a topological characterization of $\mathbb{H}$ in §5.3.

We shall give our main constructions of the bidual space $C_0(K)''$ in §5.4; this space has the form $C(\widetilde{K})$ for a hyper-Stonean space $\widetilde{K}$ that we shall call the 'hyper-Stonean envelope' of K. Indeed, we shall first construct $\widetilde{K}$ as the character space of an explicit commutative, unital C^*-algebra. Next we shall give a new construction of $\widetilde{K}$ as βS_K, where S_K is the Stone space of the Boolean ring $M(K)^+/\sim$. Further, we shall continue to show that $\widetilde{K}$ is homeomorphic to the Stone space of two other (mutually isomorphic) Boolean algebras; the first of these is the Boolean algebra of complemented faces of $P(K)$ and the second, described in §5.5, is the Boolean algebra of L-projections in $\mathscr{B}(M(K))$.

We shall conclude the chapter by summarizing in §5.6 the analogous theory for general C^*-algebras.

5.1 Hyper-Stonean spaces

We define hyper-Stonean spaces and discuss their basic properties. Normal measures and the space $N(K)$ were defined and discussed in §4.7.

H.G. Dales et al., *Banach Spaces of Continuous Functions as Dual Spaces*, CMS Books in Mathematics, DOI 10.1007/978-3-319-32349-7_5

Definition 5.1.1. Let K be a non-empty, compact space. Then

$$W_K = \bigcup\{\operatorname{supp}\mu : \mu \in N(K)\}.$$

The space K is *hyper-Stonean* if K is Stonean and W_K is dense in K. A Boolean algebra is *hyper-Stonean* if its Stone space is a hyper-Stonean topological space.

We note the comment of John Kelley in a footnote in [156]: 'The term "hyper-stonian" seems unfortunate. In spite of my affection and admiration for Marshall Stone, I find the notion of a Hyper-Stone downright appalling'. Pace Kelley's remark, the term has continued to be used, albeit in variant spellings.

It follows from Corollary 4.7.10(i) that W_K is open in K whenever K is a Stonean space.

Let K be a hyper-Stonean space. Since the restriction of a normal measure to a Borel set is a normal measure, for each non-empty, open subset U of K, there exists $\mu \in N(K) \cap P(K)$ with $\operatorname{supp}\mu \subset U$.

For example, let K be a Stonean space such that the set D_K of isolated points of K is dense in K. Then K is certainly hyper-Stonean; it is homeomorphic to βD_K. In particular, $\beta\mathbb{N}$ is hyper-Stonean.

Proposition 5.1.2. (i) *Let K be a hyper-Stonean space, and suppose that L is a clopen subspace of K. Then L is hyper-Stonean.*

(ii) *Let K be a hyper-Stonean space, and suppose that $N(K)$ is separable. Then K satisfies* CCC.

(iii) *Let K be a Stonean space. Then a subset L of K is the support of a normal measure if and only if $L \in \mathfrak{C}_K$ and $L \subset W_K$.*

(iv) *Let K be a Stonean space. Then there are disjoint, clopen subspaces K_1 and K_2 of K such that $K = K_1 \cup K_2$ and such that K_1 is hyper-Stonean and $N(K_2) = \{0\}$.*

Proof. (i) The restriction of a normal measure on K to L is a normal measure on L.

(ii) Let $\{U_\alpha : \alpha \in A\}$ be a family of pairwise-disjoint, non-empty, open subsets of K. For each $\alpha \in A$, there exists $\mu_\alpha \in N(K) \cap P(K)$ with $\operatorname{supp}\mu_\alpha \subset U_\alpha$. The family $\{\mu_\alpha : \alpha \in A\}$ is singular, and so, by Proposition 4.6.2(i), A is countable. Hence K satisfies CCC.

(iii) Suppose that $\mu \in N(K)$. By Corollary 4.7.10(i), $\operatorname{supp}\mu \in \mathfrak{C}_K$, and certainly $\operatorname{supp}\mu \subset W_K$.

Conversely, suppose that $L \in \mathfrak{C}_K$ and $L \subset W_K$. Since L is compact, there are finitely many measures $\mu_1, \dots, \mu_n \in N(K)^+$ such that $L \subset \bigcup\{\operatorname{supp}\mu_i : i \in \mathbb{N}_n\}$. Set

$$\mu = \mu_1 + \cdots + \mu_n \in N(K)^+ .$$

Then $L = \operatorname{supp}(\mu \mid L)$ and $\mu \mid L \in N(K)^+$.

(iv) Set $K_1 = \overline{W_K}$, so that K_1 is clopen, and set $K_2 = K \setminus K_1$. Then K_1 and K_2 have the required properties. □

The seminal paper of Dixmier [91] contains the following decomposition theorem, which extends clause (iv) of the above proposition; see also [234, Theorem III.1.17]. Earlier expositions were given in [23, §8, Theorem 10] and [24, Theorem 8.10]. We shall not use this theorem.

Theorem 5.1.3. *Let K be a Stonean space. Then there are three pairwise-disjoint, clopen subsets K_1, K_2, K_3 of K such that:*

(i) *K_1 is hyper-Stonean;*

(ii) *K_2 contains a dense, meagre subset, and $N(K_2) = \{0\}$;*

(ii) *in K_3, every meagre subset is nowhere dense, and the support of every measure on K_3 is nowhere dense, so that $N(K_3) = \{0\}$.* □

Example 5.1.4. (i) Let S be an infinite set, so that βS is hyper-Stonean. As in Example 1.7.14, the closed subspace S^* of βS is not even basically disconnected, and hence we cannot replace 'clopen' by 'closed' in clause (i) of Proposition 5.1.2.

(ii) The Gleason cover, $G_{\mathbb{I}}$, of $\mathbb{I}$ is Stonean by Theorem 1.6.5, but $N(G_{\mathbb{I}}) = \{0\}$ by Corollary 4.7.21, and hence $G_{\mathbb{I}}$ is not hyper-Stonean. □

Proposition 5.1.5. *Let K be a non-empty, compact F-space, and suppose that $\mu \in N(K) \cap P(K)$. Then* supp μ *contains a non-empty, clopen, hyper-Stonean subspace of K.*

Proof. Set $S = \operatorname{supp} \mu$ and $U = \operatorname{int}_K S$. By Corollary 4.7.5, (a) $\Rightarrow$ (b), $|\mu|(S \setminus U) = 0$, and so $U \neq \emptyset$. Also $|\mu|(K \setminus \overline{U}) = 0$, and hence $(K \setminus \overline{U}) \cap S = \emptyset$. Thus $\overline{U} = S$. By Proposition 4.7.9, S is Stonean, and so U is extremely disconnected. Thus there is a non-empty, clopen subspace V of U. Let $\nu = |\mu| \mid V$. Then $\nu \in N(V)$ and $\operatorname{supp} \nu = V$, and so the compact space V is hyper-Stonean. □

The terminology of the next definition is taken from [50]; the term 'perfect measure' is used in [31].

Definition 5.1.6. A positive measure μ on the Borel sets of a Stonean space K is a *category measure* if μ is regular on closed subsets of finite measure, if every non-empty, clopen set in K contains a clopen set A_0 with $0 < \mu(A_0) < \infty$, and if every nowhere dense Borel set has measure zero.

Proposition 5.1.7. *Let K be a Stonean space. Then K is hyper-Stonean if and only if there exists a category measure on K.*

Proof. Suppose that K is hyper-Stonean. Consider a maximal family $\{\mu_\alpha\}$ of measures in $N(K)^+$ with pairwise-disjoint supports, and set $\mu = \sum_\alpha \mu_\alpha$, so that μ is a positive measure on $\mathfrak{B}_K$. Take A to be a clopen subset of K. Then

$$A_0 := A \cap \operatorname{supp} \mu_{\alpha_0} \neq \emptyset$$

for some α_0 because of the maximality of the family $\{\mu_\alpha\}$ and the assumption that K is hyper-Stonean. Since K is Stonean, supp μ_{α_0} is clopen by Corollary 4.7.10(i), and so A_0 is a clopen subset of A with

$$0 < \mu(A_0) = \mu_{\alpha_0}(A_0) < \infty .$$

Clearly $\mu(B) = 0$ for each nowhere dense Borel set B because $\mu_\alpha(B) = 0$ for each such B and each α. Thus μ is a category measure.

Conversely, suppose that μ is a category measure on K. For an arbitrary clopen set A in K, take some clopen $A_0 \subset A$ with $0 < \mu(A_0) < \infty$, and set $\mu_A = \mu \mid A_0$. By Theorem 4.7.4(i), $\mu_A \in N(K)^+$ and supp $\mu_A \subset A$. Since A was arbitrary, K is hyper-Stonean. □

Let K be a Stonean space such that $N(K) \neq \{0\}$, and take $\mathscr{F}$ to be a maximal singular family in $N(K) \cap P(K)$, say

$$\mathscr{F} = \{\mu_\alpha : \alpha \in A\},$$

where the measures μ_α are distinct. For each $\alpha \in A$, set $S_\alpha = \operatorname{supp} \mu_\alpha$, so that, by Corollary 4.7.10(i), each S_α is Stonean, and hence, by Corollary 4.7.6, Φ_{μ_α} is homeomorphic to S_α. By Corollary 4.7.11(iii), $\{S_\alpha : \alpha \in A\}$ is a pairwise-disjoint family of clopen subsets of K. As in Definition 4.6.4, we set

$$U_{\mathscr{F}} = \bigcup\{S_\alpha : \alpha \in A\} = \bigcup\{\operatorname{supp} \mu_\alpha : \alpha \in A\} .$$

Then $U_{\mathscr{F}}$ is an open subset of K. In the case where K is hyper-Stonean, $U_{\mathscr{F}}$ is dense in K, and so, by Corollary 1.5.8, $\beta U_{\mathscr{F}} = K$.

The following result is now clear.

Theorem 5.1.8. *Let K be a hyper-Stonean space, and let $\mathscr{F} = \{\mu_\alpha : \alpha \in A\}$ be a maximal singular family in $N(K) \cap P(K)$. Then the map*

$$f \mapsto (f \mid S_\alpha), \quad C(K) \to \bigoplus_\infty \{L^\infty(S_\alpha, \mu_\alpha) : \alpha \in A\} = \bigoplus_\infty \{C(S_\alpha) : \alpha \in A\},$$

is a unital C^*-isomorphism, where $S_\alpha = \operatorname{supp} \mu_\alpha$ $(\alpha \in A)$.

Proof. Since K is Stonean and $U_{\mathscr{F}}$ is dense in K, the map

$$f \mapsto f \mid U_{\mathscr{F}}, \quad C(K) \to C^b(U_{\mathscr{F}}),$$

is a unital C^*-isomorphism. The map

$$g \mapsto (g \mid S_\alpha), \quad C^b(U_{\mathscr{F}}) \to \bigoplus_\infty C(S_\alpha),$$

is clearly a unital C^*-isomorphism. For each $\alpha \in A$, the measure μ_α is normal, and so $L^\infty(S_\alpha, \mu_\alpha) = C(S_\alpha)$ by Corollary 4.7.6. Thus the result follows. □

It may seem strange that, in Definition 5.1.1, which defined a 'hyper-Stonean space' K, we required in advance that K be Stonean. However, without this constraint, the key theorem, Theorem 6.4.1, to be given in the next chapter, would not hold. In fact, it is quite easy to find compact spaces K for which the union of the supports of the normal measures on K is dense, but K is not Stonean; a characterization of the compact spaces K with this property will be given in Corollary 5.1.11, but first we offer the following specific examples.

Example 5.1.9. (i) Take L_1 and L_2 to be two hyper-Stonean spaces, for example, take $L_1 = L_2 = \beta\mathbb{N}$, and form their disjoint union as a topological space. Then identify a non-isolated point in L_1 with a non-isolated point in L_2, and call the quotient space K, so that K is compact. The union of the supports of the normal measures on K is dense in K.

In the particular case where $L_1 = L_2 = \beta\mathbb{N}$, the isolated points in K are dense. The two copies of $\mathbb{N}$ in K are disjoint, countable, open sets whose closures have non-empty intersection, and so K is not even basically disconnected and K is not an F-space.

(ii) Let K be the one-point compactification of the disjoint union of an infinite family of hyper-Stonean spaces without isolated points. (The space $\mathbb{H}$, to be discussed in §5.3, is hyper-Stonean spaces without isolated points; see Proposition 5.3.3(i).) Then K is totally disconnected, but not an F-space. The union of the supports of the normal measures on K is obviously dense in K.

In fact, c is a clearly complemented subspace of $C(K)$, and so, by Proposition 2.5.7, $C(K)$ is not injective; in particular, since K is not an F-space, K is not Stonean.

(iii) In Example 6.5.5, below, we shall exhibit a compact F-space which is not Stonean, but is such that the union of the supports of the normal measures is dense in K. □

Theorem 5.1.10. *Suppose that K and L are two non-empty, compact spaces such that $\mathfrak{R}_K$ and $\mathfrak{R}_L$ are isomorphic Boolean algebras. Then W_K is dense in K if and only if W_L is dense in L.*

Proof. Let $\rho : \mathfrak{R}_L \to \mathfrak{R}_K$ be an isomorphism.

Suppose that W_K is dense in K, and take $U \in \mathfrak{R}_L \setminus \{\emptyset\}$. Since $\rho(U) \in \mathfrak{R}_K \setminus \{\emptyset\}$, there is a normal measure $\mu \geq 0$ on $\mathfrak{R}_K$ with $\mu(\rho(U)) > 0$. Then $\mu \circ \rho$ is a normal measure on $\mathfrak{R}_L$ with $(\mu \circ \rho)(U) > 0$. It follows from Theorem 4.7.27 that $\mu \circ \rho = R\nu$ for some $\nu \in N(L)$ and that $\nu(U) > 0$, where R is as in Theorem 4.7.27. Thus $U \cap \operatorname{supp} \nu \neq \emptyset$. This proves that W_L is dense in L. □

Corollary 5.1.11. *Let K be a non-empty, compact space. Then the union of the supports of the normal measures on K is dense in K if and only if the Gleason cover of K is hyper-Stonean.*

Proof. This follows from Theorem 5.1.10 because G_K is Stonean, and hence it is hyper-Stonean if and only if the union of the supports of the normal measures on G_K is dense in G_K. □

5.2 Some commutative C^*-algebras

We shall now introduce some particular commutative C^*-algebras. These examples will arise in greater generality in a main theorem, Theorem 6.4.1, and so the present results can be seen as a precursor of that theorem.

Let K be a non-empty, locally compact space, and take $\mu \in P(K)$. Recall from Definition 4.4.5 that the compact character space of the unital C^*-algebra $L^\infty(K,\mu)$ is denoted by Φ_μ and that $\mathscr{G}_\mu : L^\infty(K,\mu) \to C(\Phi_\mu)$ is a C^*-isomorphism, where $\mathscr{G}_\mu$ is the corresponding Gel'fand transform; as mentioned on page 130, Φ_μ is a Stonean space.

Note that, for each $\lambda \in L^\infty(\mu)$, the continuous linear functional $\lambda\mu$ on $C_0(K)$, defined by

$$\langle f, \lambda\mu \rangle = \int_K \lambda f \, \mathrm{d}\mu \quad (f \in C_0(K)),$$

is a measure on K, with $\|\lambda\mu\| \leq \|\lambda\|_\infty \|\mu\|$.

Since $L^\infty(\mu) = L^1(\mu)'$, there is a canonical embedding

$$\kappa_\mu : L^1(\mu) \to L^1(\mu)'' = L^\infty(\mu)' = C(\Phi_\mu)' = M(\Phi_\mu). \tag{5.1}$$

The measure μ defines a (well-defined) positive, continuous linear functional, called $\widehat{\mu}$, on $L^\infty(\mu)$. Indeed, essentially as in equation (4.3),

$$\widehat{\mu}(\lambda) = \int_K \lambda \, \mathrm{d}\mu \quad (\lambda \in L^\infty(\mu)).$$

Hence, we may regard $\widehat{\mu}$ as an element of $M(\Phi_\mu)^+$.

For each $B \in \mathfrak{B}_K$, the function $\chi_B \in B^b(K)$ (or, more precisely, the equivalence class $[\chi_B]$) is an idempotent in the C^*-algebra $L^\infty(\mu)$, and so $\mathscr{G}_\mu(\chi_B)$ is an idempotent in $C(\Phi_\mu)$, and hence it takes values in $\{0,1\}$; we set

$$K_{B,\mu} = \{\varphi \in \Phi_\mu : \mathscr{G}_\mu(\chi_B)(\varphi) = 1\}, \tag{5.2}$$

so that $\{K_{B,\mu} : B \in \mathfrak{B}_K\} = \mathfrak{C}_{\Phi_\mu}$. In particular, suppose that $B \in \mathfrak{B}_K$ and $\mu(B) = 0$. Then $K_{B,\mu} = \emptyset$. It is clear that the family $\{K_{B,\mu} : B \in \mathfrak{B}_K\}$ is a base for the topology of Φ_μ.

Let $B, C \in \mathfrak{B}_K$. Then

$$\chi_{B\cap C} = \chi_B \cdot \chi_C \quad \text{and} \quad \chi_{B\cup C} = \chi_B + \chi_C - \chi_B \cdot \chi_C\,,$$

and so

$$K_{B,\mu} \cap K_{C,\mu} = K_{B\cap C,\mu} \quad \text{and} \quad K_{B,\mu} \cup K_{C,\mu} = K_{B\cup C,\mu}\,. \tag{5.3}$$

In particular, $K_{B,\mu} \cap K_{C,\mu} = \emptyset$ if and only if $\mu(B\cap C) = 0$.

For each $B \in \mathfrak{B}_K$, we have

$$\kappa_\mu(\chi_B)(K_{B,\mu}) = \mu(B) \quad \text{and} \quad \mathscr{G}_\mu(\chi_B) = \chi_{K_{B,\mu}}\,. \tag{5.4}$$

Further,

$$\widehat{\mu}(K_{B,\mu}) = \langle \chi_{K_{B,\mu}}, \widehat{\mu} \rangle = \langle \mathscr{G}_\mu(\chi_B), \widehat{\mu} \rangle = \langle \chi_B, \mu \rangle = \mu(B), \tag{5.5}$$

so that $\widehat{\mu}(\Phi_\mu) = \mu(K)$, supp $\widehat{\mu} = \Phi_\mu$, and $\widehat{\mu} \in P(\Phi_\mu)$. It follows that the range of the map $\kappa_\mu : L^1(\mu) \to M(\Phi_\mu)$ is exactly $L^1(\Phi_\mu, \widehat{\mu})$.

Theorem 5.2.1. *Let K be a non-empty, locally compact space, and take $\mu \in P(K)$. Then Φ_μ and $St(\mathfrak{B}_\mu)$ are homeomorphic.*

Proof. Take $\varphi \in \Phi_\mu$, and set

$$p_\varphi = \{B \in \mathfrak{B}_K : \varphi([\chi_B]) = 1\} = \{B \in \mathfrak{B}_K : \varphi \in K_{B,\mu}\}.$$

Then p_φ is a filter on the Boolean algebra $\mathfrak{B}_\mu$, and it is an ultrafilter because, for each $B \in \mathfrak{B}_K$, either $B \in p_\varphi$ or $K \setminus B \in p_\varphi$. Thus p_φ is a point of $St(\mathfrak{B}_\mu)$.

Conversely, given an ultrafilter $p \in St(\mathfrak{B}_\mu)$, define φ_p by setting $\varphi_p([\chi_B]) = 1$ for a Borel set B if $B \in p$ and $\varphi_p([\chi_B]) = 0$ if $K \setminus B \in p$ and then extending φ_p by linearity and continuity to $L^\infty(\mu)$; it is clear that $\varphi_p \in \Phi_\mu$.

The two correspondences given above are inverses of each other, and the map

$$\varphi \to p_\varphi, \quad \Phi_\mu \to St(\mathfrak{B}_\mu),$$

is a homeomorphism with respect to the specified topologies on the two spaces Φ_μ and $St(\mathfrak{B}_\mu)$. □

In the next two results, we shall write σ for the weak* topology $\sigma(L^\infty(\mu), L^1(\mu))$ on $L^\infty(\mu)$.

Lemma 5.2.2. *Let K be a non-empty, locally compact space, and suppose that $\mu \in P(K)$. Consider a decreasing net $(\lambda_\alpha : \alpha \in A)$ in $L^\infty(\mu)^+$. Then $\mathscr{G}_\mu(\lambda_\alpha) \searrow 0$ in $(C(\Phi_\mu), \leq)$ if and only if $\lambda_\alpha \to 0$ in $(L^\infty(\mu), \sigma)$.*

Proof. Set $\Lambda_\alpha = \mathscr{G}_\mu(\lambda_\alpha)$ $(\alpha \in A)$, so that $(\Lambda_\alpha : \alpha \in A)$ is a decreasing net in $C(\Phi_\mu)^+$.

Suppose that $\lambda_\alpha \to 0$ in $(L^\infty(\mu), \sigma)$, and assume towards a contradiction that $\bigwedge_\alpha \Lambda_\alpha \neq 0$. Then there exist a Borel set $B \in \mathfrak{B}_K$ with $\mu(B) > 0$ and $\delta > 0$ such that $\delta\chi_{K_{B,\mu}} \leq \Lambda_\alpha$ $(\alpha \in A)$. But now

$$0 < \delta\mu(B) = \langle \chi_B, \delta\chi_B \rangle \leq \langle \chi_B, \lambda_\alpha \rangle \quad (\alpha \in A),$$

a contradiction of the fact that $\lim_\alpha \langle \chi_B, \lambda_\alpha \rangle = 0$. Hence, $\bigwedge_\alpha \Lambda_\alpha = 0$, and so $\Lambda_\alpha \searrow 0$ in $(C(\Phi_\mu), \leq)$.

Suppose that $\Lambda_\alpha \searrow 0$ in $(C(\Phi_\mu), \leq)$. We can suppose that $\|\lambda_\alpha\|_\infty \leq 1$, and so there is a subnet (λ_{α_β}) such that $\lim_\beta \lambda_{\alpha_\beta} = \lambda_0$ in $(L^\infty(\mu), \sigma)$ for some $\lambda_0 \in L^\infty(\mu)^+$. For each $f \in L^1(\mu)^+$, the net $(\langle f, \lambda_\alpha \rangle)$ decreases and a subnet has limit $\langle f, \lambda_0 \rangle$, and so $\langle f, \lambda_\alpha \rangle \geq \langle f, \lambda_0 \rangle$ $(\alpha \in A)$, whence $\lambda_\alpha \geq \lambda_0$ $(\alpha \in A)$. Set $\Lambda_0 = \mathscr{G}_\mu(\lambda_0) \in C(\Phi_\mu)$, so that $\Lambda_\alpha \geq \Lambda_0 \geq 0$ $(\alpha \in A)$. Assume towards a contradiction that $\lambda_0 \neq 0$. Then $\Lambda_0 \neq 0$ and $\bigwedge_\alpha \Lambda_\alpha \geq \Lambda_0$, a contradiction. Thus $\lambda_\alpha \to 0$ in $(L^\infty(\mu), \sigma)$. □

Theorem 5.2.3. *Let K be a non-empty, locally compact space, and suppose that $\mu \in P(K)$. Then $N(\Phi_\mu) = \kappa_\mu(L^1(\mu)) = L^1(\Phi_\mu, \widehat{\mu})$ and Φ_μ is a hyper-Stonean space.*

Proof. By a previous remark, $\kappa_\mu(L^1(\mu)) = L^1(\Phi_\mu, \widehat{\mu})$, and so we must show that $L^1(\Phi_\mu, \widehat{\mu}) = N(\Phi_\mu)$.

Take $f \in L^1(\mu)$, and suppose that $\Lambda_\alpha \searrow 0$ in $(C(\Phi_\mu), \leq)$, say $\Lambda_\alpha = \mathscr{G}_\mu(\lambda_\alpha)$, where $\lambda_\alpha \searrow 0$ in $L^\infty(\mu)$. By Lemma 5.2.2, $\lambda_\alpha \to 0$ in $(L^\infty(\mu), \sigma)$, and so

$$\langle \kappa_\mu(f), \Lambda_\alpha \rangle = \langle f, \lambda_\alpha \rangle \to 0,$$

which shows that $\kappa_\mu(f) \in N(\Phi_\mu)$. Hence, $L^1(\Phi_\mu, \widehat{\mu}) \subset N(\Phi_\mu)$.

Now suppose that $\mathrm{M} \in N(\Phi_\mu)$. We first *claim* that $\mathrm{M} \ll \widehat{\mu}$. Indeed, take $E \in \mathfrak{B}_{\Phi_\mu}$ with $\widehat{\mu}(E) = 0$. Let L be a compact subset of E, and assume that $\mathrm{int}_{\Phi_\mu} L \neq \emptyset$. Then there exists $B \in \mathfrak{B}_\mu$ such that $\mu(B) > 0$ and $K_{B,\mu} \subset L$, and so $\widehat{\mu}(E) > 0$, a contradiction. Thus $L \in \mathscr{K}_{\Phi_\mu}$, and hence $\mathrm{M}(L) = 0$ by Theorem 4.7.4(i), giving the claim.

By the Radon–Nikodým theorem, Theorem 4.4.9, $\mathrm{M} \in L^1(\Phi_\mu, \widehat{\mu})$, and hence $N(\Phi_\mu) \subset L^1(\Phi_\mu, \widehat{\mu})$.

Since $\widehat{\mu}$ is a normal measure on Φ_μ and $\mathrm{supp}\,\widehat{\mu} = \Phi_\mu$, the space Φ_μ is certainly hyper-Stonean. □

Let K be a non-empty, locally compact space, and take $\mu \in P(K)$. It is clear that, for each $x \in K_\infty$ such that $\mu(U) > 0$ for each $U \in \mathscr{N}_x$, there exists $\varphi \in \Phi_\mu$ with $\mathscr{N}_x \subset \varphi$. In particular, for each $x \in \mathrm{supp}\,\mu$, there exists $\varphi \in \Phi_\mu$ with $\mathscr{N}_x \subset \varphi$. It is also clear that, for each $\varphi \in \Phi_\mu$, there exists a unique point $x \in \mathrm{supp}\,\mu \cup \{\infty\}$ with $\mathscr{N}_x \subset \varphi$. Thus we can define a map

$$\pi_\mu : \varphi \mapsto x, \quad \Phi_\mu \to K_\infty, \tag{5.6}$$

such that $\mathrm{supp}\,\mu \subset \pi_\mu(\Phi_\mu)$. We see from the definition of the topology on the Stone space Φ_μ that π_μ is continuous. This gives the next result.

Proposition 5.2.4. *Let K be a non-empty, locally compact space, and suppose that $\mu \in P(K)$. Then $|\Phi_\mu| \geq |\mathrm{supp}\,\mu|$.* □

Proposition 5.2.5. *Let K be an infinite, locally compact space, and suppose that $\mu \in P(K)$. Then:*

(i) *$w(\Phi_\mu) = \kappa$ and $|\mathrm{supp}\,\mu| \leq |\Phi_\mu| \leq 2^\kappa$, where $\kappa = |\mathfrak{B}_\mu|$;*
(ii) *Φ_μ satisfies* CCC;
(iii) *Φ_μ has no isolated points if and only if μ is continuous.*

Proof. (i) Since the family $\{K_{B,\mu} : B \in \mathfrak{B}_\mu\}$ is a base for Φ_μ, we have $w(\Phi_\mu) \leq \kappa$, and so $|\Phi_\mu| \leq 2^\kappa$. Let $\mathscr{B}$ be a base for Φ_μ. Then each set $K_{B,\mu}$ for $B \in \mathfrak{B}_\mu$ is a finite union of sets in $\mathscr{B}$, and so $|\mathscr{B}| \geq \kappa$. Thus $w(\Phi_\mu) = \kappa$.

By Proposition 5.2.4, $|\Phi_\mu| \geq |\mathrm{supp}\,\mu|$.

(ii) Let $\{U_i : i \in I\}$ be a pairwise-disjoint family of non-empty, open subsets of Φ_μ. For each $i \in I$, choose a non-empty, clopen set $K_i \subset U_i$. Then there exists

$B_i \in \mathfrak{B}_K \setminus \mathfrak{N}_\mu$ with $K_{B_i,\mu} = K_i$. We see using (5.3) that $\mu(B_i) > 0$ $(i \in I)$ and that $\mu(B_i \cap B_j) = 0$ $(i, j \in I, i \neq j)$. Thus the index set I is countable. This shows that Φ_μ satisfies CCC.

(iii) Suppose that μ is not continuous, so that there exists $x \in K$ with $\mu(\{x\}) > 0$. Then $\varphi := \{B \in \mathfrak{B}_K : x \in B\}$ is an ultrafilter in $St(\mathfrak{B}_\mu)$, and clearly φ is an isolated point of Φ_μ.

Conversely, suppose that φ is an isolated point of Φ_μ. Then there exists $B \in \mathfrak{B}_K$ with $\mu(B) > 0$ such that $\{\psi \in \Phi_\mu : B \in \psi\} = \{\varphi\}$. Since μ is regular, we may suppose that B is compact. Thus there is a unique point $x \in B$ such that

$$\mu(U \cap B) = \mu(B) \quad (U \in \mathscr{N}_x).$$

Clearly $\mu(\{x\}) = \mu(B) > 0$, and so μ is not continuous. □

Corollary 5.2.6. *Let K be a hyper-Stonean space. Then there exists a measure $\mu \in N(K) \cap P(K)$ with $\operatorname{supp} \mu = K$ such that K is homeomorphic to Φ_μ if and only if K satisfies* CCC.

Proof. Take $\mu \in P(K)$ with $\operatorname{supp} \mu = K$. By Proposition 5.2.5(ii), Φ_μ satisfies CCC, and so K satisfies CCC whenever K is homeomorphic to Φ_μ.

Conversely, suppose that K satisfies CCC, and take $\mathscr{N}$ to be a maximal singular family in $N(K) \cap P(K)$. By Corollaries 4.7.10(i) and 4.7.11(iii), the supports of the measures in $\mathscr{N}$ are clopen and pairwise disjoint, and so $\mathscr{N}$ is countable, and hence can be enumerated as (μ_n), say. Define $\mu = \sum_{n=1}^{\infty} \mu_n / 2^n$. Then $\mu \in N(K) \cap P(K)$, and $\operatorname{supp} \mu = K$ because K is hyper-Stonean. By Corollary 4.7.6, K and Φ_μ are homeomorphic. □

Proposition 5.2.7. *Let K be a non-empty, locally compact space, and let $\mathscr{F}$ and $\mathscr{G}$ be two infinite, maximal singular families in $P_c(K)$. Then $|\mathscr{F}| = |\mathscr{G}|$.*

Proof. Suppose that $\mathscr{F} = \{\mu_i : i \in I\}$ and $\mathscr{G} = \{\nu_j : j \in J\}$, where $\mu_i, \nu_j \in M_c(K)^+$. We *claim* that $|I| = |J|$.

Assume towards a contradiction that $|I| < |J|$. For each $i \in I$, consider the set

$$H_i = \{j \in J : \Phi_{\nu_j} \cap \Phi_{\mu_i} \neq \emptyset\}.$$

By Proposition 5.2.5(ii), Φ_{μ_i} satisfies CCC, and so $|H_i| \leq \aleph_0$. Also $\bigcup\{H_i : i \in I\} = J$ because $\mathscr{F}$ is a maximal family in $P_c(K)$. Thus $|J| \leq \aleph_0 \cdot |I| = |I|$, a contradiction.

We conclude that $|I| = |J|$. □

Corollary 5.2.8. *Let K be an uncountable, compact, metrizable space. Then every maximal singular family in $P(K)$ that contains all the point masses consists of $\mathfrak{c}$ point masses and $\mathfrak{c}$ continuous measures.* □

5.3 The hyper-Stonean space of the unit interval

We now describe a particular instance of the above situation.

Definition 5.3.1. The character space of the commutative, unital C^*-algebra $L^\infty(\mathbb{I})$ is the *hyper-Stonean space of the unit interval*, and is denoted by $\mathbb{H}$.

Thus $L^\infty(\mathbb{I}) = C(\mathbb{H})$. In the notation of Definition 4.4.5, we have $\mathbb{H} = \Phi_m$, where $m \in P(\mathbb{I})$ is Lebesgue measure. The space $\mathbb{H}$ was earlier defined in [72, Definition 2.18]; see also [106, A7H].

Theorem 5.3.2. *Let K be a non-empty, locally compact space, and take $\mu \in P_c(K)$. Suppose that $(L^1(K,\mu), \|\cdot\|_1)$ is separable. Then the Stonean spaces Φ_μ and $\mathbb{H}$ are homeomorphic.*

Proof. By Theorem 4.3.6, the Boolean algebras $\mathfrak{B}_\mu$ and $\mathfrak{B}_m$ are isomorphic, and so $St(\mathfrak{B}_\mu)$ and $St(\mathfrak{B}_m)$ are homeomorphic. By Theorem 5.2.1, $St(\mathfrak{B}_\mu)$ and $St(\mathfrak{B}_m)$ are homeomorphic to Φ_μ and $\mathbb{H}$, respectively. Hence the topological spaces Φ_μ and $\mathbb{H}$ are homeomorphic. □

Proposition 5.3.3. (i) *$\mathbb{H}$ is a hyper-Stonean space, $\mathbb{H}$ satisfies* CCC *and has no isolated points, and $N(\mathbb{H}) \cong L^1(\mathbb{I})$, so that $N(\mathbb{H})$ is separable;*

(ii) *$|\mathbb{H}| = 2^{\mathfrak{c}}$ and $w(\mathbb{H}) = \mathfrak{c}$;*

(iii) *the space $C(\mathbb{H})_{[1]}$ is metrizable in the weak* topology $\sigma(L^\infty(\mathbb{I}), L^1(\mathbb{I}))$.*

Proof. (i) These are special cases of parts of Theorem 5.2.3 and Proposition 5.2.5.

(ii) By Corollary 1.4.15, $|\mathfrak{B}_\mathbb{I}| = \mathfrak{c}$, and so, by Proposition 5.2.5(i), $w(\mathbb{H}) = \mathfrak{c}$ and $|\mathbb{H}| \leq 2^{\mathfrak{c}}$.

For $n \in \mathbb{N}$, set $F_n = [t_{2n+1}, t_{2n}]$, where (t_n) is a sequence in $\mathbb{I}$ such that $t_n \searrow 0$. For each $S \subset \mathbb{N}$, set $B_S = \bigcup\{F_n : n \in S\}$, and, for each $p \in \mathbb{N}^*$, set

$$C_p = \bigcap\{K_{B_S} : S \in p\}.$$

Then C_p is a non-empty, closed subset of Φ_μ, and $C_p \cap C_q = \emptyset$ whenever p and q are distinct points of $\mathbb{N}^*$. By Proposition 1.5.4, $|\mathbb{N}^*| = 2^{\mathfrak{c}}$, and so it follows that $|\mathbb{H}| \geq 2^{\mathfrak{c}}$.

(iii) This follows from Theorem 2.1.4(iii) because $L^1(\mathbb{I})$ is separable. □

The space $\mathbb{H}$ was characterized topologically in [72, Corollary 2.22] as follows.

Theorem 5.3.4. *A topological space K is homeomorphic to $\mathbb{H}$ if and only if K has the following properties:*

(i) *K is a hyper-Stonean space;*

(ii) *K has no isolated points;*

(iii) *the space $N(K)$ is separable.*

Proof. By Proposition 5.3.3(i), the space $\mathbb{H}$ has the specified properties.

Now suppose that K is a topological space with the specified properties. By Proposition 5.1.2(ii), K satisfies CCC, and so, by Corollary 5.2.6, there is a measure $\mu \in N(K) \cap P(K)$ with $\operatorname{supp} \mu = K$ such that K is homeomorphic to Φ_μ. By (ii), $\operatorname{supp} \mu$ has no isolated points, and so, by Proposition 5.2.5(iii), $\mu \in N_c(K)$. By Corollary 4.7.6, $L^\infty(K,\mu) = C(K)$, and, by Theorem 5.2.3, $N(K) \cong L^1(K,\mu)$, so that $N(K)' = C(K)$. By (iii), $L^1(K,\mu)$ is separable, and so it follows from Theorem 5.3.2 that K and $\mathbb{H}$ are homeomorphic. □

Example 5.3.5. We note that clause (iii) of the above characterization of $\mathbb{H}$ is necessary: there are a compact space L and a measure $\mu \in M_c(L)^+$ such that $K = \Phi_\mu$ is a hyper-Stonean space with no isolated points, but such that $N(K)$ is not separable.

Indeed, set $L = \mathbb{Z}_2^{\mathfrak{c}}$, the Cantor cube of weight $\mathfrak{c}$; this space was described on page 16. Let m_L be the Haar measure on L, so that $m_L \in P_c(L)$, and then set $K = \Phi_{m_L}$, so that K is a hyper-Stonean space that has no isolated points and satisfies CCC.

Take $B \in \mathfrak{B}_L$. Since m_L is regular and the space L is totally disconnected, for each $n \in \mathbb{N}$, there exists $C_n \in \mathfrak{C}_L$ with $m_L(B \,\Delta\, C_n) < 1/n$. Hence there exists $C \in \sigma(\mathfrak{C}_L)$, the σ-algebra generated by $\mathfrak{C}_L$, with $m_L(B \,\Delta\, C) = 0$. By equation (1.5), $|\mathfrak{C}_L| = \mathfrak{c}$; as on page 4, $|\sigma(\mathfrak{C}_L)| = \mathfrak{c}$. Hence $|\mathfrak{B}_{m_L}| \leq \mathfrak{c}$. For each $\sigma < \mathfrak{c}$, set $B_\sigma = \{\varepsilon = (\varepsilon_\tau) \in L : \varepsilon_\sigma = 0\}$, a clopen set in L. Then

$$m_L(B_\sigma \,\Delta\, B_\tau) = \|\chi_{B_\sigma} - \chi_{B_\tau}\|_1 = \frac{1}{2} \quad (\sigma, \tau < \mathfrak{c}, \sigma \neq \tau). \tag{5.7}$$

This shows that $|\mathfrak{B}_{m_L}| \geq \mathfrak{c}$, and so $|\mathfrak{B}_{m_L}| = \mathfrak{c}$. Further, by (5.7), the Banach space $(L^1(L,m_L), \|\cdot\|_1)$ is not separable. By Theorem 5.2.3, $N(K) \cong L^1(L,m_L)$, and so $N(K)$ is not separable.

We note that, by Proposition 5.2.5(i), $w(K) = \mathfrak{c}$ and $2^{\mathfrak{c}} = |L| \leq |K| \leq 2^{\mathfrak{c}}$, and hence $w(K) = w(\mathbb{H})$ and $|K| = 2^{\mathfrak{c}} = |\mathbb{H}|$. By Theorem 2.1.7(ii), $d(C(K)) = \mathfrak{c}$. It follows that no conditions on the cardinality and weight of K can rescue the uniqueness statement in Theorem 5.3.4 in the absence of clause (iii). □

5.4 The bidual of $C_0(K)$

Let K be a non-empty, locally compact space. We shall now obtain a representation of the bidual space $C_0(K)''$ by applying Theorem 2.2.30.

Indeed, to apply this theorem, we take E to be the Banach space $C_0(K)$, so that $E' = M(K)$. The subset S of $S_{E'}$ is taken to be a maximal singular family $\mathscr{F}$ in $P(K)$, as in §4.6, and the closed subspace of $M(K)$ corresponding to $\mu \in \mathscr{F}$ is $F_\mu = L^1(K,\mu)$, so that $F'_\mu = L^\infty(K,\mu)$, a commutative, unital C^*-algebra identified with $C(\Phi_\mu)$. Let $U_{\mathscr{F}}$ be as in Definition 4.6.4. Thus, as in equation (2.4),

$$F = \bigoplus_1 \{L^1(K,\mu) : \mu \in \mathscr{F}\}$$

and

$$F' = \bigoplus_{\infty}\{L^{\infty}(K,\mu) : \mu \in \mathscr{F}\} = \bigoplus_{\infty}\{C(\Phi_\mu) : \mu \in \mathscr{F}\}$$

is a commutative, unital C^*-algebra. By Theorem 4.6.5, equation (2.3) is satisfied in the present situation.

Let $\mu \in P(K)$. Suppose that $f_\alpha \to 0$ in $L^{\infty}(K,\mu)$ and that $g \in L^{\infty}(K,\mu)$. Then

$$\int_K f_\alpha \lambda \,\mathrm{d}\mu \to 0 \quad (\lambda \in L^1(K,\mu)).$$

Since $g\lambda \in L^1(\mu)$ $(\lambda \in L^1(K,\mu))$ and

$$\int_K (gf_\alpha)\lambda \,\mathrm{d}\mu = \int_K f_\alpha(g\lambda)\,\mathrm{d}\mu \to 0 \quad (\lambda \in L^1(K,\mu)),$$

it follows that the product $\cdot$ in F' is separately continuous with respect to the weak* topology, $\sigma(F',F)$. Also, the involution $*$ is similarly continuous on F'.

Theorem 5.4.1. *Let K be a non-empty, locally compact space, and suppose that $\mathscr{F}$ is a maximal singular family in $P(K)$. Then $(C_0(K)'',\Box)$ is a commutative, unital C^*-algebra, and the map*

$$T : \Lambda \mapsto (\Lambda \mid L^1(K,\mu) : \mu \in \mathscr{F}), \quad (C_0(K)'',\Box) \to (F',\cdot),$$

where

$$F' = \bigoplus_{\infty}\{L^{\infty}(K,\mu) : \mu \in \mathscr{F}\} = \bigoplus_{\infty}\{C(\Phi_\mu) : \mu \in \mathscr{F}\} \cong C^b(U_{\mathscr{F}}) \cong C(\beta U_{\mathscr{F}}),$$

is a C^*-algebra isomorphism. Further, $C_0(K)$ is Arens regular, the extended linear involution $*$ on $C_0(K)''$ is an involution, and $(C_0(K)'',\Box,*)$ is a commutative, unital C^*-algebra.

Proof. As in Theorems 2.2.30 and 4.6.5, the map T is a $*$-linear isometry, and T is weak*-weak*-continuous.

Now take $\mathrm{M},\mathrm{N} \in C_0(K)''$, say $\mathrm{M} = \lim_\alpha f_\alpha$ and $\mathrm{N} = \lim_\beta g_\beta$ for nets (f_α) and (g_β) in $C_0(K)$, where the limits are taken in the weak* topology. Then

$$\begin{aligned} T(\mathrm{M}\Box\mathrm{N}) &= T(\lim_\alpha \lim_\beta f_\alpha g_\beta) = \lim_\alpha \lim_\beta T(f_\alpha g_\beta) \\ &= \lim_\alpha \lim_\beta T(f_\alpha)T(g_\beta) = T(\mathrm{M}) \cdot T(\mathrm{N}). \end{aligned}$$

Similarly, $T(\mathrm{M}\Diamond\mathrm{N}) = T(\mathrm{M}) \cdot T(\mathrm{N})$ for $\mathrm{M},\mathrm{N} \in C_0(K)''$, and so $C_0(K)$ is Arens regular and $T : (C_0(K)'',\Box) \to (F',\cdot)$ is an algebra isomorphism.

It follows from Theorem 3.1.10 that the linear involution $*$ on $(C_0(K)'',\Box)$ is an involution. □

The above representation of $C_0(K)''$ is related to the theory of generalized functions of Wong [244], which is itself based on an earlier discussion of a special case in Šreĭder [227]; see also [120]. In the terminology of [244], an element $(f_\mu : \mu \in P(K))$ in the space

$$\bigoplus_\infty \{L^\infty(K,\mu) : \mu \in P(K)\}$$

is a *generalized function* if $f_\nu = f_\mu$ as elements of $L^\infty(K,\nu)$ whenever $\mu, \nu \in P(K)$ and $\nu \ll \mu$. Wong's theorem obtains a representation of $C_0(K)''$ as the commutative C^*-algebra consisting of generalized functions.

The bidual $C_0(K)''$ of $C_0(K)$ is also studied by Kaplan in [153].

We have shown in some detail that the bidual of a commutative C^*-algebra is itself a commutative, unital C^*-algebra of the form $C(\beta U_{\mathscr{F}})$; indeed, $\beta U_{\mathscr{F}}$ is the character space of the unital, commutative C^*-algebra $(C_0(K)'', \Box)$.

Definition 5.4.2. Let K be a non-empty, locally compact space. Then the character space, called $\widetilde{K}$, of the unital, commutative C^*-algebra $C_0(K)''$ is the *hyper-Stonean envelope* of K.

Thus $\widetilde{K}$ is homeomorphic to $\beta U_{\mathscr{F}}$. Clearly $\widetilde{K}$ is a Stonean space. Since $\widehat{\mu} \in N(\widetilde{K})$ and $\operatorname{supp} \widehat{\mu} = \Phi_\mu$ for each $\mu \in P(K)$, we see that $W_{\widetilde{K}} \supset U_{\mathscr{F}}$, and so $\widetilde{K}$ is certainly hyper-Stonean. The space $\widetilde{K}$ will be studied in §6.5.

The following is the most elementary realization of the above theory.

Example 5.4.3. Take $K = \mathbb{N}$, so that $C_0(K) = c_0$. Then $c_0' = \ell^1$. The relevant maximal singular family is $\{\delta_n : n \in \mathbb{N}\}$, and $(c_0)'' = \ell^\infty = C(\beta\mathbb{N})$. The hyper-Stonean envelope of $\mathbb{N}$ is $\beta\mathbb{N}$. In this example, the subspace $\mathbb{N} = W_{\beta\mathbb{N}}$ consisting of the isolated points of $\beta\mathbb{N}$ is dense in $\beta\mathbb{N}$.

We shall later calculate the cardinalities of some hyper-Stonean envelopes; note here that $|\beta\mathbb{N}| = 2^{\mathfrak{c}}$ by Proposition 1.5.4. □

The next most elementary realization of the above theory gives the space $\widetilde{\mathbb{I}}$; this space will be studied and characterized in §6.5.

Let K be a non-empty, locally compact space. We recall from Theorem 4.3.11 that we denote by S_K the Stone space of the Dedekind complete Boolean ring

$$(M(K)^+/\sim, \leq),$$

so that S_K is an extremely disconnected, locally compact space and

$$S_K = \bigcup\{\Phi_\mu : \mu \in P(K)\};$$

the family $\mathfrak{C}_{S_K}$ of compact–open subspaces of S_K, together with the partial order of subset inclusion, is isomorphic as a partially ordered set to $(M(K)^+/\sim, \leq)$.

Theorem 5.4.4. *Let K be a non-empty, locally compact space, and suppose that $\mathscr{F}$ is a maximal singular family in $P(K)$. Then $U_{\mathscr{F}}$ is a dense, open subspace of S_K with the property that*

$$C^b(U_{\mathscr{F}}) \cong C^b(S_K) \cong C(\widetilde{K}),$$

and so the space $\widetilde{K}$ *is homeomorphic to* βS_K *and is independent of the choice of the family* $\mathscr{F}$.

Proof. Let $f \in C^b(U_{\mathscr{F}})$. For each $V \in \mathfrak{C}_{S_K}$, the set $V \cap U_{\mathscr{F}}$ is dense in V. The space V is Stonean, and so, by Lemma 1.5.7, $f \mid (V \cap U_{\mathscr{F}})$ has an extension to $f_V \in C(V)$ with $|f_V|_V \leq |f|_{U_{\mathscr{F}}}$. Suppose that also $W \in \mathfrak{C}_{S_K}$. Then $f_V \mid (V \cap W) = f_W \mid (V \cap W)$ because the two functions coincide on the subspace $V \cap W \cap U_{\mathscr{F}}$, which is dense in $V \cap W$. Define $F \in C^b(S_K)$ by setting $F \mid V = f_V$ $(V \in \mathfrak{C}_{S_K})$. Then $F \in C^b(S_K)$ with $|F|_{S_K} = |f|_{U_{\mathscr{F}}}$. The result follows. □

We shall now give a further way of representing the space $\widetilde{K}$. The Boolean algebra of complemented faces of a simplex was introduced in Example 1.7.15, and $\mathrm{Comp}_{P(K)}$ was mentioned on page 111.

Theorem 5.4.5. *Let K be a non-empty, locally compact space. Then the Boolean algebra* $\mathrm{Comp}_{P(K)}$ *is isomorphic to* $\mathfrak{C}_{\widetilde{K}}$, *so that* $\widetilde{K}$ *is homeomorphic to* $St(\mathrm{Comp}_{P(K)})$, *and* $\mathrm{Comp}_{P(K)}$ *is a complete Boolean algebra.*

Proof. Set $\kappa = \kappa_{M(K)} : M(K) \to M(\widetilde{K})$. As shown later on page 202, κ is an isometry from $M(K)$ onto $N(\widetilde{K})$.

Let U be a clopen subset of $\widetilde{K}$, and set $F_U = \{\mu \in P(K) : \operatorname{supp} \kappa(\mu) \subset U\}$. Then F_U is a face in $P(K)$; this face is complemented, with $F_U^{\perp} = F_V$, where $V = \widetilde{K} \setminus U$.

Now suppose that F is a complemented face of $P(K)$, and define

$$U = \overline{\bigcup \{\operatorname{supp} \kappa(\mu) : \mu \in F\}}.$$

Each set $\operatorname{supp} \kappa(\mu)$ is a clopen subset of $\widetilde{K}$, and so U is a clopen subset of $\widetilde{K}$ with $F \subset F_U$.

We *claim* that $F_U \subset F$. Indeed, take $\mu \in P(K)$ with $\operatorname{supp} \kappa(\mu) \subset U$. Then $\mu = t\mu_1 + (1-t)\mu_2$, where $\mu_1 \in F$, $\mu_2 \in F^{\perp}$, and $t \in \mathbb{I}$. Assume toward a contradiction that $t < 1$. For each $\nu \in F$, $\nu \perp \mu_2$ by Proposition 4.2.13, so that $\kappa(\nu) \perp \kappa(\mu_2)$ by Corollary 4.2.6, and $\operatorname{supp} \kappa(\nu) \cap \operatorname{supp} \kappa(\mu_2) = \emptyset$ by Corollary 4.7.11(iii). Hence $\operatorname{supp} \kappa(\mu_2) \cap U = \emptyset$. But $\kappa(\mu) = t\kappa(\mu_1) + (1-t)\kappa(\mu_2)$, so that $\operatorname{supp} \kappa(\mu_2) \subset \operatorname{supp} \kappa(\mu) \subset U$. Thus $\operatorname{supp} \kappa(\mu_2) = \emptyset$, so $\kappa(\mu_2) = 0$ and $\mu_2 = 0$, a contradiction. Thus $t = 1$ and $\mu = \mu_1 \in F$. We have shown that $F = F_U$.

Finally, suppose that W is a clopen subset of $\widetilde{K}$ with $F_W = F_U$. Set $A = U \setminus W$, a clopen set. If $A \neq \emptyset$, then, since $\widetilde{K}$ is hyper-Stonean, there exists $\mu \in P(K)$ with $\operatorname{supp} \kappa(\mu) \subset A$, hence $\mu \in F_U \setminus F_W$. Thus $A = \emptyset$. Similarly, $W \setminus U = \emptyset$, and $W = U$.

We have established that the map $U \mapsto F_U$ is a bijection from $\mathfrak{C}_{\widetilde{K}}$ onto the family of complemented faces of $P(K)$; the map preserves the Boolean operations. □

A generalization of the above is outlined in a series of exercises in the text of Lacey [166, p. 203], without reference to any original literature; we have not found it treated elsewhere.

5.5 *L*-decompositions

We now present yet another slightly different way of recognizing the space $\widetilde{K}$; for this, we require the notions of *L*-decompositions and *L*-projections.

Definition 5.5.1. Let E be a normed space, and suppose that F and G are closed subspaces of E such that $E = F \oplus_1 G$. Then the decomposition is an *L-decomposition*; the corresponding projections onto F and G are *L-projections*.

Each *L*-projection on a normed space E is a bounded projection and a contraction. Suppose that P is the *L*-projection of E onto F. Then it is clear that $I_E - P$ is the *L*-projection onto G and that

$$\|x\| = \|Px\| + \|(I_E - P)x\| \quad (x \in E). \tag{5.8}$$

See [133] for an account of *L*-decompositions of a Banach space.

In the mid-1950s, F. Cunningham, a student of L. Loomis at Havard, wrote a Ph.D. thesis [66] which investigated the algebraic structure of the operator algebra generated by the *L*-projections on a Banach space; this work was later published in [67]. This established that any two *L*-projections commute and that the closed linear span of these operators is in fact a commutative, unital C^*-algebra whose idempotent elements, which always form a Boolean algebra, are just the *L*-projections. The Boolean algebra of *L*-projections is complete as a Boolean algebra. But more is true: it is even *Bade complete*, in the sense of [22]; see also Definition 5.5.5, below. This means essentially that infinite joins are continuous in the strong operator topology. (Details are explained in Dunford–Schwartz, Part III [95, §XVII.3], which is an account of the work of Bade. A more recent exposition of Bade complete algebras is given in [207, Chapters IV–VII].) It turns out that the Bade completeness of a Boolean algebra of *L*-projections is precisely the condition that the Boolean algebra be hyper-Stonean; we shall now provide some details of this result, simplifying the proofs from the original; we shall begin with some standard results on *L*-projections.

Proposition 5.5.2. *Let E be a Banach space. Then any two L-projections on E commute.*

Proof. Let P and Q be *L*-projections on E, and take $x \in E$. By (5.8), we see that

$$\begin{aligned}
\|Qx\| &= \|PQx\| + \|(I_E - P)Qx\| \\
&= \|QPQx\| + \|(I_E - Q)PQx\| + \|Q(Qx - PQx)\| + \|(I_E - Q)(Qx - PQx)\| \\
&= \|QPQx\| + 2 \cdot \|PQx - QPQx\| + \|Qx - QPQx\| \\
&\geq \|Qx\| + 2 \cdot \|PQx - QPQx\| \quad \text{(by the triangle inequality for } \|Qx\|\text{)},
\end{aligned}$$

so $PQ = QPQ$. By replacing Q by $I_E - Q$, we obtain $P(I_E - Q) = (I_E - Q)P(I_E - Q)$, which is equivalent to $QP = QPQ$. Hence $PQ = QP$, as required. □

Definition 5.5.3. Let E be a Banach space. The collection of L-projections on E is denoted by Proj_E. The closed linear span of Proj_E in $\mathscr{B}(E)$ is the *Cunningham algebra* of E.

Thus the Cunningham algebra is a commutative Banach algebra; we shall see shortly that it is a C^*-algebra.

Proposition 5.5.4. *Let E be a Banach space. Then* Proj_E *forms a Boolean algebra with respect to the operations defined by*

$$P \wedge Q = PQ\,, \quad P \vee Q = P + Q - PQ\,, \quad P' = I_E - P \quad (P, Q \in \mathrm{Proj}_E)\,. \tag{5.9}$$

Proof. We have noted above that $I_E - P$ is an L-projection for each L-projection P and that $PQ = QP$ whenever P and Q are L-projections.

Let P and Q be L-projections. We shall show that PQ is an L-projection by the following calculation: for each $x \in E$, we have

$$\begin{aligned} \|x\| &= \|Qx\| + \|(I_E - Q)x\| \\ &= \|PQx\| + \|Qx - PQx\| + \|x - Qx\| \\ &\geq \|PQx\| + \|x - PQx\| \geq \|x\|\,. \end{aligned}$$

Therefore, also $P + Q - PQ = I_E - (I_E - P)(I_E - Q)$ is an L-projection.

It is now straightforward to verify that the set of L-projections on E is a Boolean algebra with respect to the stated operations. □

Definition 5.5.5. Let E be a Banach space, and let A be a Boolean subalgebra of Proj_E. Then A is *Bade complete* if A is complete as an abstract Boolean algebra and, moreover, for any increasing net $\{P_\alpha\}$ in A, the supremum $\bigvee P_\alpha$ exists in A with

$$\left(\bigvee P_\alpha\right) x = \lim_\alpha P_\alpha(x) \quad (x \in E)\,.$$ [1]

An abstract Boolean algebra is *Bade complete* if it is isomorphic to some Bade complete Boolean algebra of L-projections on a Banach space.

Proposition 5.5.6. *Let E be a Banach space. Then the Boolean algebra* Proj_E *is Bade complete.*

Proof. Let $\mathscr{M}$ be an increasing net of L-projections in Proj_E (using the induced order: $P \leq Q$ if and only if $PQ = P$). For $x \in E$ and $P \leq Q$ in $\mathscr{M}$, we have

$$\|Qx\| - \|Px\| = \|Qx - Px\| \geq 0\,,$$

[1] This condition is not the original definition, but it is equivalent to the original by virtue of [95, Lemma XVII.3.4]; see also [207, Theorem IV.1].

so that $\{\|Px\| : P \in \mathscr{M}\}$ is increasing and bounded in $\mathbb{R}^+$ for each $x \in E$. For $\varepsilon > 0$, choose $P_0 \in \mathscr{M}$ with $\|P_0 x\| \geq \sup\{\|Px\| : P \in \mathscr{M}\} - \varepsilon$. Then, for $P_1, P_2 \geq P_0$ in $\mathscr{M}$, we have

$$\|P_1 x - P_2 x\| \leq \|P_1 x - P_0 x\| + \|P_2 x - P_0 x\| = \|P_1 x\| - \|P_0 x\| + \|P_2 x\| - \|P_0 x\| < 2\varepsilon,$$

so that $\{Px : P \in \mathscr{M}\}$ is a convergent net in E. Consider the pointwise limit S of this net, defined by

$$Sx = \lim\{Px : P \in \mathscr{M}\} \quad (x \in E).$$

Clearly $S : E \to E$ is linear, and $\|x\| = \|Sx\| + \|(I_E - S)x\|$ $(x \in E)$, so that S is bounded with $\|S\| \leq 1$. Thus $S \in \mathscr{B}(E)_{[1]}$.

To see that $S^2 = S$, take $x \in E$, $\varepsilon > 0$, and some sufficiently large $P_0 \in \mathscr{M}$ such that $\|PSx - S^2 x\| < \varepsilon$ and $\|Px - Sx\| < \varepsilon$ whenever $P \geq P_0$ in $\mathscr{M}$, so that we have $\|Px - PSx\| = \|P(Px - Sx)\| < \varepsilon$. Hence,

$$\|Sx - S^2 x\| \leq \|Sx - Px\| + \|Px - PSx\| + \|PSx - S^2 x\| < 3\varepsilon,$$

so that indeed $S = S^2$. We now see that $S \in \mathrm{Proj}_E$.

Finally, for $Q \in \mathscr{M}$ and $x \in E$, we have

$$QSx = Q\left(\lim_{P \in \mathscr{M}} Px\right) = \lim_{P \in \mathscr{M}} QPx = Qx$$

since $QP = Q$ for sufficiently large P, and hence $QS = Q$, i.e., $Q \leq S$. Moreover, if $Q \geq P$ for all $P \in \mathscr{M}$, then $SQx = \lim_{P \in \mathscr{M}} PQx = \lim_{P \in \mathscr{M}} Px = Sx$, i.e., $S \leq Q$, so that $S = \bigvee\{P : P \in \mathscr{M}\}$.

It follows that Proj_E forms a Bade complete Boolean algebra. □

Theorem 5.5.7. *Let B be an abstract Boolean algebra.*

(i) *Suppose that B is isomorphic to a Bade complete Boolean algebra consisting of L-projections on a Banach space. Then the Stone space St(B) is hyper-Stonean.*

(ii) *Suppose that St(B) is hyper-Stonean. Then B is isomorphic to the (Bade complete) Boolean algebra of all L-projections on the Banach space N(St(B)) of normal measures on St(B).*

Proof. (i) Since B is complete as a Boolean algebra, $St(B)$ is a Stonean space by Corollary 1.7.5.

Take a Banach space E such that B is isomorphic to a Bade complete Boolean subalgebra of Proj_E, and regard B as a subalgebra of Proj_E. For each $x \in E$, define $\mu_x : B \to \mathbb{R}^+$ by $\mu_x(P) = \|Px\|$. Take $P, Q \in B$ with $P \wedge Q = 0$. Then

$$\mu_x(P \vee Q) = \|Px + Qx\| = \|Px\| + \|Qx\|$$

because P and Q are orthogonal L-projections, and so μ_x is a positive measure on the Boolean algebra B in the sense of Definition 1.7.12; μ_x is bounded because

$$\mu_x(P) = \|Px\| \leq \|x\| \quad (x \in E).$$

Each such measure has a unique extension to a measure $\widehat{\mu_x} \in M(St(B))$, and in fact each $\widehat{\mu_x}$ is a normal measure because, by the definition of Bade completeness, $\widehat{\mu_x}(F) = 0$ for every nowhere dense, closed subset F of $St(B)$. Finally, for every non-empty, clopen set D in $St(B)$, there is some $x \in E$ with $\widehat{\mu_x}(D) \neq 0$ because, for every non-zero $P \in B$, necessarily $Px \neq 0$ for some $x \in E$. Thus the supports of the measures $\widehat{\mu_x}$ form a dense subset of $St(B)$, which must therefore be hyper-Stonean.

(ii) Suppose that $X = St(B)$ is hyper-Stonean for some Boolean algebra B. By Theorem 1.7.2(iv), we can suppose that B is the Boolean algebra $\mathfrak{C}_X$ of clopen sets in X. For each $D \in B$, define $P_D : N(X) \to N(X)$ by

$$P_D(\mu)(B) = \mu(D \cap B) \quad (\mu \in N(X), B \in \mathfrak{B}_X).$$

Then P_D is easily seen to be an L-projection on $N(X)$. Moreover, all L-projections on $N(X)$ are of this form: for an L-projection P, take D to be the largest clopen set L such that $L \cap \operatorname{supp} P\mu \neq \emptyset$ for some $\mu \in N(X)$. The map $D \mapsto P_D$ is a Boolean isomorphism of B onto $\{P_D : D \in B\}$. □

Theorem 5.5.8. *Let E be a Banach space, and let $\mathscr{A} \subset \mathscr{B}(E)$ be a Boolean subalgebra of* Proj_E *with $0, I_E \in \mathscr{A}$. Then the norm-closure of* $\operatorname{lin}\mathscr{A}$ *is a Banach subalgebra of $\mathscr{B}(E)$ which is isometrically and algebraically isomorphic to the space $C(St(\mathscr{A}))$.*

Suppose that $\mathscr{A} = \mathrm{Proj}_E$. Then $St(\mathscr{A})$ is hyper-Stonean.

Proof. Set $X = St(\mathscr{A})$, and let $\rho : \mathfrak{C}_X \to \mathscr{A}$ be a Boolean isomorphism realizing the Stone duality. Define S to be the set of simple functions in $C(X)$, so that S is a subalgebra of $C(X)$. Each $f \in S$ is represented by a unique set of disjoint elements $D_1, \dots, D_n \in \mathfrak{C}_X$ with distinct coefficients $a_1, \dots, a_n$, namely,

$$D_i = \{x \in X : f(x) = a_i\} \quad (i \in \mathbb{N}_n),$$

so that $f = \sum_{i=1}^n a_i \chi_{D_i}$. Define $\widehat{\rho} : S \to \operatorname{lin}\mathscr{A}$ by

$$\widehat{\rho}(f) = \sum_{i=1}^n a_i \rho(D_i) \quad \left(f = \sum_{i=1}^n a_i \chi_{D_i}\right),$$

so that $\widehat{\rho}$ is well defined and is easily seen to be an algebra homomorphism onto the subalgebra $\operatorname{lin}\mathscr{A}$ of $\mathscr{B}(E)$.

We shall now show that the map $\widehat{\rho}$ is in fact an isometry. It suffices to show that, for each $P_1, \dots, P_n \in A$ with $P_iP_j = 0$ for $i, j \in \mathbb{N}_n$ with $i \neq j$ and scalars $\alpha_1, \dots, \alpha_n$ such that $|\alpha_1| = \max\{|\alpha_i| : i \in \mathbb{N}_n\}$, we have

$$\left\| \sum_{i=1}^n \alpha_i P_i \right\| = |\alpha_1| .$$

We first observe that $\|\sum_{i=1}^n y_i\| = \sum_{i=1}^n \|y_i\|$ whenever $y_i \in P_i(E)$ $(i \in \mathbb{N}_n)$. This is easily seen for $n = 2$: indeed,

$$\|y_1 + y_2\| = \|P_1(y_1 + y_2)\| + \|(I_E - P_1)(y_1 + y_2)\| = \|y_1\| + \|y_2\|,$$

and the case of general $n \in \mathbb{N}$ follows by induction. For $x \in E$, we have

$$\begin{aligned}\left\|\sum_{i=1}^n \alpha_i P_i(x)\right\| &= \sum_{i=1}^n |\alpha_i| \, \|P_i(x)\| \leq |\alpha_1| \sum_{i=1}^n \|P_i(x)\| \\ &= |\alpha_1| \left\|\sum_{i=1}^n P_i(x)\right\| \leq |\alpha_1| \left\|\sum_{i=1}^n P_i\right\| \|x\| = |\alpha_1| \|x\|\end{aligned}$$

because $\sum_{i=1}^n P_i = \bigvee_{i=1}^n P_i \in \mathscr{A}$, and hence $\|\sum_{i=1}^n P_i\| = 1$. Thus

$$\left\|\sum_{i=1}^n \alpha_i P_i\right\| \leq |\alpha_1| .$$

On the other hand, for each $x \in P_1(E)$ with $x \neq 0$, we have

$$\left\|\sum_{i=1}^n \alpha_i P_i(x)\right\| = \|\alpha_1 P_1(x)\| = |\alpha_1| \, \|x\| ,$$

whence $\|\sum_{i=1}^n \alpha_i P_i\| \geq |\alpha_1|$. Thus $\widehat{\rho}$ is an isometry.

Since S is norm-dense in $C(X)$, the map $\widehat{\rho}$ extends to an isometry of $C(X)$ onto the norm-closed operator algebra in $\mathscr{B}(E)$ generated by $\mathscr{A}$, as desired.

Finally, in the case where $\mathscr{A}$ consists of all the L-projections on E, we have seen that $\mathscr{A}$ is Bade complete, and so its Stone space is hyper-Stonean. □

Theorem 5.5.9. *Let K be a non-empty, compact space. Then the Boolean algebra* $\mathrm{Comp}_{P(K)}$ *is isomorphic to the Boolean algebra* $\mathrm{Proj}_{M(K)}$ *of L-projections in* $\mathscr{B}(M(K))$.

Proof. Let F be a complemented face of $P(K)$. Then

$$M(K) = \operatorname{lin} F \oplus_1 \operatorname{lin} F^{\perp}$$

is an L-decomposition of $M(K)$.

Conversely, suppose that $M(K) = G \oplus_1 H$ is an L-decomposition of $M(K)$. We first *claim* that $G \cap P(K)$ is a face of $P(K)$. To see this, take $\lambda \in G \cap P(K)$, and suppose that $\lambda = t\mu + (1-t)\nu$ for some $\mu, \nu \in P(K)$ and some $t \in (0,1)$. Then $\mu = \mu_1 + \mu_2$ and $\nu = \nu_1 + \nu_2$, where $\mu_1, \nu_1 \in G$ and $\mu_2, \nu_2 \in H$. Furthermore,

$$\|\mu\| = \|\mu_1\| + \|\mu_2\| = 1 \quad \text{and} \quad \|\nu\| = \|\nu_1\| + \|\nu_2\| = 1 .$$

This implies that $\mu_1, \mu_2, \nu_1, \nu_2 \in M(K)^+$. Since $\lambda \in G$, we have $t\mu_2 + (1-t)\nu_2 = 0$, and so $\mu_2 = \nu_2 = 0$. Thus $\mu, \nu \in G \cap P(K)$, giving the claim.

Set $F = G \cap P(K)$ and $F^\dagger = H \cap P(K)$, so that F and $F^\dagger$ are faces in $P(K)$. It is clear that $F^\dagger$ is the complementary face to F.

The result follows. □

Corollary 5.5.10. *Let K be a non-empty, compact space. Then $\widetilde{K}$ is homeomorphic to $St(\mathrm{Proj}_{M(K)})$.*

Proof. This follows from Theorems 5.4.5 and 5.5.9. □

Let K be a non-empty, compact space. Then our programme is to understand the L-projections on $M(K)$ (equivalently, the L-decompositions of $M(K)$). The bidual of $C(K)$ is $C(\widetilde{K})$, and we have identified the hyper-Stonean envelope $\widetilde{K}$ with the character space of the Cunningham algebra on $M(K)$ and the Stone space of the Boolean algebra $\mathrm{Proj}_{M(K)}$; the latter is isomorphic to the Boolean algebra $\mathrm{Comp}_{P(K)}$ of complemented faces of $P(K)$.

5.6 Biduals of C^*-algebras

We now summarize, without proof, various generalizations of the above identification of $C_0(K)''$ as a commutative C^*-algebra to arbitrary C^*-algebras.

The following theorem is given in [68, Theorem 3.2.36] and [234, Theorem III.2.4]. Recall that every C^*-algebra has a universal $*$-representation.

Theorem 5.6.1. *Let A be a C^*-algebra, and suppose that $\pi : A \to \mathscr{B}(H)$ is a universal $*$-representation. Then A is Arens regular, $(A'', \Box)$ is a von Neumann algebra, and there is an isometric, unital $*$-representation $\widetilde{\pi} : A'' \to \mathscr{B}(H)$ such that $\widetilde{\pi} \mid A = \pi$ and $\widetilde{\pi}(A'') = \pi(A)^{cc}$.* □

Definition 5.6.2. Let A be a C^*-algebra. Then $(A'', \Box)$ is the *enveloping von Neumann algebra* of A.

Thus the enveloping von Neumann algebra of a C^*-algebra A can be identified with the algebra $\widetilde{\pi}(A'') = \pi(A)^{cc}$, where the map $\pi : A \to \mathscr{B}(H)$ is a universal $*$-representation of A. Let K be a non-empty, locally compact space. Then we have identified the bidual of the commutative C^*-algebra $C_0(K)$ with $C(\widetilde{K})$, where $\widetilde{K}$ is the hyper-Stonean envelope of K: the algebra $C(\widetilde{K})$ is the enveloping von Neumann algebra of $C_0(K)$.

There is an attractive and somewhat different approach to the proof of the above theorem; it is less well known than the one indicated, and we wish to sketch it.

Let A be a unital Banach algebra, and take $h \in A$. Then h is *hermitian* if

$$\|\exp(\mathrm{i}th)\| = 1 \quad (t \in \mathbb{R});$$

this is equivalent to requiring that $\langle h, \lambda \rangle$ belongs to $\mathbb{R}$ for each $\lambda \in K_A$. See [42, §10]. This generalizes the standard notion for C^*-algebras: in the latter case, an element h is hermitian if and only if it is self-adjoint, so that $h^* = h$. The set of hermitian elements in A is denoted by $H(A)$; $H(A)$ is a closed, real-linear subspace of A, but, in general, it is not a subalgebra of the underlying real-linear algebra of A. (For a simple counter-example, see [41, Example 1, §6].) Next, define

$$J(A) = H(A) \oplus \mathrm{i}H(A),$$

so that $J(A)$ is the (complex) linear span of $H(A)$ and $J(A)$ is a closed linear subspace of A. Each element of $J(A)$ has a unique representation in the form $h + \mathrm{i}k$, where $h, k \in H(A)$, and the map

$$* : h + \mathrm{i}k \mapsto h - \mathrm{i}k, \quad J(A) \to J(A),$$

is a linear involution on $J(A)$. In general, $J(A)$ is not a subalgebra of A; this is so if and only if $h^2 \in H(A)$ for each $h \in H(A)$, and, in this case, $*$ is an (algebra) involution on $J(A)$ and $(J(A), *)$ is a Banach algebra with an isometric involution. The algebra A is a *V-algebra* if $J(A) = A$, and now A has an isometric involution from $J(A)$. See [42, §38] and [194, §2.6]. It is clear that each unital C^*-algebra is a V-algebra; the converse is the following *Vidav–Palmer theorem*, from the beautiful paper [193]. See also [41, §6], [42, Theorem 38.14], and [195, Theorem 9.59].

Theorem 5.6.3. *Let A be a unital Banach algebra which is a V-algebra. Then A, with the involution from $J(A)$, is a C^*-algebra.* □

Although the space $J(\mathscr{B}(E))$, which is the linear span of the hermitian operators on a Banach space E, is not necessarily a subalgebra of $\mathscr{B}(E)$, it is a remarkable result of Kalton [152] that, in the case where E is a (complex) Banach lattice, $J(\mathscr{B}(E))$ is a closed subalgebra of $\mathscr{B}(E)$ and hence isometrically a C^*-algebra.

Again take A to be a unital Banach algebra. We set $H(A') = \mathrm{lin}_{\mathbb{R}} K_A$, so that $H(A')$ consists of the *hermitian functionals* on A. We also set

$$Z(A) = \mathrm{co}\,(K_A \cup -\mathrm{i}K_A),$$

so that $Z(A)$ is a convex subset of $A'_{[1]}$. The subset K_A is a closed face of $Z(A)$, and $-\mathrm{i}K_A$ is a complementary face, but, in general, K_A is not a split face (as defined on page 8) of $Z(A)$. We note that the definition of K_A and $Z(A)$ depends on only the norm on the algebra A and the existence of an element e_A, and apparently does not depend on the algebra structure of A, and so we can define $K_{A''}$ and $Z(A'')$ with respect to A'' without knowing anything about the product in A''.

The following theorem of Asimov and Ellis is proved in [17, Theorem 4]; it uses the above Vidav–Palmer theorem. See also [18, Chapter 4].

Theorem 5.6.4. *Let A be a unital Banach algebra. Then the following conditions on A are equivalent:*

(a) *K_A is a split face of $Z(A)$;*

(b) *$H(A') \cap iH(A') = \{0\}$;*

(c) *$H(A)$ separates the points of K_A;*

(d) *A is a C^*-algebra;*

(e) *$A = J(A)$.* □

It is rather easy to see that K_A is a split face of $Z(A)$ if and only if $K_{A''}$ is a split face of $Z(A'')$ [98, Theorem 2]. Thus we obtain the following theorem [98, Corollary 3].

Theorem 5.6.5. *Let A be a unital Banach algebra.*

(i) *Suppose that A is a C^*-algebra. Then A is Arens regular, the extended linear involution $*$ is an involution on $(A'', \Box)$, and A'' is a von Neumann algebra with respect to the product $\Box$ and the involution $*$.*

(i) *Suppose that A'' is a C^*-algebra with respect to the product $\Box$ and an involution. Then A is a C^*-subalgebra of A''.* □

Chapter 6
The Banach Space $C(K)$

The main aim of this chapter is to determine when a space of the form $C(K)$ for a compact space K is a dual space or a bidual space, either isometrically or isomorphically. However, we shall first discuss when two spaces $C(K)$ and $C(L)$ are isomorphic and when they are isometrically isomorphic. Some results come from rather elementary considerations, but some require more sophisticated background.

In §6.1, we shall show that $|K| = |L|$ whenever $C(K) \sim C(L)$, so that the cardinality of K is an isomorphic invariant of $C(K)$-spaces, and, in §6.2, we shall give various examples of compact spaces K such that $C(K)$ is not (isometrically or isomorphically) a dual space. In §6.3, we shall show that the Banach space $M(K) = C(K)'$ has a unique predual only in very special circumstances.

The question when a space $C(K)$ is isometrically a dual space, that is, when $C(K)$ is a von Neumann algebra, is fully determined in Theorem 6.4.1 in §6.4. For example, this is the case if and only if the compact space K is hyper-Stonean, and then the unique isometric predual $C(K)_*$ of $C(K)$ is identified with the space $N(K)$ of normal measures on K. This result shows why we have described hyper-Stonean spaces and normal measures at some length in earlier chapters.

In §6.5 and §6.6, we shall discuss the topological properties of the hyper-Stonean envelope $\widetilde{K}$ of a compact space K, in particular giving results on the cardinality of certain subsets of $\widetilde{K}$, and, in §6.7, we shall discuss the Baire classes $B_\alpha(\mathbb{I})$, in particular proving in Theorem 6.7.5 that $B_1(\mathbb{I})$ is not isomorphic to any Baire class $B_\beta(\mathbb{I})$ for $2 \leq \beta < \omega_1$. In §6.8, we shall show that $C(K)$ is 1-injective if and only if K is Stonean, and we shall characterize 1-injective Banach spaces as those spaces that are isometrically isomorphic to $C(K)$ for some Stonean space K.

In §6.9, we shall consider when $C(K)$ is isomorphically a dual space. Here it seems that a topological characterization of such spaces K is not known, and may be inaccessible, but we do give various examples and conditions that show when $C(K)$ is or is not isomorphically a dual space. In particular, we should like to show that K must be totally disconnected whenever $C(K)$ is a dual space: the closest that we come to this is Theorem 6.9.6, which shows that K must contain a dense, open, extremely disconnected subset, and hence that K cannot be connected.

H.G. Dales et al., *Banach Spaces of Continuous Functions as Dual Spaces*,
CMS Books in Mathematics, DOI 10.1007/978-3-319-32349-7_6

In §6.10, we shall return to a question raised in the Introduction: we wish to show that X is homeomorphic to a space of the form $\widetilde{K}$ for some locally compact space K whenever $C(X)$ is isometrically a bidual space. We do show that X is always homeomorphic to a clopen subspace of a space $\widetilde{K}$ and that X is homeomorphic to either $\beta\mathbb{N} = \widetilde{\mathbb{N}}$ or to $\widetilde{\mathbb{I}}$ whenever $C(X)$ is isometrically isomorphic to the bidual of a separable Banach space. We shall also obtain results close to a classical theorem of Lindenstrauss in the setting of complex-valued functions.

In §6.11, we shall summarize some results that we have mentioned on the questions when $C(K)$ is injective and when it is a dual or bidual space, and, in §6.12, we shall repeat some open questions that we have raised in the text.

6.1 Isomorphisms of the spaces $C(K)$

Let K and L be two non-empty, locally compact spaces. We shall recall the well-known theory showing when the Banach spaces $C_0(K)$ and $C_0(L)$ are isometrically isomorphic.

The first, elementary example shows that two such spaces can be isomorphic, but not isometrically isomorphic.

Example 6.1.1. The spaces c_0 and c are not isometrically isomorphic: the unit ball $c_{[1]}$ has many extreme points, but the unit ball $(c_0)_{[1]}$ has no extreme points. However, c_0 and c are isomorphic. To see this, set

$$T\alpha = (2\alpha_\infty, \alpha_1 - \alpha_\infty, \alpha_2 - \alpha_\infty, \dots)$$

for $\alpha = (\alpha_n) \in c$ with $\lim_{n\to\infty} \alpha_n = \alpha_\infty$. Then $T : c \to c_0$ is an isomorphism with $\|T\| = 2$ and $\|T^{-1}\| = 3/2$, and so $c_0 \sim c$, with $d(c_0, c) \leq 3$. In fact, this latter estimate is sharp: $d(c_0, c) = 3$. For this, we follow Cambern [49]; see [51, 121] for more general results.

Assume to the contrary that $T : c \to c_0$ is an isomorphism with $\|T\| \|T^{-1}\| < 3$. We may suppose that $\|T^{-1}\| = 1$, so that $1 \leq \|T\| < 3$. We choose ε in $(0,1)$ and $p \in \mathbb{N}$ with

$$(1+\varepsilon)\|T\| < 3 - \varepsilon \quad \text{and} \quad \frac{1}{p} < \frac{3 - (\|T\| + \varepsilon)}{2},$$

so that $\|T\| < 3 - 2/p - \varepsilon$ and $p \geq 2$.

For notational convenience, we set $e = (1,1,1,\dots) \in c$ and $h_n = T\delta_n \in c_0$ for $n \in \mathbb{N}$, so that $|h_n|_{\mathbb{N}} \geq 1$ $(n \in \mathbb{N})$. We fix $K \in \mathbb{N}$ such that $|(Te)(n)| < \varepsilon$ $(n > K)$, and then define $f, g \in c_0$ by setting $f(n) = (Te)(n)$ $(n \leq K)$, $f(n) = 0$ $(n > K)$, and $g = Te - f$, so that $g \in c_0$ and $|g|_{\mathbb{N}} < \varepsilon$, and hence $T^{-1}g \in c$ with $|T^{-1}g|_{\mathbb{N}} < \varepsilon$.

Next, for each $n \in \mathbb{N}$, we define

$$S_n = \{k \in \mathbb{N} : |h_n(k)| = |h_n|_{\mathbb{N}}\} \quad \text{and} \quad T_n = \{k \in \mathbb{N} : |h_n(k)| > 1/p\},$$

so that $S_n \neq \emptyset$ and $S_n \subset T_n$.

We *claim* that $|\{n \in \mathbb{N} : T_n \cap \mathbb{N}_K \neq \emptyset\}| < 3p$. For otherwise, there exists $m \in \mathbb{N}_K$ belonging to $3p$ of the sets T_n, say to $T_{n_1}, \dots, T_{n_{3p}}$, where $n_1 < n_2 < \cdots < n_{3p}$. Choose $\zeta_1, \dots, \zeta_{3p} \in \mathbb{T}$ such that $\zeta_i h_{n_i}(m) \in \mathbb{R}^+$ $(i \in \mathbb{N}_{3p})$, and set $h = \sum_{i=1}^{3p} \zeta_i \delta_{n_i}$, so that $|h|_{\mathbb{N}} = 1$ and

$$|Th|_{\mathbb{N}} \geq |(Th)(m)| = \left|\sum_{i=1}^{3p} \zeta_i h_{n_i}(m)\right| = \sum_{i=1}^{3p} |h_{n_i}(m)| \geq 3p \cdot \frac{1}{p} = 3,$$

a contradiction because $|Th|_{\mathbb{N}} \leq \|T\| \, |h|_{\mathbb{N}} < 3$. Thus the claim holds.

By the claim, we can choose $n_0 \in \mathbb{N}$ with $T_{n_0} \cap \mathbb{N}_K = \emptyset$. Consider the element $\widetilde{f} := f + 2h_{n_0}$ in c_0. Then

$$\left|\widetilde{f}\right|_{\mathbb{N}} \geq \left|T^{-1}\widetilde{f}\right|_{\mathbb{N}} = \left|e - T^{-1}g + 2\delta_{n_0}\right|_{\mathbb{N}} > 3 - \varepsilon.$$

Suppose that there exists $k \in \mathbb{N}_K$ with $\left|\widetilde{f}(k)\right| > 3 - \varepsilon$. Then $g(k) = 0$ and

$$3 - \frac{2}{p} - \varepsilon > \|T\| \geq |Te|_{\mathbb{N}} \geq |Te(k)| \geq \left|\widetilde{f}(k)\right| - 2\left|h_{n_0}(k)\right| > 3 - \varepsilon - \frac{2}{p},$$

a contradiction.

Suppose that there exists $k > K$ with $\left|\widetilde{f}(k)\right| > 3 - \varepsilon$, so that $\widetilde{f}(k) = 2h_{n_0}(k)$. Consider the element $u = e - T^{-1}g - 2\delta_{n_0} \in c$, so that $Tu = f - 2h_{n_0}$. Then

$$|u|_{\mathbb{N}} \leq \left|e - 2\delta_{n_0}\right|_{\mathbb{N}} + \left|T^{-1}g\right|_N < 1 + \varepsilon$$

and $3 - \varepsilon < 2\left|h_{n_0}(k)\right| = |(Tu)(k)| \leq \|T\| \, |u|_{\mathbb{N}} < (1+\varepsilon)\|T\|$, also a contradiction.

We conclude that $\|T\| \left\|T^{-1}\right\| \geq 3$, and so $d(c_0, c) = 3$. □

The question when two spaces $C(K)$ and $C(L)$ are isometrically isomorphic is answered by the following classic *Banach–Stone theorem*; see [94, V.8.8], for example. We recall one of the very standard proofs.

Theorem 6.1.2. *Let K and L be two non-empty, compact spaces, and suppose that $T : C(L) \to C(K)$ is an isometric isomorphism. Then there are a homeomorphism $\eta : K \to L$ and a function $h \in C(K, \mathbb{T})$ such that*

$$(Tf)(x) = h(x)(f \circ \eta)(x) \quad (f \in C(L), x \in K).$$

In the case where $T(1_L) = 1_K$, we have $Tf = f \circ \eta$ $(f \in C(L))$, so that $T = \eta^\circ$.

Proof. The map $T' : M(K) \to M(L)$ is also an isometric isomorphism, and so T' maps $\operatorname{ex} M(K)_{[1]}$ onto $\operatorname{ex} M(L)_{[1]}$. By the identification of these sets of extreme points in Proposition 4.4.15(i), we see that there are a bijection $\eta : K \to L$ and a function $h : K \to \mathbb{T}$ such that $T'(\delta_x) = h(x)\delta_{\eta(x)}$ $(x \in K)$.

Let $x_\alpha \to x$ in K. Then $\delta_{x_\alpha} \to \delta_x$ weak* in $M(K)$. Since T' is weak*-weak*-continuous, $h(x_\alpha)\delta_{\eta(x_\alpha)} \to h(x)\delta_{\eta(x)}$ weak* in $M(L)$. By considering the action of these measures on 1_L, we see that $h(x_\alpha) \to h(x)$, and so $h \in C(K, \mathbb{T})$. Assume that there is a subnet (x_β) of (x_α) such that $\eta(x_\beta) \to y \neq \eta(x)$ for some $y \in K$. By considering a function $g \in C(L)$ with $g(y) = 0$ and $g(\eta(x)) = 1$, we obtain a contradiction. Thus $\eta(x_\alpha) \to \eta(x)$, η is continuous, and hence η is a homeomorphism. □

Corollary 6.1.3. *Let K and L be two non-empty, compact spaces, and suppose that there is an isometry from $C_{\mathbb{R}}(K)$ onto $C_{\mathbb{R}}(L)$. Then K and L homeomorphic.*

Proof. By Theorem 2.2.10, $C_{\mathbb{R}}(K)$ and $C_{\mathbb{R}}(L)$ are isometrically isomorphic as real-linear spaces, and so $C(K) \cong C(L)$. By the theorem, K and L homeomorphic. □

In fact, it is a theorem of Amir [10] that the conclusions of the above Banach–Stone theorem are valid whenever $d(C(K), C(L)) < 2$; see [3, Exercise 4.11]. The same conclusions hold when $d(C(K), C(L)) = 2$ provided that at least one of K and L is totally disconnected [57, p. 6], but in general there are non-homeomorphic spaces K and L with $d(C(K), C(L)) = 2$ [56].

Theorem 6.1.4. *Let K and L be two non-empty, compact spaces. Then the following are equivalent:*

(a) *the spaces K and L are homeomorphic;*
(b) *there is a Banach-lattice isometry from $C(L)$ onto $C(K)$;*
(c) *there is a Riesz isomorphism from $C(L)$ onto $C(K)$;*
(d) *$C(L)$ and $C(K)$ are C^*-isomorphic;*
(e) *there is an algebra isomorphism from $C(L)$ onto $C(K)$;*
(f) *$C(L) \cong C(K)$;*
(g) *there is an isometry from $C_{\mathbb{R}}(L)$ onto $C_{\mathbb{R}}(K)$.*

Proof. The equivalence of (a)–(e) is given in Theorem 3.2.6, and clearly we have (b) $\Rightarrow$ (f), (g). By Theorem 6.1.2, (f) $\Rightarrow$ (a), and (g) $\Rightarrow$ (a) by Corollary 6.1.3. □

We shall soon consider the much deeper question of when $C(K)$ and $C(L)$ are isomorphic. However, we first discuss the following related question: *Suppose that $C(K)$ and $C(L)$ are isomorphic. What properties do K and L share?* As in Definition 2.2.5, such a property is called an isomorphic invariant of $C(K)$-spaces.

Towards this, we actually give consequences concerning the relation of K to L supposing only that $C(K)$ is isomorphic to a closed subspace of $C(L)$. It was observed by Cengiz [53] that K and L have the same cardinality when $C(K) \sim C(L)$. Our statement in clause (iii) of the next theorem is somewhat more delicate: $|K| \leq |L|$ under our hypotheses. To our knowledge, this fact has not appeared in the literature; it is a direct consequence of the key assertion (6.1) below, which was used in [53] with the stronger hypothesis. Our proof is a simplification and slight modification of that of Cengiz to accommodate the weaker assumption of an *into* isomorphism. A quantitative version of (6.1) has recently been given in [205].

Theorem 6.1.5. *Let K and L be non-empty, compact spaces such that $C(K)$ is isomorphic to a closed subspace of $C(L)$. Then:*

(i) *K is metrizable if L is metrizable;*

(ii) $w(K) \leq w(L)$;

(iii) $|K| \leq |L|$.

Proof. We may suppose that K and L are infinite, for otherwise the results are trivial.

Let $T : C(K) \to C(L)$ be an isomorphism onto a closed subspace, say E, of $C(L)$, so that $T' : E' \to M(K)$ is an isomorphism. We may suppose that

$$m|g|_K \leq |Tg|_L \leq |g|_K \quad (g \in C(K))$$

for some $m > 0$. In particular, $T'(\delta_y)$ and $|T'(\delta_y)|$ belong to $M(K)$ for each $y \in L$; for $x \in K$ and $y \in L$, we set

$$\nu_y = |T'(\delta_y)| \in M(K) \quad \text{and} \quad \nu_y(x) = \nu_y(\{x\}),$$

so that $\|\nu_y\| \leq \|T\| \leq 1$.

(i) and (ii) Since T preserves the density character of $C(K)$, these follow immediately from Theorem 2.1.7, (i) and (ii).

(iii) The key step in the proof is to establish the following claim:

$$\textit{for each } x \in K, \textit{ there exists } y \in L \textit{ such that } \nu_y(x) \neq 0. \tag{6.1}$$

Suppose that (6.1) has been established. For $y \in L$, set $K_y = \{x \in K : \nu_y(x) \neq 0\}$. Then each set K_y is countable and, by (6.1), $K = \bigcup\{K_y : y \in L\}$. Thus

$$|K| \leq \aleph_0 \cdot |L| = |L|,$$

giving the result.

Assume towards a contradiction that (6.1) is false. Then there exists $x \in K$ such that $\nu_y(x) = 0$ for all $y \in L$. Let $\mathcal{N}_x = \{U_\alpha : \alpha \in A\}$ be the family of all open neighbourhoods of x, and fix $\varepsilon > 0$ with $\varepsilon(m+1) < m$. For each $\alpha \in A$, define

$$W_\alpha = \{y \in L : \nu_y(\overline{U_\alpha}) < \varepsilon\}, \quad V_\alpha = K \setminus \overline{U_\alpha}.$$

For each $y \in L$, there exists $\alpha \in A$ with $\nu_y(U_\alpha) < \varepsilon$. Hence $\bigcup W_\alpha = L$. Moreover, for each fixed $\alpha \in A$, the function

$$y \mapsto \nu_y(\overline{U_\alpha}) = \nu_y(K) - \nu_y(V_\alpha), \quad L \to \mathbb{R},$$

being the difference of two lower semi-continuous functions on L (see Proposition 4.1.13), is a Borel function on L, so that $W_\alpha \in \mathfrak{B}_L$.

Take a fixed $\mu \in M(L)^+$. We *claim* that

$$\mu(L) = \sup\{\mu(W_\alpha) : \alpha \in A\}. \tag{6.2}$$

Observe that $W_{\alpha_1} \subset W_{\alpha_2}$ when $U_{\alpha_2} \subset U_{\alpha_1}$, and so the family $\{W_\alpha : \alpha \in A\}$ is upwards directed by inclusion. Hence we can choose an increasing sequence (W_{α_n}) such that

$$\mu(W) = \sup\{\mu(W_\alpha) : \alpha \in A\},$$

where $W = \bigcup_{n=1}^\infty W_{\alpha_n}$. Set $B = L \setminus W$, and observe that $\mu(B \bigcap W_\alpha) = 0$ $(\alpha \in A)$. Indeed, for $\alpha \in A$ and $n \in \mathbb{N}$, choose $\beta \in A$ with $W_\beta \supset W_\alpha \cup W_{\alpha_n}$. Then

$$\begin{aligned}\mu(B \cap W_\alpha) \leq \mu(B \cap W_\beta) = \mu(W_\beta) - \mu(W_\beta \setminus B) \leq \mu(W_\beta) - \mu(W_{\alpha_n}) \\ \leq \mu(W) - \mu(W_{\alpha_n}) \to 0,\end{aligned}$$

whence $\mu(B \cap W_\alpha) = 0$. To prove the claim that (6.2) holds, it suffices to show $\mu(B) = 0$, i.e., that $\mu(F) = 0$ for each compact subset F of B.

Assume to the contrary that, for some compact $F \subset B$, we have $\mu(F) > 0$. By replacing μ by $\mu \mid \mathrm{supp}\,(\mu \mid F)$, we may suppose that $\mathrm{supp}\,\mu = F$. Now set

$$M = \sup\{\|\nu_y\| : y \in F\},$$

so that $0 \leq M \leq 1$. For each open subset U of K, the function $y \mapsto \nu_y(U)$, $F \to \mathbb{R}$, is lower semi-continuous (see Proposition 4.1.13), and so each set

$$P_\alpha := \{y \in F : \nu_y(V_\alpha) > M - \varepsilon\} \quad (\alpha \in A)$$

is an open subset of F. For each $\alpha \in A$ and $y \in P_\alpha$, we have

$$M \geq \|\nu_y\| = \nu_y(V_\alpha) + \nu_y(\overline{U_\alpha}) > M - \varepsilon + \nu_y(\overline{U_\alpha}),$$

and so $\nu_y(\overline{U_\alpha}) < \varepsilon$. This shows that $P_\alpha \subset F \cap W_\alpha \subset B \cap W_\alpha$, and so $\mu(P_\alpha) = 0$. Take $y \in F$ with $\|\nu_y\| > M - \varepsilon$. Since $\|\nu_y\| = \sup_{\alpha \in A} \nu_y(V_\alpha)$, there exists $\alpha_0 \in A$ such that $y \in P_{\alpha_0}$. Hence P_{α_0} is a non-empty, open subspace of F such that $\mu(P_{\alpha_0}) = 0$, a contradiction of the defining property of the support. Thus $\mu(F) = 0$, and the claim is established.

We shall now prove statement (6.1). By the Hahn–Banach theorem, there exists a measure $\mu \in M(L)_{[1/m]}$ with

$$\langle Tf, \mu \rangle = f(x) \quad (f \in C(K)).$$

Choose $\alpha \in A$ such that $|\mu|\,(L \setminus W_\alpha) < \varepsilon$, and take $f \in C(K)$ with $\chi_{\{x\}} \leq f \leq \chi_{\overline{U_\alpha}}$. Then

$$|(Tf)(y)| = |\langle f, T'(\delta_y) \rangle| \leq \int_K f \,\mathrm{d}\nu_y \leq \nu_y(\overline{U_\alpha}) \quad (y \in L),$$

and so we have $|(Tf)(y)| < \varepsilon$ $(y \in W_\alpha)$. It follows that

$$1 = f(x) = \langle Tf, \mu \rangle < \varepsilon|\mu|(W_\alpha) + |\mu|(L \setminus W_\alpha) < \varepsilon(1/m + 1) < 1,$$

a contradiction.

Thus (6.1) holds, and the proof is complete. □

Clause (iv) of following result is a theorem of Cengiz; for (iii), see page 63.

Corollary 6.1.6. *Let K and L be non-empty, compact spaces such that $C(K) \sim C(L)$. Then:*

(i) L is metrizable if and only if K is metrizable;

(ii) $w(L) = w(K)$;

(iii) $c(K) = c(L)$;

(iv) $|L| = |K|$. □

Thus, in particular, the cardinality, weight, and Souslin number of K are isomorphic invariants of $C(K)$-spaces.

The question of the isomorphism of $C(K)$ and $C(L)$ in the case where K and L are uncountable, compact, metrizable spaces is answered by the following classic theorem of Milutin; for a fine modern exposition of this result and some extensions, see [3, §4.4]. We like the 'conceptual, but indirect' proof that Ditor gives of 'Milutin's lemma'; this was given in [88, 89], and a somewhat simplified account was given by William Bade in [24]. This approach is explained in some detail in [215, §2*A*].

Theorem 6.1.7. *Let K and L be uncountable, compact, metrizable spaces. Then the Banach spaces $C(K)$ and $C(L)$ are isomorphic.* □

It follows that the two spaces $C(\mathbb{I})$ and $C(\Delta)$ are isomorphic. However, it seems that no explicit formula for an isomorphism between the spaces is known, and the value of the Banach–Mazur distance $d(C(\mathbb{I}),C(\Delta))$ is not known; the above result and an argument from Theorem 2.4.9 shows that $2 < d(C(\mathbb{I}),C(\Delta)) \leq 12$ (see page 186). For some results on the Banach–Mazur distance between c and other $C(K)$ spaces, see [51].

Thus we see that there is a large difference between the isometric and the isomorphic theories: the Cantor set Δ, the closed unit interval $\mathbb{I}$, and the closed unit square $\mathbb{I} \times \mathbb{I}$ are each uncountable, compact metric spaces, and so the Banach spaces $C(\Delta)$, $C(\mathbb{I})$, and $C(\mathbb{I} \times \mathbb{I})$ are pairwise mutually isomorphic. However Δ is totally disconnected and $\mathbb{I}$ is connected, and so Δ and $\mathbb{I}$ are certainly not homeomorphic; also $\mathbb{I}$ and $\mathbb{I} \times \mathbb{I}$ are not homeomorphic. Thus no two of $C(\Delta)$, $C(\mathbb{I})$, and $C(\mathbb{I} \times \mathbb{I})$ are isometrically isomorphic. The fact that the Banach spaces $C(\mathbb{I})$ and $C(\mathbb{I} \times \mathbb{I})$ are not isometrically isomorphic to each other was already proved by Banach himself in 1932 [30, pp. 172/173].

The situation when K and L are countable and compact is as follows: By a theorem of Mazurkiewicz and Sierpiński [225, Theorem 8.6.10], each countable, compact space is homeomorphic to the space $\alpha + 1 = [0, \alpha]$ of all ordinals up to and including the ordinal α for some countable α (taking the order topology on $\alpha + 1$). Two such spaces $C(\alpha + 1)$ and $C(\beta + 1)$, where $\alpha < \beta$, are isomorphic if and only if $\beta < \alpha^{\omega}$ [225, §21.5.14]; for each infinite, countable, compact space K, there is a countable ordinal α such that $C(K)$ is isomorphic to $C(\omega^{\omega^{\alpha}} + 1)$, and the latter spaces are pairwise non-isomorphic. See [131, §2.6] and [215, Theorem 2.14].

For example, consider the compact spaces

$$\omega+1, \quad 2\cdot\omega+1, \quad \text{and} \quad \omega^{\omega}+1,$$

where ω is the first infinite ordinal, and the spaces are taken with the order topology; these three spaces are countable and compact, but no two of them are mutually homeomorphic. The two Banach spaces $C(\omega+1)$ and $C(2\cdot\omega+1)$ are isomorphic, but $C(\omega+1)$ and $C(\omega^{\omega}+1)$ are not isomorphic. The dual of each of these three spaces is isometrically isomorphic to ℓ^1, and so the bidual is isometrically isomorphic to $C(\beta\mathbb{N})$.

We now consider compact, non-metrizable spaces, for which we do have one result. By Theorem 4.4.7, $\ell^\infty = C(\beta\mathbb{N})$ is isomorphic to $L^\infty(\mathbb{I}) = C(\mathbb{H})$. However, it is certainly not true that $\beta\mathbb{N}$ is homeomorphic to $\mathbb{H}$ – the set of isolated points is dense in $\beta\mathbb{N}$, but $\mathbb{H}$ has no isolated points – and so $C(\beta\mathbb{N})$ is not isometrically isomorphic to $C(\mathbb{H})$. This gives a further example of two compact spaces K and L such that $C(K)\sim C(L)$, but $C(K)\not\cong C(L)$.

However, as stated by Johnson and Lindenstrauss in [148]: 'It seems a hopeless task to get an isomorphic classification of general $C(K)$ spaces, but some information is contained in [248]'.

We see that, for each uncountable, compact, metrizable space K, the Banach space $C(K)$ is isomorphic to $C(\Delta)$ for the totally disconnected, compact space Δ. For a long time, the following question was open [215, p. 1594]:

Is it true that, for each compact space K, the Banach space $C(K)$ is isomorphic to $C(L)$ for some totally disconnected space L?

However, this question was recently resolved in a deep and impressive paper of Piotr Koszmider [160]. Indeed, this paper presents two constructions of infinite, separable, compact spaces K. In the first construction, K is totally disconnected and $C(K)$ is not isomorphic to any of its proper, closed subspaces; in particular, $C(K)$ is not isomorphic to any subspace of codimension one. This gives a very strong counter-example, for 'natural' Banach spaces, to the 'hyperplane problem'; see also [204]. (Earlier, the first counter-example to the hyperplane problem was given by Gowers in [122]; here the author produced a separable Banach space that is not isomorphic to any proper, closed linear subspace of itself.) In particular, there is a locally compact, non-compact space L such that $C_0(L)\not\sim C(L_\infty)$. However, we always have $M(L)\cong M(L_\infty)$, as observed in Proposition 4.2.14.

In the second construction of Koszmider, K is a connected, compact space and $C(K)$ is not isomorphic to $C(L)$ for any totally disconnected, compact space L. This gives a negative answer to the above question; the result is proved in [160, Lemma 5.3]. In [21], the authors give another example of a compact space K such that $C(K)$ is not isomorphic to $C(L)$ for any totally disconnected, compact space L, but now K is 'relatively nice': for example, K is sequentially compact.

There is a valuable summary of spaces $C(K)$ with exotic properties in [161]; for further results on the above question, see [205].

6.2 Elementary theory

We now consider when $C_0(K)$ is injective and when it is a dual space. We already know from Corollary 2.4.17 that $C_0(K)$ is not injective and not isomorphically a dual space whenever K is a locally compact space that contains a convergent sequence of distinct points. In particular, this holds whenever K is an infinite, compact, metrizable space, such as $\mathbb{I}$ or Δ. Also, by Theorem 2.4.12, $C_0(K)$ is not injective whenever K is a locally compact space that is not pseudo-compact. On the other hand, of course, the space $\ell^\infty = C(\beta\mathbb{N}) \cong (\ell^1)' = c_0''$ is isometrically a bidual space.

Here we shall prove rather easily that a compact space with only finitely many clopen subsets is not isometrically a dual space. In §6.4, we shall characterize the compact spaces K such that $C(K)$ is isometrically a dual space, and so this result will be subsumed in a later theorem; however, the later theorem requires more substantial preliminaries.

The following theorem concerning the space of real-valued, continuous functions uses no more than the Krein–Milman theorem. A stronger result will be given in Corollary 6.9.7.

Theorem 6.2.1. *Let K be an infinite, compact space that has only finitely many clopen subsets. Then $C_\mathbb{R}(K)$ is not isometrically a dual space.*

Proof. The extreme points of $C_\mathbb{R}(K)_{[1]}$ are exactly the functions $f \in C(K)$ with $f(x) = \pm 1$ $(x \in K)$. Thus in our case the linear span, say F, of $\operatorname{ex} C_\mathbb{R}(K)_{[1]}$ is finite dimensional.

Assume that $C_\mathbb{R}(K) \cong E'$ for a Banach space E. Then F is $\sigma(E', E)$-closed, and so, by the Krein–Milman theorem, Theorem 2.6.1, $C_\mathbb{R}(K) = F$ is finite dimensional, a contradiction. Thus $C_\mathbb{R}(K)$ is not isometrically a dual space. □

We remark that it is an easy exercise to show that $C_\mathbb{R}(K)_{[1]} = \overline{\text{co}}(\operatorname{ex}(C_\mathbb{R}(K)_{[1]}))$ if and only if K is totally disconnected.

One might think that a straightforward modification of the above argument would give the same result for $C(K)$ itself. However, the following result suggests that this is not the case.

Theorem 6.2.2. *Let K be a non-empty, compact space. Then*

$$C(K)_{[1]} = \overline{\text{co}}(\operatorname{ex}(C(K)_{[1]}))\,.$$

Proof. Set $B = C(K)_{[1]}$ and $S = \{f \in C(K) : |f(t)| = 1 \ (t \in K)\}$. By Proposition 2.6.13, $S = \operatorname{ex} B$, and clearly S is circled, and so, by Corollary 2.1.3, it suffices to show that

$$\|\mu\| \leq \sup\left\{\left|\int_K f \,\mathrm{d}\mu\right| : f \in S\right\} \quad (\mu \in M(K))\,. \tag{6.3}$$

Fix $\mu \in M(K)$ and $\varepsilon > 0$, and take $f \in B$ with $|\langle f, \mu\rangle| > \|\mu\| - \varepsilon$, so that

$$\int_K 1 \, \mathrm{d}|\mu| = \|\mu\| < \left| \int_K f \, \mathrm{d}\mu \right| + \varepsilon \leq \int_K |f| \, \mathrm{d}|\mu| + \varepsilon .$$

Let $B_1 = \{x \in K : \Im f(x) \geq 0\} \setminus \mathbf{Z}(f)$ and $B_2 = \{x \in K : \Im f(x) < 0\}$. Then B_1 and B_2 are disjoint Borel sets in K with $B_1 \cup B_2 \cup \mathbf{Z}(f) = K$. Take L_1, L_2 to be compact sets with $L_1 \subset B_1$, $L_2 \subset B_2$, and $|\mu|((B_1 \cup B_2) \setminus (L_1 \cup L_2)) < \varepsilon$. Then there exists $g \in C_{\mathbb{R}}(K)$ (with $|g|_K \leq \pi$) such that

$$f(x) = |f|(x) \exp(\mathrm{i}g(x)) \quad (x \in L_1 \cup L_2 \cup \mathbf{Z}(f)),$$

and in this case

$$\begin{aligned}\left| \int_K (\exp(\mathrm{i}g) - f) \, \mathrm{d}\mu \right| &\leq \int_K |\exp(\mathrm{i}g) - f| \, \mathrm{d}|\mu| < \int_{L_1 \cup L_2 \cup \mathbf{Z}(f)} (1 - |f|) \, \mathrm{d}|\mu| + \varepsilon \\ &\leq \int_K (1 - |f|) \, \mathrm{d}|\mu| + \varepsilon < 2\varepsilon .\end{aligned}$$

Thus,

$$\|\mu\| < \left| \int_K f \, \mathrm{d}\mu \right| + \varepsilon \leq \left| \int_K \exp(\mathrm{i}g) \, \mathrm{d}\mu \right| + 3\varepsilon .$$

Since $\exp(\mathrm{i}g) \in S$, equation (6.3) follows. □

The above result is a special case of the *Russo–Dye* theorem [68, Theorem 3.2.18], which asserts that, for each unital C^*-algebra A, we have $A_{[1]} = \overline{\mathrm{co}}(\mathscr{U}(A))$, where $\mathscr{U}(A)$ denotes the set of unitary elements of A. For a proof of a slightly stronger result, see [78, Theorem I.8.4].

It follows that, for each compact space K such that $C(K) \cong E'$ for a Banach space E, it is indeed true that $\mathrm{co}(\mathrm{ex}\, C(K)_{[1]})$ is $\sigma(E', E)$-dense in $C(K)_{[1]}$.

Nevertheless, we shall obtain (in Theorem 6.2.6) a version of Theorem 6.2.1 for $C(K)$ without using anything beyond the Krein–Milman theorem. We require some preliminary results.

Lemma 6.2.3. *Let K be a compact space, and let $\mu \in M(K)^+$. Take $f, g \in L^1_{\mathbb{R}}(\mu)$ and $\varepsilon > 0$. Suppose that $\|f + \mathrm{i}g\|_1 = 1$ and $1 - \varepsilon < \|f\|_1 \leq 1$. Then $\|g\|_1 \leq \sqrt{2\varepsilon}$.*

Proof. Take $a, b > 0$. Since $(1+t)^{1/2} \leq 1 + t/2$ $(t \geq 0)$, we have

$$a^2 + b^2 \geq a^2 \left(1 + \frac{b^2}{a^2}\right)^{1/2} + \frac{b^2}{2} = a(a^2 + b^2)^{1/2} + \frac{b^2}{2},$$

and so $(a^2 + b^2)^{1/2} \geq a + b^2/2(a^2 + b^2)^{1/2}$.

Set $h = g^2/(f^2 + g^2)^{1/2}$. It follows that

$$1 = \int_K (f^2 + g^2)^{1/2} \, \mathrm{d}\mu \geq \int_K |f| \, \mathrm{d}\mu + \frac{1}{2} \int_K h \, \mathrm{d}\mu ,$$

and so $\int_K h \, d\mu < 2\varepsilon$. We then have

$$\int_K |g| \, d\mu = \int_K \frac{|g|}{(f^2+g^2)^{1/4}} (f^2+g^2)^{1/4} \, d\mu$$
$$\leq \left(\int_K h \, d\mu \right)^{1/2} \left(\int_K (f^2+g^2)^{1/2} \, d\mu \right)^{1/2}$$

by Cauchy–Schwarz, and so $\|g\|_1 \leq \sqrt{2\varepsilon}$. □

Corollary 6.2.4. *Let K be a compact space, and let $\mu, \nu \in M_{\mathbb{R}}(K)$. Take $\varepsilon > 0$, and suppose that $\|\mu + \mathrm{i}\nu\| = 1$ and that $1 - \varepsilon < \|\mu\| \leq 1$. Then $\|\nu\| \leq \sqrt{2\varepsilon}$.*

Proof. Consider the measure $\lambda = |\mu| + |\nu| \in M(K)^+$. Then $\mu = f\lambda$ and $\nu = g\lambda$ for some $f, g \in L^1_{\mathbb{R}}(\lambda)$ such that $\|\mu\| = \|f\|_1$ and $\|\nu\| = \|g\|_1$. Thus the result follows from the lemma. □

Proposition 6.2.5. *Let K be a non-empty, compact space. Then the Banach space $C(K)$ is isometrically the dual of a Banach space if and only if the real Banach space $C_{\mathbb{R}}(K)$ is isometrically the dual of a real Banach space.*

Proof. First, suppose that $C_{\mathbb{R}}(K)$ is isometrically the dual of a real Banach space. Then $C(K)$ is isometrically a dual space by Proposition 2.3.6.

Second, suppose that $C(K) \cong E'$, where E is a Banach space; we regard E as a closed subspace of $E'' = M(K)$. Define

$$F = \{\Re\mu \in M_{\mathbb{R}}(K) : \mu \in E\}.$$

Then F is a real-linear subspace of $M_{\mathbb{R}}(K)$, and $\Re\mu, \Im\mu \in F$ whenever $\mu \in E$, so that $E = F \oplus \mathrm{i}F$.

For each (real-linear) $\lambda \in F'$, define $\widetilde{\lambda}$ on E by

$$\widetilde{\lambda}(\mu + \mathrm{i}\nu) = \lambda(\mu) + \mathrm{i}\lambda(\nu) \quad (\mu, \nu \in F).$$

Then $\widetilde{\lambda}$ is a continuous, complex-linear functional on E with $\|\lambda\| \leq \left\|\widetilde{\lambda}\right\| \leq \sqrt{2}\|\lambda\|$. Thus there exist unique elements $f, g \in C_{\mathbb{R}}(K)$ with

$$\widetilde{\lambda}(\mu + \mathrm{i}\nu) = \langle f + \mathrm{i}g, \mu + \mathrm{i}\nu \rangle \quad (\mu + \mathrm{i}\nu \in E).$$

It follows that $\lambda(\mu) = \langle f, \mu \rangle - \langle g, \nu \rangle$ and $\lambda(\nu) = \langle f, \nu \rangle + \langle g, \mu \rangle$ when $\mu + \mathrm{i}\nu \in E$. Define $T : \lambda \mapsto f$ in this case. Then $T : F' \to C_{\mathbb{R}}(K)$ is a continuous, real-linear map (with $\|T\| \leq \sqrt{2}$) such that $|T(\lambda)|_K \geq \|\lambda\| \ (\lambda \in F')$.

Take $f \in C_{\mathbb{R}}(K)$, and define $\lambda(\mu) = \langle f, \mu \rangle \ (\mu \in F)$. Then $\lambda \in F'$ is such that $\|\lambda\| \leq |f|_K$, and clearly $T(\lambda) = f$. This shows that T is a surjection.

Take $\lambda \in F'$ with $T(\lambda) = 0$, and assume towards a contradiction that $\lambda \neq 0$. Then $\widetilde{\lambda} \neq 0$, and so we may suppose that $\left\|\widetilde{\lambda}\right\| = 1$. Thus there exists $g \in C_{\mathbb{R}}(K)$ with $|g|_K = 1$ such that $\lambda(\mu) = -\langle g, \nu\rangle$ and $\lambda(\nu) = \langle g, \mu\rangle$ when $\mu + \mathrm{i}\nu \in E$. Choose $t \in K$ with $|g(t)| = 1$; by replacing λ by $-\lambda$, if necessary, we may suppose that $g(t) = 1$. The closed unit ball $E_{[1]}$ is weak*-dense in $M(K)_{[1]}$, and so, for each $\varepsilon > 0$, there exists $\mu_0 + \mathrm{i}\nu_0 \in E_{[1]}$ with $|\langle g, \delta_t - (\mu_0 + \mathrm{i}\nu_0)\rangle| < \varepsilon$. Thus,

$$|1 - \langle g, \mu_0\rangle| \leq |1 - \langle g, \mu_0 + \mathrm{i}\nu_0\rangle| < \varepsilon .$$

Since $1 - \varepsilon < \|\mu_0\| \leq 1$, it follows from Corollary 6.2.4 that $\|\nu_0\| \leq \sqrt{2\varepsilon}$, and so

$$1 - \varepsilon \leq |\langle g, \mu_0\rangle| = |\lambda(\nu_0)| \leq \|\nu_0\| \leq \sqrt{2\varepsilon},$$

a contradiction for some $\varepsilon > 0$. Thus $\lambda = 0$ and so T is an injection.

We have shown that $T : F' \to C_{\mathbb{R}}(K)$ is an isomorphism. To show that T is an isometry, it remains to show that $|T(\lambda)|_K \leq \|\lambda\|$ $(\lambda \in F')$. Take $f \in C_{\mathbb{R}}(K)$, and then take $t \in K$ with $|f(t)| = |f|_K$. For each $\varepsilon > 0$, there exists $\mu + \mathrm{i}\nu \in E_{[1]}$ with $|f(t) - \langle f, \mu + \mathrm{i}\nu\rangle| < \varepsilon$. We have $\mu \in F$ with $\|\mu\| \leq 1$. Take the unique λ with $T(\lambda) = f$, so that, as above, $\lambda(\mu) = \langle f, \mu\rangle$. Then

$$\|\lambda\| \geq |\langle f, \mu\rangle| > |f(t)| - \varepsilon = |f|_K - \varepsilon = |T(\lambda)|_K - \varepsilon,$$

and so $|T(\lambda)|_K \leq \|\lambda\| + \varepsilon$. This holds true for each $\varepsilon > 0$, and so $|T(\lambda)|_K \leq \|\lambda\|$. But this is true for each $\lambda \in F'$, and so T is an isometry. □

Theorem 6.2.6. *Let K be an infinite, compact space that has only finitely many clopen subsets. Then $C(K)$ is not isometrically a dual space.*

Proof. Assume towards a contradiction that $C(K)$ is isometrically a dual space. By Proposition 6.2.5, $C_{\mathbb{R}}(K)$ is isometrically the dual of a real Banach space. But, by Theorem 6.2.1, this is not the case. □

We shall note in Corollary 6.9.7 that, in the above situation, $C(K)$ is not even isomorphically a dual space. The above theorem is also an immediate consequence of Theorem 6.4.1, below.

6.3 Uniqueness of preduals of $M(K)$

Let K be a non-empty, locally compact space. Then $C_0(K)' = M(K)$. Now suppose that E and F are Banach spaces such that $E' \cong F' \cong M(K)$. Can we say that there is any uniqueness about E and F? Is it the case that necessarily $E \cong F$ or that $E \sim F$? The short answer to these questions is 'no'.

Let us consider the situation when $K = \mathbb{N}$, and so $M(K) = \ell^1$.

It is standard that $(c_0)' = \ell^1$, with the duality

$$(\alpha, \beta) \mapsto \sum_{n=1}^{\infty} \alpha_n \beta_n, \quad c_0 \times \ell^1 \to \mathbb{C}.$$

Similarly, the dual of c is $\ell^1(\mathbb{N} \cup \{\infty\}) \cong \ell^1$. Thus ℓ^1 has two preduals which, by Example 6.1.1, are not isometrically isomorphic; in the terminology of Definition 2.2.29, the Banach space ℓ^1 does not have a unique predual. However, as in Example 6.1.1, the two spaces c_0 and c are isomorphic. We now seek examples of Banach spaces E and F such that $E' \cong F' \cong \ell^1$, but such that E and F are not even isomorphic to each other.

In fact, there is a multitude of isometric preduals of ℓ^1, no two of which are mutually isomorphic. For example, suppose that $E = C(K)$ for a countable, compact space K. Then $E' \cong \ell^1$. As we noted in §6.1, there are many countable, compact spaces of the form $\omega^{\omega^\alpha} + 1$ such that no two of the Banach spaces $C(\omega^{\omega^\alpha} + 1)$ are pairwise isomorphic. The 'smallest' countable, compact space K such that $C(K)$ is not isomorphic to c is $K = [0, \omega^\omega]$.

There is a famous example of Benyamini and Lindenstrauss given in [32] (see also [215, p. 1599]) which exhibits a Banach space F such that $F' \cong \ell^1$, but such that F is not isomorphic to any complemented subspace of a space of the form $C(K)$ for a compact space K; further, the space F is not isomorphic to any Banach lattice. The space constructed by Alspach in [7] is also an isometric predual of ℓ^1 that is not isomorphic to a complemented subspace of any $C(K)$ space; a further example of such a space, with additional properties, is given by Gasparis in [109].

In [44], Bourgain and Delbaen showed that there is an isomorphic predual E of ℓ^1 such that E has the Radon–Nikodým property and each infinite-dimensional subspace of E contains a further infinite-dimensional subspace that is reflexive. The underlying construction used in this paper has become a standard technique. Also it is shown in [105] that, given any Banach space F with F' separable, there is a predual E of ℓ^1 which contains an isomorphic copy of F.

A further remarkable predual of ℓ^1 is constructed by Argyros and Haydon in [16]. This is an infinite-dimensional Banach space E that is a closed subspace of ℓ^∞ and such that $E' \sim \ell^1$. The space E is such that every bounded operator on it is a scalar multiple of the identity I_E plus a compact operator. That there exists an infinite-dimensional Banach space E such that $\mathscr{B}(E) = \mathbb{C}I_E + \mathscr{K}(E)$ answered a very famous question that had been open for at least 40 years, and was probably known to Banach. The space E, which is built by using a development of the Bourgain–Delbaen construction, has several other striking properties. This space E clearly has the property that the only closed subspaces that are complemented in E are those that are either of finite dimension or of finite codimension.

The paper [80] considers (isometric and isomorphic) preduals E of $\ell^1_{\mathbb{R}}(\mathbb{Z})$ which are invariant under the bilateral shift when they are regarded as subspaces of $\ell^\infty_{\mathbb{R}}(\mathbb{Z})$; many exotic preduals of this form are constructed. We note that, in the case where K is a countable, compact space, the canonical duality between $C_{\mathbb{R}}(K)$ and $\ell^1_{\mathbb{R}}(\mathbb{Z})$ does not make $C_{\mathbb{R}}(K)$ invariant under the bilateral shift. It is shown in [80, Theorem 5.8] that there is a Banach subspace E of $\ell^\infty_{\mathbb{R}}(\mathbb{Z})$ such that $E' \cong \ell^1_{\mathbb{R}}(\mathbb{Z})$ and E is invariant under the bilateral shift, but such that E is not isomorphic to $c_{0,\mathbb{R}}$. It is also shown that, given $\zeta \in \overline{\mathbb{D}}$, there is an isometric predual of $\ell^\infty(\mathbb{Z})$ such that $\zeta\delta_0$ belongs to the $\sigma(\ell^1(\mathbb{Z}),E)$-closure of the set $\{\delta_n : n \in \mathbb{Z}\}$. This implies that neither $\ell^1_{\mathbb{R}}(\mathbb{Z})$ nor $\ell^1(\mathbb{Z})^+$ is closed in $(\ell^1(\mathbb{Z}),\sigma(\ell^1(\mathbb{Z}),E))$, a fact that prevents some 'soft' proofs of some later results.

Thus the only positive result that we can hope for is that the compact spaces K and L share some property whenever $M(K) \cong M(L)$.

First we note that $|L| = |K|$ in this case. For suppose that $T : M(K) \to M(L)$ is an isometric isomorphism, and take $x \in K$. As in the proof of Theorem 6.1.2, $T\delta_x = \zeta\delta_y$ for some $\zeta \in \mathbb{T}$ and $y \in L$. Set $\theta(x) = y$. Then $\theta : K \to L$ is a bijection, and so $|L| = |K|$. However, the next example, taken from [71, Example 4.26], shows that metrizability of K is not necessarily preserved.

Example 6.3.1. Let $K = \mathbb{I}$ with the usual topology, and take

$$L = ((0,1] \times \{0\}) \cup ([0,1) \times \{1\})$$

as a set; L is ordered lexicographically, and then we assign the order topology to L, so that a typical basic set has the form

$$U = \{(c,0),(c,1) : a < c < b\},$$

where $0 < a < b < 1$; each other basic open set differs from such a set by at most 2 points. This space is called the *two-arrows space* [99, 3.10.C].

The relative topology from L on $(0,1]$, identified with $(0,1] \times \{0\}$, coincides with the Sorgenfrey topology on $(0,1]$; the latter topology is generated by intervals of the form $(a,b]$, where $0 < a < b \leq 1$. The space L is compact, but it is not metrizable because the Sorgenfrey topology on $(0,1]$ is not metrizable.

Clearly K and L have the same cardinality, so the spaces $M_d(K)$ and $M_d(L)$ of discrete measures can be identified.

We *claim* that it is also true that the spaces $M_c(K)$ and $M_c(L)$ of continuous measures can be identified. To see this, consider the family $\mathscr{F}$ of subsets S of L such that $S\Delta(B \times \{0,1\})$ is countable for some $B \in \mathfrak{B}_K$. The family $\mathscr{F}$ is a σ-algebra, and $\mathscr{F}$ contains all basic open sets of the above form U, and hence all basic open sets in L. It is easy to see that each open set in L is a countable union of basic open sets, and so $\mathscr{F}$ contains all the open sets in L. Hence $\mathscr{F}$ contains $\mathfrak{B}_L$, so that, in fact, $\mathscr{F} = \mathfrak{B}_L$. Take $\mu \in M_c(L)$, and define $T\mu \in M_c(K)$ by

$$(T\mu)(B) = \mu(B \times \{0,1\}) \quad (B \in \mathfrak{B}_K),$$

so that $T\mu \in M_c(L)$, $T \in \mathscr{B}(M_c(L),M_c(K))$, and $\|T\mu\| = \|\mu\|$. For each $\nu \in M_c(K)$ and $S \in \mathfrak{B}_L$, define $\mu(S) = \nu(B)$, where $B \in \mathfrak{B}_K$ is such that $S\Delta(B \times \{0,1\})$ is countable. Then $\mu(S)$ is well-defined, $\mu \in M_c(L)$, and $T\mu = \nu$. Thus T is a surjection. It is now clear that $T : M_c(L) \to M_c(K)$ is an isometric isomorphism, giving the claim.

It follows that $C(K)' = M(K) \cong C(L)' = M(L)$. However K is metrizable and $C(K)$ is separable, whereas L is not metrizable and $C(L)$ is not separable. Here we are using Theorem 2.1.7(i).

In particular, we see that we have two Banach spaces E and F such that $E' \cong F'$, but $d(E) \neq d(F)$. □

6.4 The isometric theory

Let K be a non-empty, compact space. We first recall the well-known theory that explains when $C(K)$ is isometrically a dual space and hence when $C(K)$ is a von Neumann algebra, equivalently, a W^*-algebra.

The theory involves hyper-Stonean spaces, and there is some dispute on the most appropriate definition of this term among several competing, equivalent, properties. Here, we have chosen a 'topological–measure theoretic' definition (see Definition 5.1.1), following the seminal paper of Dixmier [91]; in [72], we started with a functional-analytic approach. We now record the following theorem, which shows that several different definitions of 'hyper-Stonean' in the literature are equivalent.

We note that we have no purely topological characterization of hyper-Stonean spaces.

Part of the following theorem is contained in [234, Chapter III, §1], but we have avoided reference to representations on Hilbert spaces.

Recall that the set W_K was defined in Definition 5.1.1 and that K is hyper-Stonean if and only if K is Stonean and W_K is dense in K.

Theorem 6.4.1. *Let K be a non-empty, compact space. Then the following are equivalent:*

(a) *$C(K)$ is isometrically a dual space, so that $C(K)$ is a von Neumann algebra;*

(b) *there is a C^*-isomorphism*

$$T : f \mapsto (f \mid S_\alpha), \quad C(K) \to \bigoplus_\infty L^\infty(S_\alpha, \mu_\alpha),$$

where $\{\mu_\alpha : \alpha \in A\}$ is a maximal singular family in $N(K) \cap P(K)$ and we are setting $S_\alpha = \operatorname{supp} \mu_\alpha$ $(\alpha \in A)$;

(c) *the map $T : C(K) \to N(K)'$ defined by*

$$(Tf)(\mu) = \langle f, \mu \rangle = \int_K f \, d\mu \quad (f \in C(K), \mu \in N(K))$$

is an isometric isomorphism, and so $C(K) \cong N(K)'$;

(d) *K is Stonean and, for each $f \in C(K)^+$ with $f \neq 0$, there exists $\mu \in N(K)^+$ with $\langle f, \mu \rangle \neq 0$;*

(e) *K is hyper-Stonean;*

(f) *K is Stonean and there exists a category measure on K;*

(g) *there is a locally compact space Γ and a decomposable measure ν on Γ such that $C(K)$ is C^*-isomorphic to $L^\infty(\Gamma,\nu)$.*

Proof. (c) $\Rightarrow$ (a) and (b) $\Rightarrow$ (a) These are trivial.

(a) $\Rightarrow$ (d) By Proposition 6.2.5, there exists a real-linear subspace E of $M_{\mathbb{R}}(K)$ with $E' = C_{\mathbb{R}}(K)$. We write σ for the weak* topology $\sigma(C_{\mathbb{R}}(K),E)$ on the space $C_{\mathbb{R}}(K)$.

The space $(C_{\mathbb{R}}(K)_{[1]},\sigma)$ is compact. Since the map

$$f \mapsto \frac{1}{2}(1+f), \quad C_{\mathbb{R}}(K)_{[1]} \to C(K)^+_{[1]},$$

is a bijection which is a homeomorphism with respect to σ, the space $(C(K)^+_{[1]},\sigma)$ is also compact. By the Krein–Šmulian theorem, Theorem 2.1.4(vi), the positive cone $C(K)^+$ is closed in $(C_{\mathbb{R}}(K),\sigma)$. Take $f \in C_{\mathbb{R}}(K) \setminus C(K)^+$. Then the Hahn–Banach theorem, Theorem 2.1.8(ii), applies to show that there exists $\lambda \in (C_{\mathbb{R}}(K),\sigma)' = E$ with

$$\langle f,\lambda\rangle < \inf\{\langle g,\lambda\rangle : g \in C(K)^+\}.$$

It cannot be that $\langle g,\lambda\rangle < 0$ for some $g \in C(K)^+$: indeed, this would imply that $\langle ng,\lambda\rangle < \langle f,\lambda\rangle$ for some $n \in \mathbb{N}$, a contradiction, and so

$$\inf\{\langle g,\lambda\rangle : g \in C(K)^+\} = 0.$$

Thus $\lambda \in E^+$. It follows that, for each $f \in C_{\mathbb{R}}(K)$, we have $f \geq 0$ if and only if $\langle f,\lambda\rangle \geq 0$ $(\lambda \in E^+)$.

Let $(f_\alpha : \alpha \in A)$ be an increasing net in $C(K)^+_{[1]}$. Then $(f_\alpha : \alpha \in A)$ has an accumulation point, say f_0, in $(C(K)^+_{[1]},\sigma)$; by passing to a subnet, we may suppose that $\lim_\alpha f_\alpha = f_0$ with respect to σ. For each $\lambda \in E^+$, the net $(\langle f_\alpha,\lambda\rangle : \alpha \in A)$ is increasing with limit $\langle f_0,\lambda\rangle$, and so $\langle f_\alpha,\lambda\rangle \leq \langle f_0,\lambda\rangle$ $(\alpha \in A)$. It follows that $f_\alpha \leq f_0$ $(\alpha \in A)$.

Suppose that $g \in C(K)^+$ with $f_\alpha \leq g$ $(\alpha \in A)$. Then $\langle f_\alpha,\lambda\rangle \leq \langle g,\lambda\rangle$ $(\lambda \in E^+)$ for each $\alpha \in A$, and so $\langle f_0,\lambda\rangle \leq \langle g,\lambda\rangle$. This implies that $f_0 \leq g$ and hence that $f_0 = \bigvee\{f_\alpha : \alpha \in A\}$. Thus $C(K)$ is Dedekind complete, and so it now follows from Theorem 2.3.3 that K is a Stonean space.

Next, suppose that $\mu \in E$ and $g_\alpha \searrow 0$ in $C_{\mathbb{R}}(K)$. Then $1 = \bigvee(1-g_\alpha)$, and so $\lim_\alpha \langle g_\alpha,\mu\rangle = 0$. This shows that μ is normal. Thus $E \subset N(K)$.

For each $f \in C(K)^+$ with $f \neq 0$, there exists $\mu \in E^+$ with $\langle f,\mu\rangle \neq 0$; since $E^+ \subset N(K)^+$, (d) follows.

(d) $\Rightarrow$ (e) Let U be a non-empty, open subset of the Stonean space K. Then there exists $f \in C(K)^+$ with $f \neq 0$ such that $\operatorname{supp} f \subset U$. By (d), there exists $\mu \in N(K)^+$ with $\langle f,\mu\rangle \neq 0$. Clearly $(\operatorname{supp}\mu) \cap U \neq \emptyset$. This shows that W_K is dense in K, and so K is hyper-Stonean.

(e) $\Rightarrow$ (b) This follows from Theorem 5.1.8.

(b) $\Rightarrow$ (c) Since (b) $\Rightarrow$ (a) $\Rightarrow$ (d), the space K is Stonean, and hence, by Corollary 4.7.11(iii), the spaces S_α are pairwise disjoint.

Set $E = \bigoplus_1 L^1(S_\alpha, \mu_\alpha)$, so that $E' = \bigoplus_\infty L^\infty(S_\alpha, \mu_\alpha)$. The map $T' : E'' \to M(K)$ is an isometric isomorphism. We *claim* that T' maps E onto $N(K)$.

Indeed, take $h = (h_\alpha)$ in E (so that only countably many of the h_α are non-zero), and set $\nu = T'h \in M(K)$. Take $f \in C(K)$, and, for each α, set $f_\alpha = f \mid S_\alpha$, so that

$$\int_K f \, d\nu = \langle f, \nu \rangle = \langle Tf, h \rangle = \sum_{\alpha \in A} \int_{S_\alpha} f_\alpha h_\alpha \, d\mu_\alpha , \tag{6.4}$$

where we note that

$$\int_{S_\alpha} f_\alpha h_\beta \, d\mu_\alpha = 0 \quad (\alpha, \beta \in A, \alpha \neq \beta).$$

Take $L \in \mathscr{K}_K$. Then, for each $\alpha \in A$, we have $L \cap S_\alpha \in \mathscr{K}_K$ and $\mu_\alpha \in N(K)$, and so $\mu_\alpha(L \cap S_\alpha) = 0$. By (6.4) with $f = \chi_L$, we have $\nu(L) = 0$. By Theorem 4.7.4(i), $\nu \in N(K)$.

Conversely, take $\nu \in N(K)$. Then $|\nu|(K \setminus \bigcup_\alpha S_\alpha) = 0$. For each $\alpha \in A$, it follows from Corollary 4.7.11(i) that $\nu \mid S_\alpha \ll \mu_\alpha$, and so, by the Radon–Nikodým theorem, Theorem 4.4.9, there exists $h_\alpha \in L^1(S_\alpha, \mu_\alpha)$ with $\nu \mid S_\alpha = h_\alpha \mu_\alpha$ and $\|h_\alpha\|_1 = \|\nu \mid S_\alpha\|$. Set $h = (h_\alpha)$. Then

$$\sum_\alpha \|h_\alpha\|_1 = \sum_\alpha \|\nu \mid S_\alpha\| = \|\nu\| ,$$

so that $h \in E$, and

$$\int_K f \, d\nu = \sum_\alpha \int_{S_\alpha} f_\alpha h_\alpha \, d\mu_\alpha ,$$

whence $T'h = \nu$. This establishes the claim.

It follows that $C(K) \cong N(K)'$. When we identify E and $N(K)$, we obtain the formula in (c).

(e) $\Leftrightarrow$ (f) This follows from Proposition 5.1.7.

(g) $\Rightarrow$ (a) This follows because $L^\infty(\Gamma, \nu) \cong L^1(\Gamma, \nu)'$ by Theorem 4.4.10(ii).

(b) $\Rightarrow$ (g) We take Γ to be the (pairwise-disjoint) union of the family $\{S_\alpha\}$, and set $\nu = \sum_\alpha \mu_\alpha$, so that ν is a decomposable measure. It is clear that $C(K)$ is C^*-isomorphic to $L^\infty(\Gamma, \nu)$. $\square$

The implication (e) $\Rightarrow$ (a) of the above theorem is due to Dixmier [91]; the converse implication, (a) $\Rightarrow$ (e), is due to Grothendieck [125].

The Bade 1957 Notes [23] were first developed in a 1957 seminar on $C(K)$ at Berkeley and were mimeographed and distributed to many of Bade's students and colleagues. He held similar seminars in 1964 and 1967. A (somewhat faded) copy of these notes is in the holdings of the Mathematics–Statistics library at Berkeley. The 1957 Notes prove (Grothendieck's theorem) that, if $C(K)$ has an isometric predual, then K is hyper-Stonean, as in our definition. No proof of this direction was included in the 1971 Aarhus Notes [24], which retain much of the earlier material and a few additional topics. The converse direction – that, if K is hyper-Stonean, then $C(K)$ has an isometric predual (Dixmier's theorem) – is mentioned without proof in [23] and proved in [24]. So the 1957 and 1971 Notes of Bade contained the only treatment of the main results of the above theorem available in English for many years until the topic found its way into several texts (for example, [166, 223, 234]) in the mid-1970s. Thus, these notes were, for a long period, the only place where some basic facts about $C(K)$ were unified and clearly expounded.

Theorem 6.4.2. *Let K be a non-empty, compact space. Then $C(K)$ is a von Neumann algebra if and only if K is hyper-Stonean. In this case, an isometric predual of $C(K)$ is $N(K)$, and this predual is strongly unique.*

Proof. By the equivalence of (a) and (e) in Theorem 6.4.1, $C(K)$ is a von Neumann algebra if and only if K is hyper-Stonean. Further, by (c), $N(K)' = C(K)$. It remains to show that $N(K)$ is strongly unique in this case.

Suppose that E is an isometric predual of $C(K)$. Then we can regard E as a closed linear subspace of $M(K)$, and we have noted in the proof of the implication (a) $\Rightarrow$ (d) in Theorem 6.4.1 that $E \subset N(K)$. By the Hahn–Banach theorem, $E = N(K)$.

Suppose that G is a Banach space and that $T : C(K) \to G'$ is an isometric operator. Equip the Banach space G' with the product and the involution transferred from $C(K)$. Then $T' : G'' \to M(K)$ is continuous when G'' has the topology $\sigma(G'', G')$ and $M(K)$ has the topology $\sigma(M(K), C(K))$. By Theorem 2.1.4(iv), T' maps G onto $N(K)$. Thus $N(K)$ is the strongly unique predual of $C(K)$. □

Thus, in the case where K is hyper-Stonean, equivalently, where $C(K)$ is isometrically a dual space, the strongly unique isometric predual of $C(K)$ is the space $N(K)$ of normal measures on K; we write $N(K) = C(K)_*$.

Let K be a non-empty, compact space, and suppose that E is a Banach sublattice of $M(K)$, so that E is an AL-space and, by Theorem 2.3.8, E' is an AM-space. The map $\mu \mapsto \mu(K)$, $E \to \mathbb{C}$, is an AM-unit in E', and so, by Theorem 2.3.9(iii), E' is Banach-lattice isometric to $C(X)$ for a compact space X.

Example 6.4.3. In view of the above uniqueness result, one could ask the following 'isomorphic' question. Let K be a hyper-Stonean space, so that $C(K)$ is isometrically a dual space, with predual $N(K)$. Suppose also that E is a Banach space such that $E' \sim C(K)$. Can we conclude that $E \sim N(K)$? In fact, this is easily seen not to be the case. For example, the isometric preduals of $\ell^\infty = C(\beta\mathbb{N})$ and $L^\infty(\mathbb{I}) = C(\mathbb{H})$ are ℓ^1 and $L^1(\mathbb{I})$, respectively. By Theorem 4.4.7, $L^\infty(\mathbb{I}) \sim \ell^\infty$, and so $L^1(\mathbb{I})$ and ℓ^1 are isomorphic preduals of both ℓ^∞ and $C(\mathbb{H})$. However $L^1(\mathbb{I})$ and ℓ^1 are not isomorphic: $\ell^1 = c_0'$ is isometrically a dual space, but, by Theorem 4.4.19, $L^1(\mathbb{I})$ is not even isomorphically a dual space. Hence we see that both ℓ^∞ and $C(\mathbb{H})$ have isometric preduals and also isomorphic preduals that are not isomorphic to the unique isometric predual. □

Proposition 6.4.4. *Let K be a non-empty, compact space. Suppose that $C(K)$ is isometrically a dual space and that L is a non-empty, clopen subspace of K. Then $C(L)$ is isometrically a dual space, and $N(L) \neq \{0\}$.*

Proof. By Theorem 6.4.1, (a) $\Rightarrow$ (e), K is hyper-Stonean. By Proposition 5.1.2(i), L is hyper-Stonean, and so, by Theorem 6.4.1, (e) $\Rightarrow$ (a), $C(L)$ is isometrically a dual space. Since $C(L) \neq \{0\}$, necessarily $N(L) \neq \{0\}$. □

Proposition 6.4.5. *Let K be a non-empty, compact space. Then $C(K)$ is not isometrically a dual space whenever K contains a non-empty, clopen subspace from which there is a continuous, open surjection onto a compact space L that has one of the following properties:*

(i) *L is a separable, compact space without isolated points;*

(ii) *L is a non-discrete, locally compact group;*

(iii) *L is a locally connected, compact space without isolated points;*

(iv) *$L = S^*$, where S is an infinite set.*

Proof. By Propositions 4.7.17 and 6.4.4, it suffices to show that $N(L) = \{0\}$ in each of the four cases. This follows from Proposition 4.7.20, Corollary 4.7.22, Theorem 4.7.23, and Corollary 4.7.14, respectively. □

We shall see in Example 6.9.10 that there is a separable, compact space K without isolated points such that $C(K)$ is isomorphically a dual space, despite the fact that $N(K) = \{0\}$. On the other hand, $C(K)$ and $C(S^*)$ are not even isomorphically dual spaces for any locally connected, compact space K without isolated points or for any infinite set S; see Corollary 6.9.7.

There are generalizations of parts of Theorems 6.4.1 and 6.4.2 to non-commutative C^*-algebras. We sketch these without giving proofs, which can be found in the texts [149, 150] of Kadison and Ringrose, [222, §1.13] of Sakai, and [234] of Takesaki; the key points are due to Sakai [221] in 1956.

By [150, Definition 7.1.11], a positive linear functional λ on a von Neumann algebra A is *normal* if it is order-continuous, in the sense that $\lambda(a_\alpha) \to \lambda(a)$ for each monotone increasing net (a_α) in A^+ with supremum a; a continuous linear functional is *normal* if it is a linear combination of normal, positive linear functionals. By [150, Theorem 7.1.12], a state on A is normal if and only if it is weak operator continuous on $A_{[1]}$ (and several other equivalences are given in this reference).

Now suppose that A is a von Neumann algebra and that A has an isometric predual E. By [150, Theorem 7.4.2], the $\sigma(A,E)$ topology on $A_{[1]}$ coincides with the weak operator topology on $A_{[1]}$. Alternatively, [222, Theorem 1.13.2] shows directly that a positive linear functional λ on A is normal if and only if it is $\sigma(A,E)$-continuous. In each case it follows that the Banach space of normal functionals is the unique isometric predual, called A_*, of A. An argument similar to the one that we gave then shows that the predual E is strongly unique.

6.5 The space $\widetilde{K}$

We first summarize some results that we have already obtained related to the hyper-Stonean envelope $\widetilde{K}$ of a locally compact space K. The main theorems that contribute to the following summary are Theorem 5.4.1, in which we identify the bidual, $(C_0(K)'', \Box)$, of $C_0(K)$ with $C(\widetilde{K})$, and Theorem 6.4.2, where we identify the unique isometric predual of a space $C(L)$, where L is a hyper-Stonean space.

Theorem 6.5.1. *Let K be a non-empty, locally compact space. Then:*

(i) *the dual of $C_0(K)$ is the space $M(K)$ of complex-valued, regular Borel measures on K;*

(ii) *the bidual of $C_0(K)$ and the dual of $M(K)$ is the space $C(\widetilde{K})$, where $\widetilde{K}$ is the hyper-Stonean envelope of K;*

(iii) *the strongly unique predual $C(\widetilde{K})_*$ of $C(\widetilde{K})$ is the space $N(\widetilde{K})$ of normal measures on $\widetilde{K}$, and $N(\widetilde{K})$ is Banach-lattice isometric to $M(K)$;*

(iv) *the third dual $C_0(K)'''$ of $C_0(K)$ and bidual of $M(K)$ and the dual of $C(\widetilde{K})$ is the space $M(\widetilde{K})$ of complex-valued, regular Borel measures on $\widetilde{K}$;*

(v) *the canonical embedding $\kappa_{C_0(K)} : C_0(K) \to C_0(K)''$ is a C^*-algebra embedding, and the pointwise product on $C(\widetilde{K})$ is equal to the Arens product $\Box$ on $C_0(K)''$ when the spaces $C_0(K)''$ and $C(\widetilde{K})$ are identified.* □

Thus, with appropriate identifications, we have:

$$C_0(K)' = M(K) = N(\widetilde{K}), \quad C_0(K)'' = M(K)' = C(\widetilde{K}), \tag{6.5}$$

$$C_0(K)''' = M(K)'' = C(\widetilde{K})' = M(\widetilde{K}). \tag{6.6}$$

Since $M(K) \cong M(K_\infty)$ (by Proposition 4.2.14), we also have

$$\widetilde{K} \quad \text{is homeomorphic to} \quad \widetilde{K_\infty}. \tag{6.7}$$

We now describe the topology of the hyper-Stonean envelope $\widetilde{K}$ of K in more detail, and calculate the cardinalities of some subsets; most of the results are taken from [72].

Let (K, τ) be a non-empty, locally compact space, with hyper-Stonean envelope $\widetilde{K}$. The topology on the space $\widetilde{K}$ is the weak* topology $\sigma = \sigma(M(\widetilde{K}), C(\widetilde{K}))$ restricted to $\widetilde{K}$, the character space of $C(\widetilde{K})$.

The dual of the canonical embedding of $C_0(K)$ into $C_0(K)'' = C(\widetilde{K})$ is a bounded projection $\pi : M(\widetilde{K}) \to M(K)$, and the restriction of π to $\widetilde{K}$ (where $\widetilde{K}$ is identified with the point masses in $M(\widetilde{K})$) is a continuous projection $\pi : (\widetilde{K}, \sigma) \to (K_\infty, \tau)$. The canonical embedding is

$$\kappa = \kappa_{M(K)} : M(K) \to M(K)'' = M(\widetilde{K}),$$

with $\kappa(M(K)) = N(\widetilde{K})$, and $\pi \circ \kappa$ is the identity on $M(K)$; the pair $(\widetilde{K}, \pi)$ is a cover of K_∞ in the sense of Definition 1.4.20. We see that

$$N_d(\widetilde{K}) = \kappa(M_d(K)) \quad \text{and} \quad N_c(\widetilde{K}) = \kappa(M_c(K)).$$

Since $M(K) = M_d(K) \oplus_1 M_c(K)$, we have $N(\widetilde{K}) = N_d(\widetilde{K}) \oplus_1 N_c(\widetilde{K})$.

For each $x \in K$, the *fibre* $K_{\{x\}}$ is defined to be $\pi^{-1}(\{x\})$; the canonical embedding of $C(K)$ in $C(\widetilde{K})$ identifies $C(K)$ with the functions in $C(\widetilde{K})$ that are constant on each fibre.

Let E be a $C_0(K)$-submodule of $M(K)$. Then

$$E^\circ = \{F \in C(\widetilde{K}) : F \mid E = 0\}$$

is a σ-closed subspace of $C(\widetilde{K})$. Further, it is clear that E° is a closed ideal in $C(\widetilde{K})$, and so, by Proposition 3.1.6, it has the form $\{F \in C(\widetilde{K}) : F \mid L_E = 0\}$ for some compact subspace L_E of $\widetilde{K}$. In particular, we have noted that $M_d(K)$ and $M_c(K)$ are $C_0(K)$-submodules of $M(K)$; the corresponding compact subspaces of $\widetilde{K}$ are denoted by $\widetilde{K}_d$ and $\widetilde{K}_c$, respectively. Since $N(\widetilde{K}) = N_d(\widetilde{K}) \oplus_1 N_c(\widetilde{K})$, it follows easily that $\{\widetilde{K}_d, \widetilde{K}_c\}$ is a partition of $\widetilde{K}$ into two clopen subspaces; further,

$$N_d(\widetilde{K})^\circ = \{F \in C(\widetilde{K}) : F \mid \widetilde{K}_d = 0\},$$

identified with $C(\widetilde{K}_c)$; similarly, $N_c(\widetilde{K})^\circ$ is identified with $C(\widetilde{K}_d)$. It follows that

$$C(\widetilde{K}) = C(\widetilde{K}_d) \oplus_\infty C(\widetilde{K}_c).$$

Definition 6.5.2. Let K be a non-empty, locally compact space. Then

$$U_K = \bigcup\{\Phi_\mu : \mu \in P(K)\} \quad \text{and} \quad U_{K,c} = \bigcup\{\Phi_\mu : \mu \in P_c(K)\}.$$

In fact, U_K is exactly $W_{\widetilde{K}}$, in the notation of Definition 5.1.1, and U_K contains the set $U_{\mathscr{F}}$ that arose in Theorem 5.4.1; the sets U_K and $U_{K,c}$ are open subsets of $\widetilde{K}$. The following summary uses Theorem 5.4.1 and Corollary 5.2.6.

Theorem 6.5.3. *Let (K, τ) be a non-empty, locally compact space. Then $\widetilde{K}$ is a hyper-Stonean space for the specified topology σ. The canonical embedding of K in $\widetilde{K}$ identifies K with the set $D_{\widetilde{K}}$ of isolated points of $\widetilde{K}$; the space (K, σ) is discrete; there is a continuous surjection $\pi : (\widetilde{K}, \sigma) \to (K_\infty, \tau)$; the closure of K in $(\widetilde{K}, \sigma)$ is identified with $\beta K_d = \widetilde{K_d}$.*

For each $\mu \in P(K)$, the character space of $L^\infty(K, \mu)$ is identified with the non-empty, clopen subset Φ_μ of $\widetilde{K}$, and $L^\infty(K, \mu)$ and $C(\Phi_\mu)$ are C^*-isomorphic. The non-empty, clopen subspaces of $\widetilde{K}$ that satisfy CCC *are exactly those of the form Φ_μ for some $\mu \in P(K)$.*

The sets U_K and $U_{K,c}$ are dense, open subspaces of $\widetilde{K}$ and $\widetilde{K}_c$, respectively; further, $\beta U_K = \widetilde{K}$ and $\beta U_{K,c} = \widetilde{K}_c$.

Let $\mathscr{F}$ be a maximal singular family in $P(K)$. Then

$$C(\widetilde{K}) = M(K)' = \bigoplus_\infty\{C(\Phi_\mu) : \mu \in \mathscr{F}\}.$$

The canonical image of $M(K)$ in $M(K)''$ is

$$N(\widetilde{K}) = \bigoplus_1 \{L^1(\Phi_\mu, \widehat{\mu}) : \mu \in \mathscr{F}\} = \bigoplus_1 \{N(\Phi_\mu) : \mu \in \mathscr{F}\}.$$

□

Thus the maximal singular family $\mathscr{F}$ in $P(K)$ corresponds to a maximal singular family in $N(\widetilde{K}) \cap P(\widetilde{K})$, as described on page 164.

Let K be an uncountable, compact, metrizable space. Recall from Corollary 5.2.8 that every maximal singular family in $P(K)$ consists of $\mathfrak{c}$ point masses and $\mathfrak{c}$ continuous measures.

We now describe the space $\widetilde{\mathbb{I}}$; in fact, this is equal to $\widetilde{K}$ for each uncountable, compact, metrizable space K because, for each such K, we have $M(\mathbb{I}) \cong M(K)$ by Corollary 4.6.7. The result is taken from [72, Theorem 4.16]; a picture of $\widetilde{K}$ is given on page 205.

Theorem 6.5.4. *Let K be an uncountable, compact, metrizable space. Then the hyper-Stonean envelope $X = \widetilde{K}$ has the following properties:*

(i) *X is a Stonean space;*

(ii) *the set D_X of isolated points of X has cardinality $\mathfrak{c}$;*

(iii) *$X \setminus \overline{D_X}$ contains a maximal pairwise-disjoint family of $\mathfrak{c}$ clopen subspaces, each homeomorphic to $\mathbb{H}$.*

Further, any two spaces X_1 and X_2 satisfying (i)–(iii) *are mutually homeomorphic.*

Proof. We have shown that $X = \widetilde{K}$ satisfies (i)–(iii).

Let X_1 and X_2 be two spaces satisfying (i)–(iii), and let D_{X_i} $(i = 1,2)$ be the respective sets of isolated points. Since D_{X_i} is a discrete, hence open, subspace of the Stonean space X_i, the space $\overline{D_{X_i}} = \beta D_{X_i}$ is clopen in X_i. Since the spaces βD_{X_i} are determined by the cardinality of D_{X_i}, $\overline{D_{X_1}}$ is homeomorphic to $\overline{D_{X_2}}$.

Let the families specified in (iii) corresponding to X_1 and X_2 be $\mathscr{F}_i$ $(i = 1,2)$, and let U_i denote the union of the sets in $\mathscr{F}_i$. Since each set in $\mathscr{F}_1$ is homeomorphic to each set in $\mathscr{F}_2$, and there is a bijection between $\mathscr{F}_1$ and $\mathscr{F}_2$, the open sets U_1 and U_2 are homeomorphic. Using the maximality of $\mathscr{F}_i$, it follows from Corollary 1.5.8 that $\overline{U_i} = X_i \setminus \overline{D_{X_i}} = \beta U_i$ for $i = 1,2$, and hence $X_1 \setminus \overline{D_{X_1}}$ is homeomorphic to $X_2 \setminus \overline{D_{X_2}}$. This implies that X_1 and X_2 are mutually homeomorphic. □

Example 6.5.5. There is a compact F-space K without isolated points such that K is not Stonean, but is such that the union of the supports of the normal measures on K is dense in K.

Indeed, let L be an uncountable, compact, metrizable space, and let $V = U_{L,c}$, a locally compact space. We *claim* that ∞ is a P-point in V_∞. Indeed, every compact subset of V is contained in a finite union of sets of the form Φ_μ, and so every σ-compact subset of V is contained in a countable union of the sets Φ_μ, and hence in

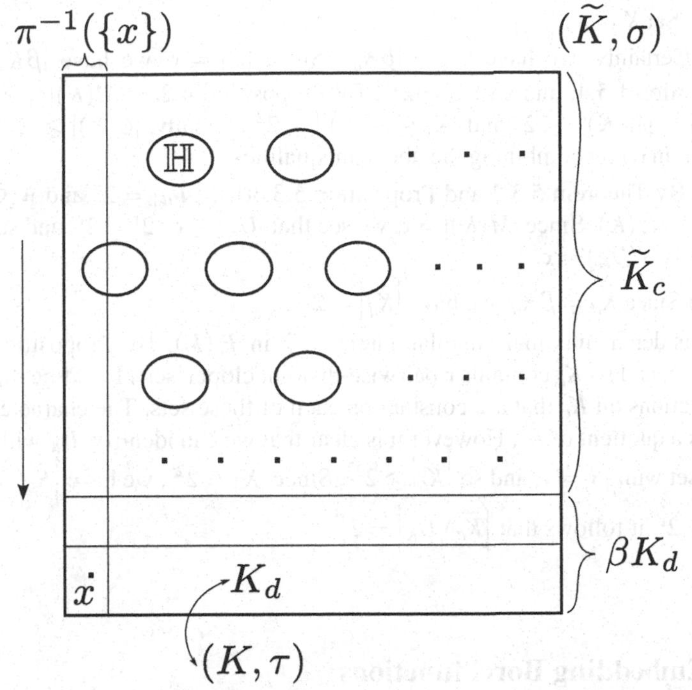

Fig. 6.1 The hyper-Stonean envelope $\widetilde{K}$ of an uncountable, compact, metrizable space (K,τ).

a set Φ_ν for some $\nu \in M_c(L)^+$. Suppose that (W_n) is a sequence of open neighbourhoods of ∞. Then there is a compact subset L of V such that $\bigcup\{V_\infty \setminus W_n : n \in \mathbb{N}\} \subset L$. This shows that $V_\infty \setminus L$ is an open neighbourhood of ∞ contained in $\bigcap W_n$, giving the claim.

Now take K to be the union of two disjoint copies of V_∞, with the two points at infinity identified to a point x_0. Then x_0 is not an extremely disconnected point because the two copies of V are open sets that do not contain x_0, but x_0 belongs to the closure of each set. On the other hand, it is easy to check that pairs of disjoint, cozero subsets of K have disjoint closures, and so K is an F-space. □

Let K be an uncountable, compact, metrizable space, and set $X = \widetilde{K}$. Then U_K is a dense, open subset of $\widetilde{K}$, and so it appears that U_K is 'large' in $\widetilde{K}$. However the following calculation of cardinalities shows that, in fact, there is a sense in which U_K is 'small' in $\widetilde{K}$. The result is essentially [72, Theorem 4.17].

Theorem 6.5.6. *Let K be an uncountable, compact, metrizable space. Then:*

(i) $\left|C(\widetilde{K})\right| = 2^{\mathfrak{c}}$ *and* $\left|\widetilde{K}\right| = 2^{2^{\mathfrak{c}}}$;

(ii) $|U_K| = 2^{\mathfrak{c}}$ *and* $w(U_K) = \mathfrak{c}$;

(iii) $\left|\widetilde{K}_d\right| = \left|\widetilde{K}_c\right| = \left|\widetilde{K}_c \setminus U_K\right| = 2^{2^{\mathfrak{c}}}$.

Proof. Set $X = \widetilde{K}$.

(i) Certainly, we have $|X| \geq |\beta K_d|$. Since $|K| = \mathfrak{c}$, we have $|\beta K_d| = 2^{2^{\mathfrak{c}}}$ by Proposition 1.5.4, and so $|X| \geq 2^{2^{\mathfrak{c}}}$. By Proposition 4.2.3, $|M(K)| = \mathfrak{c}$, and hence $|C(X)| = |M(K)'| \leq 2^{\mathfrak{c}}$ and $|X| \leq |C(X)'| \leq 2^{2^{\mathfrak{c}}}$. Finally, $|C(X)| \geq |\ell^\infty(K_d)| = 2^{\mathfrak{c}}$. We obtain (i) by combining the above inequalities.

(ii) By Theorem 5.3.2 and Proposition 5.3.3(ii), $|\Phi_\mu| = 2^{\mathfrak{c}}$ and $w(\Phi_\mu) = \mathfrak{c}$ for each $\mu \in P_c(K)$. Since $|M(K)| = \mathfrak{c}$, we see that $|U_K| \leq \mathfrak{c} \cdot 2^{\mathfrak{c}} = 2^{\mathfrak{c}}$, and so $|U_K| = 2^{\mathfrak{c}}$. Similarly $w(U_K) = \mathfrak{c}$.

(iii) Since $\widetilde{K}_d = \beta K_d$, we have $\left|\widetilde{K}_d\right| = 2^{2^{\mathfrak{c}}}$.

Consider a maximal singular family $\mathscr{F}_c$ in $P_c(K)$. By Proposition 4.6.2(iii), $|\mathscr{F}_c| = \mathfrak{c}$, and so $\widetilde{K}_c$ contains $\mathfrak{c}$ pairwise-disjoint clopen sets. Let $\mathfrak{A}$ be the algebra of all functions on $\widetilde{K}_c$ that are constant on each of these sets. The character space $\Phi_{\mathfrak{A}}$ of $\mathfrak{A}$ is a quotient of $\widetilde{K}_c$. However it is clear that we can identify $\Phi_{\mathfrak{A}}$ with βS, where S is a set with $|S| = \mathfrak{c}$, and so $\left|\widetilde{K}_c\right| \geq 2^{2^{\mathfrak{c}}}$. Since $|X| = 2^{2^{\mathfrak{c}}}$, we have $\left|\widetilde{K}_c\right| = 2^{2^{\mathfrak{c}}}$. Since $|U_K| = 2^{\mathfrak{c}}$, it follows that $\left|\widetilde{K}_c \setminus U_K\right| = 2^{2^{\mathfrak{c}}}$. □

6.6 Embedding Borel functions

Let K be a non-empty, locally compact space. In this section, we shall discuss the subalgebra $B^b(K)$ of the space $C(\widetilde{K})$. Take $f \in B^b(K)$. In Definition 4.1.5, we defined $\kappa(f) \in M(K)' = C_0(K)''$ by the formula

$$\langle \kappa(f), \mu \rangle = \int_K f \,\mathrm{d}\mu \quad (\mu \in M(K)).$$

Now we see that $\kappa(f) \in C(\widetilde{K})$ and $\kappa(f) \mid K = f$ for each $f \in B^b(K)$.

Theorem 6.6.1. *Let K be a non-empty, locally compact space. Then the map*

$$\kappa : B^b(K) \to C(\widetilde{K})$$

is a unital C^*-embedding identifying $B^b(K)$ as a closed, unital C^*-subalgebra of $C(\widetilde{K})$ and extending the canonical embedding $\kappa : C_0(K) \to C(K)''$.

Proof. Clearly κ is a continuous, linear operator that extends the canonical embedding. Take $f, g \in B^b(K)$ and $\mu \in M(K)$. Then

$$\langle \kappa(f)\kappa(g), \mu \rangle = \langle \kappa(f), \kappa(g) \cdot \mu \rangle = \langle \kappa(f), g\mu \rangle = \int_\Omega fg \,\mathrm{d}\mu = \langle \kappa(fg), \mu \rangle,$$

and so $\kappa(fg) = \kappa(f)\kappa(g)$. Thus κ is an algebra homomorphism. Also κ preserves complex conjugates and $\kappa(1_K) = 1_{\widetilde{K}}$, and so the above map is a unital C^*-embedding. □

Let K be a non-empty, compact space, and take $\mu \in P(K)$. We have a restriction map $\rho_\mu : C(\widetilde{K}) \to C(\Phi_\mu)$. On the other hand, there is a quotient map

$$q_\mu : B^b(K) \to L^\infty(K,\mu),$$

formed by identifying an element $g \in B^b(K)$ with its equivalence class in $L^\infty(K,\mu)$. We have

$$\langle q_\mu(g), f\rangle = \int_K fg\,\mathrm{d}\mu = \langle \kappa(g), f\mu\rangle \quad (f \in L^1(K,\mu),\, g \in B^b(K)).$$

Hence $\mathscr{G}_\mu(q_\mu(g)) = \rho_\mu(\kappa(g))$ $(g \in B^b(K))$, and so $\mathscr{G}_\mu \circ q_\mu = \rho_\nu \circ \kappa$; this shows that the diagram

$$\begin{array}{ccc} B^b(K) & \xrightarrow{\ \kappa\ } & C(\widetilde{K}) \\ {\scriptstyle q_\mu}\downarrow & & \downarrow{\scriptstyle \rho_\mu} \\ L^\infty(K,\mu) & \xrightarrow{\ \mathscr{G}_\mu\ } & C(\Phi_\mu) \end{array}$$

is commutative, and that $\kappa(B^b(K)) \mid \Phi_\mu = C(\Phi_\mu)$. Further, $\ker q_\mu$ is a closed ideal in $B^b(K)$, and $L^\infty(K,\mu) \cong B^b(K)/\ker q_\mu$.

In the case where there is a non-Borel set in K, it cannot be that $\kappa(B^b(K))$ separates the points of $\widetilde{K}$. For if this were so, we would have $\kappa(B^b(K)) = C(\widetilde{K})$ by the Stone–Weierstrass theorem, Theorem 1.4.26 (ii). However $B^b(K)$ is not a Dedekind complete lattice, but $C(\widetilde{K})$ is a Dedekind complete lattice.

As in Definition 3.3.2, page 102, the character space of $B^b(K)$ is denoted by $\Phi_b(K)$; we now see that $\Phi_b(K)$ is (homeomorphic to) the compact space $\widetilde{K}/\sim$, where, for points $\varphi, \psi \in \widetilde{K}$, we set

$$\varphi \sim \psi \quad \text{if} \quad \kappa(f)(\varphi) = \kappa(f)(\psi) \quad (f \in B^b(K)).$$

Since $\mathrm{lin}\,\{\chi_B : B \in \mathfrak{B}_K\}$ is dense in $B^b(K)$, it follows that

$$\varphi \sim \psi \quad \text{if and only if} \quad \kappa(\chi_B)(\varphi) = \kappa(\chi_B)(\psi) \quad (B \in \mathfrak{B}_K).$$

Since $C_{\mathbb{R}}(\Phi_b(K))$ is not usually a Dedekind complete lattice, $\Phi_b(K)$ is not usually a Stonean space; as we remarked in Proposition 3.3.3(i), $\Phi_b(K)$ is always a basically disconnected space.

Definition 6.6.2. Let K be a non-empty, locally compact space, and take $\varphi, \psi \in \widetilde{K}$. Then φ and ψ are *Borel equivalent* if $\varphi \sim \psi$.

The equivalence class under the relation $\sim$ that contains $\varphi \in \widetilde{K}$ is denoted by $[\varphi]$. Clearly $[\varphi]$ is contained in the fibre $\widetilde{K}_{\pi(\varphi)}$ for $\varphi \in \widetilde{K}$. Also, it is immediate that $\varphi \not\sim \psi$ whenever $\varphi, \psi \in U_K$ and $\varphi \neq \psi$. Thus each equivalence class $[\varphi]$ meets the set U_K in at most one point.

For each $\mu \in P(K)$, define

$$[\Phi_\mu] := \bigcup\{[\varphi] : \varphi \in \Phi_\mu\},$$

and then set

$$[U_K] = \bigcup\{[\Phi_\mu] : \mu \in P(K)\} = \bigcup\{[\varphi] : \varphi \in U_K\}.$$

Clearly each $[\Phi_\mu]$ is a closed subset of $\widetilde{K}$. It seemed possible that it would be the case that the subspace $[U_K]$ would be equal to the whole of $\widetilde{K}$. However the following theorem, taken from [72, Theorems 4.19 and 4.24], shows that this is far from the case whenever K is an uncountable, compact, metrizable space.

Theorem 6.6.3. *Let K be an uncountable, compact, metrizable space. Then*

$$|\beta K_d \setminus [U_K]| = 2^{2^{\mathfrak{c}}} \tag{6.8}$$

and

$$|[U_K]| = \left|[U_K] \cap \widetilde{K}_c\right| = \left|\widetilde{K}_c \setminus [U_K]\right| = 2^{2^{\mathfrak{c}}}. \tag{6.9}$$

We shall need the following definitions from [139, Definitions 3.13 and 3.60] within the proof of this theorem.

Let S be a non-empty set, and let κ be an infinite cardinal. Then a *κ-uniform ultrafilter* on S is an ultrafilter $\mathscr{U}$ on S such that each set in $\mathscr{U}$ has cardinality at least κ. Let $\mathscr{A}$ be a non-empty family of subsets of S. Then $\mathscr{A}$ has the *κ-uniform finite intersection property* if each non-empty, finite subfamily of $\mathscr{A}$ has an intersection of cardinality at least κ. Theorem 3.62 of [139] is the following.

Theorem 6.6.4. *Let S be an infinite set of cardinality κ, and let $\mathscr{A}$ be a non-empty family of at most κ subsets of S such that $\mathscr{A}$ has the κ-uniform finite intersection property. Then there are at least 2^{2^κ} κ-uniform ultrafilters on S that contain $\mathscr{A}$.* □

Proof of Theorem 6.6.3. We can suppose that the compact set K is the unit interval $\mathbb{I}$, and we shall do this.

Consider the family of G_δ-subsets B of $\mathbb{I}$ such that $B \supset \mathbb{Q} \cap \mathbb{I}$. Each such B is a Borel set, and, by Theorem 1.4.11, B is uncountable, and so $|B| = \mathfrak{c}$ by Proposition 1.4.14. The family $\mathscr{F}$ of all such sets B is a filter of Borel subsets of $\mathbb{I}$ and also $|\mathscr{F}| = \mathfrak{c}$, and so, by Theorem 6.6.4, there are $2^{2^{\mathfrak{c}}}$ $\mathfrak{c}$-uniform ultrafilters $\mathscr{U}$ on K with $\mathscr{F} \subset \mathscr{U}$. We identify these ultrafilters with points ψ of βK_d.

Let ψ be such an ultrafilter. We *claim* that, for each $\mu \in M(K)^+$, there exists $B \in \mathfrak{B}_K$ with $B \in \psi$ and such that $\mu(B) = 0$.

First, suppose that $\mu \in M_d(K)^+$, and set

$$C = \operatorname{supp} \mu \quad \text{and} \quad B = K \setminus C.$$

Since C is countable and ψ is a $\mathfrak{c}$-uniform ultrafilter, it is not true that $C \in \psi$. Thus B is a Borel set, $B \in \psi$, and $\mu(B) = 0$.

Second, suppose that $\mu \in M_c(K)^+$. By Proposition 4.2.2, there is a G_δ-subset B of K containing $\mathbb{Q}$ with $\mu(B) = 0$, and so again $B \in \mathscr{F} \subset \psi$.

Now let $\mu \in M(K)^+$. Then there exist $\mu_1 \in M_d(K)^+$ and $\mu_2 \in M_c(K)^+$ such that $\mu = \mu_1 + \mu_2$. Take subsets $B_1, B_2 \in \mathfrak{B}_K$ such that $B_1, B_2 \in \psi$ and

$$\mu_1(B_1) = \mu_2(B_2) = 0,$$

and set $B = B_1 \cap B_2$, so that $B \in \mathfrak{B}_K$ with $B \in \psi$ and $\mu(B) = 0$.

For each $\varphi \in \Phi_\mu$, we have $\kappa_E(\chi_B)(\varphi) = 0$, whereas $\kappa_E(\chi_B)(\psi) = 1$ because $B \in \psi$. This shows that $\psi \notin [\Phi_\mu]$, and hence we see that $|\beta K_d \setminus [U_K]| = 2^{2^{\mathfrak{c}}}$, giving equation (6.8).

To obtain equation (6.9), we shall require some more preliminary material; we shall first associate a certain filter of Borel sets with each $\varphi \in \widetilde{K}$.

Definition 6.6.5. Let K be a non-empty, locally compact space, and take $\varphi \in \widetilde{K}$. Then

$$\mathscr{G}_\varphi = \{B \in \mathfrak{B}_K : \varphi \in K_B\}.$$

Clearly each $\mathscr{G}_\varphi$ is a subset of $\mathfrak{B}_K$ that is closed under finite intersections. In the case where K is compact and metrizable, $|\mathscr{G}_\varphi| \leq \mathfrak{c}$.

We begin with a preliminary lemma and a corollary.

Let K be an uncountable, compact, metrizable space. By Proposition 4.6.2(iii), there is a maximal singular family of continuous measures in $P_c(K)$, say

$$\mathscr{F}_c = \{\mu_i \in P_c(K) : i \in I\},$$

with $|\mathscr{F}_c| = \mathfrak{c}$. For each $B \in \mathfrak{B}_K$, we set

$$J_B = \{i \in I : K_B \cap \Phi_i \neq \emptyset\},$$

where we write Φ_i for Φ_{μ_i}. Clearly $|J_B| \leq \mathfrak{c}$.

Lemma 6.6.6. *Let K be a compact, metrizable space, and let B be an uncountable Borel set in K. Then:* (i) $|J_B| = \mathfrak{c}$; (ii) $K_B \cap (\widetilde{K}_c \setminus U_K) \neq \emptyset$.

Proof. (i) By Proposition 1.4.17, B contains an uncountable, compact subset, say C. We *claim* that the family

$$\{\mu_i \mid C : i \in J_B\} \cup \{0\}$$

is a maximal singular family of measures in $M_c(C)^+$. Indeed, all pairs of distinct elements of this family are mutually singular. Suppose that $\nu \in M_c(C)^+$ is such that $\nu \perp (\mu_i \mid C)$ for each $i \in J_B$. Then $\nu \perp \mu_i$ for each $i \in I$, and so $\nu = 0$. This gives the claim. By Corollary 5.2.8, $|J_B| = \mathfrak{c}$.

(ii) Assume towards a contradiction that $K_B \cap \widetilde{K}_c \subset U_K$. Then

$$K_B \subset \bigcup\{\Phi_\mu : \mu \in P_c(K)\}.$$

Since the space K_B is compact, since each space Φ_μ is open in $\widetilde{K}$, and since the family $\{\Phi_\mu : \mu \in P_c(K)\}$ is closed under finite unions, there exists a measure $\mu \in P_c(K)$ such that $K_B \subset \Phi_\mu$. By (i), the set $\{i \in I : \Phi_\mu \cap \Phi_i \neq \emptyset\}$ is uncountable. But this contradicts the fact, given in Proposition 5.2.5(ii), that Φ_μ satisfies CCC. Thus $K_B \cap \widetilde{K}_c \not\subset U_K$. □

Corollary 6.6.7. *Let K be an uncountable, compact, metrizable space, and take $\varphi \in \widetilde{K}_c \cup (\beta K_d \setminus U_K)$. Then there exists $\psi \in \widetilde{K}_c \setminus U_K$ such that $\psi \sim \varphi$.*

Proof. Since $\varphi \in \widetilde{K}_c \cup (\beta K_d \setminus U_K)$, each $B \in \mathscr{G}_\varphi$ is uncountable. For $B \in \mathscr{G}_\varphi$, the set $K_B \cap (\widetilde{K}_c \setminus K)$ is closed in the compact space $\widetilde{K}_c \setminus U_K$, and, by Lemma 6.6.6(ii), the set $K_B \cap (\widetilde{K}_c \setminus K)$ is not empty. Thus

$$\bigcap\{K_B \cap (\widetilde{K}_c \setminus U_K) : B \in \mathscr{G}_\varphi\} \neq \emptyset;$$

choose ψ in the set on the left. Then $\psi \in \widetilde{K}_c \setminus U_K$, and $\psi \in K_B$ whenever $\varphi \in K_B$, and hence $\psi \sim \varphi$. □

Proposition 6.6.8. *Let K be an uncountable, compact, metrizable space, and take $\varphi \in \widetilde{K}$.*

(i) *Suppose that there exists $B \in \mathscr{G}_\varphi$ such that B is countable. Then $[\varphi] = \{\varphi\}$, and so $|[\varphi]| = 1$.*

(ii) *Suppose that each $B \in \mathscr{G}_\varphi$ is uncountable. Then $|[\varphi]| = |[\varphi] \cap \beta K_d| = 2^{2^{\mathfrak{c}}}$.*

(iii) *Suppose that $\varphi \in \widetilde{K}_c$. Then $|[\varphi]| = \left|[\varphi] \cap \widetilde{K}_c\right| = 2^{2^{\mathfrak{c}}}$.*

Proof. Recall from Theorem 6.5.6 (i) that $\left|\widetilde{K}\right| = 2^{2^{\mathfrak{c}}}$.

(i) Take $B \in \mathscr{G}_\varphi$ with B countable. Suppose that $\psi \in [\varphi]$. Since $\chi_B \in B^b(K)$ and $\varphi \in K_{\chi_B} = \beta B \subset \beta K_d$, necessarily $\psi \in \beta B$. Since $\ell^\infty(B) \subset B^b(K)$ and the functions in $\ell^\infty(B)$ separate the points of βB, it follows that $\psi = \varphi$.

(ii) We first note that $|\mathscr{G}_\varphi| \leq \mathfrak{c}$ and that each member of $\mathscr{G}_\varphi$ has cardinality $\mathfrak{c}$. Since $\mathscr{G}_\varphi$ is closed under finite intersections, it is clear that $|\mathscr{G}_\varphi|$ has the $\mathfrak{c}$-uniform finite intersection property. By Theorem 6.6.4,

$$|\{\psi \in \beta K_d : \psi \supset \mathscr{G}_\varphi\}| = 2^{2^{\mathfrak{c}}}.$$

However, for each $\psi \supset \mathscr{G}_\varphi$ and each $B \in \mathscr{G}_\varphi$, we have $\psi \in K_B$, and so $\psi \sim \varphi$. It follows that $|[\varphi] \cap \beta K_d| = 2^{2^{\mathfrak{c}}}$.

(iii) First, we consider the case where $\varphi \in \widetilde{K}_c \setminus U_K$. Again consider the above family $\mathscr{F}_c$ (from page 209), so that $\{\Phi_i : i \in I\}$ is a pairwise-disjoint family of subsets of $\widetilde{K}$ having cardinality $\mathfrak{c}$.

For each $B \in \mathscr{G}_\varphi$, define J_B as above. By Lemma 6.6.6(i), $|J_B| = \mathfrak{c}$. Certainly $\left|\{J_B : B \in \mathscr{G}_\varphi\}\right| \leq |\mathfrak{B}_K|$, and $|\mathfrak{B}_K| = \mathfrak{c}$ by Corollary 1.4.15. Thus, by Theorem 6.6.4, there are $2^{2^{\mathfrak{c}}}$ ultrafilters $\mathscr{U}$ on I each containing $\{J_B : B \in \mathscr{G}_\varphi\}$.

For each such ultrafilter $\mathscr{U}$ and each $B \in \mathscr{G}_\varphi$, define

$$C(\mathscr{U},B) = \bigcap_{U\in\mathscr{U}} \left\{\overline{\bigcup_{i\in U} K_B \cap \Phi_i}\right\} \quad \text{and} \quad C(\mathscr{U}) = \bigcap\{C(\mathscr{U},B) : B \in \mathscr{G}_\varphi\}.$$

Since each set $\overline{\bigcup_{i\in U} K_B \cap \Phi_i}$ is a non-empty, closed subset of the compact space $\widetilde{K}_c$, it follows that $C(\mathscr{U}) \neq \emptyset$ for each such $\mathscr{U}$. Suppose that $\mathscr{U}_1$ and $\mathscr{U}_2$ are distinct ultrafilters on I each containing $\{J_B : B \in \mathscr{G}_\varphi\}$ and that $B_1, B_2 \in \mathscr{G}_\varphi$. Then $C(\mathscr{U}_1,B_1) \cap C(\mathscr{U}_2,B_2) = \emptyset$, and so $C(\mathscr{U}_1) \cap C(\mathscr{U}_2) = \emptyset$. Thus there are $2^{2^{\mathfrak{c}}}$ sets of the form $C(\mathscr{U})$, and the family of these sets is pairwise disjoint.

Let $\mathscr{U}$ be an ultrafilter on I containing $\{J_B : B \in \mathscr{G}_\varphi\}$, and let $\psi \in C(\mathscr{U})$. For each $B \in \mathscr{G}_\varphi$, we have $\psi \in C(\mathscr{U},B) \subset K_B$, and so $\psi \sim \varphi$.

We have shown that $\left|[\varphi] \cap \widetilde{K}_c\right| = 2^{2^{\mathfrak{c}}}$ for this element φ.

Second, we consider the case where $\varphi \in \widetilde{K}_c \cap U_K$. By Corollary 6.6.7, there exists $\psi \in \widetilde{K}_c \setminus U_K$ such that $\psi \sim \varphi$. Thus we have $\left|[\varphi] \cap \widetilde{K}_c\right| = \left|[\psi] \cap \widetilde{K}_c\right| = 2^{2^{\mathfrak{c}}}$, as required. □

Proof of Theorem 6.6.3, continued. We can now verify equation (6.9) of Theorem 6.6.3.

Take $\varphi \in \widetilde{K}_c \cap U_K$. By Proposition 6.6.8(iii), $\|[\varphi]\| = \left|[\varphi] \cap \widetilde{K}_c\right| = 2^{2^{\mathfrak{c}}}$. Since $[\varphi] \subset [U_K]$, we have $\|[U_K]\| = \left|[U_K] \cap \widetilde{K}_c\right| = 2^{2^{\mathfrak{c}}}$.

By equation (6.8), there exists $\varphi \in \beta K_d \setminus [U_K]$. By Corollary 6.6.7, there exists an element $\psi \in \widetilde{K}_c \setminus U_K$ such that $\psi \sim \varphi$. Since $\varphi \notin [U_K]$, we have $[\psi] \cap U_K = \emptyset$, and so $\|[\psi]\| = 2^{2^{\mathfrak{c}}}$. Thus $\left|\widetilde{K}_c \setminus [U_K]\right| = 2^{2^{\mathfrak{c}}}$.

This concludes the proof of Theorem 6.6.3. □

The space $\widetilde{K} \setminus [U_K]$ has been called the 'dark matter' of $\widetilde{K}$: we can gain information about points of $[U_K]$ by using the bounded Borel functions on K and the subalgebra $\kappa(B^b(K))$ of $C(\widetilde{K})$, but these functions give little information about points of $\widetilde{K} \setminus [U_K]$.

6.7 Baire classes

As promised in §3.3, we can now provide some of the central ideas associated with Banach spaces of Baire functions. The goal is to identify Banach-space invariants which may help to classify the Baire classes.

Let K be a non-empty, compact space. We are regarding each Baire space $B_\alpha(K)$ for an ordinal α with $0 \le \alpha \le \omega_1$ as a C^*-subalgebra of $B^b(K)$, and hence as a closed subspace of $C(K)'' = C(\widetilde{K})$. In the Banach space $C(K)$, each bounded, pointwise convergent sequence (with a possibly discontinuous limit) is weakly Cauchy (by the Lebesgue dominated convergence theorem), and the limit is a function in $B_1(K)$ which is an element of $C(K)''$. The space $B_1(K)$ is in fact the Banach subalgebra of $C(\widetilde{K})$ generated by the weak*-convergent sequences in $C(K)$; $B_1(K)$ contains $C(K)$ as a closed subalgebra.

There is a parallel story for a general Banach space E. We consider the linear subspace E_w of E'' consisting of the limits of weak*-convergent sequences in E (where E is identified with its canonical image in E''). In other words, E_w consists of all the limits (in E'') of weakly Cauchy sequences in E. This linear space was first discussed by Grothendieck in [124, p. 159]; see also [97, p. 646]. It is a significant fact that E_w is always a norm-closed subspace of E''. This was first shown by McWilliams [181]; other proofs are given in [178, Corollary 4.38], [192, p. 381] and (for separable E) in [15, Theorem II.1.2(a)].

In the case where $E = C(K)$ for a compact space K, we have $E_w \cong B_1(K)$, so that, as stated in Theorem 3.3.9, E_w is a Grothendieck space. It would be interesting to know, in the light of the non-commutative generalizations mentioned on page 143, the answer to the following question.

Question 2: *Let E be a C^*-algebra. Is E_w necessarily a Grothendieck space?*

Suppose that E is a Banach space and F is a closed subspace of E. Then F_w is a closed subspace of E_w; the embedding of F_w in E_w is the restriction to F_w of the second dual of the embedding of F in E.

Example 6.7.1. Let $E = c_0(\Gamma)$ for some non-empty set Γ. Then $(c_0(\Gamma))_w = \ell_c^\infty(\Gamma)$, where $\ell_c^\infty(\Gamma)$ is the subspace of 'countably supported' elements of $\ell^\infty(\Gamma)$, so that

$$\ell_c^\infty(\Gamma) = \{f \in \ell^\infty(\Gamma) : \{s \in \Gamma : f(s) \neq 0\} \text{ is countable}\}.$$

In the case where Γ is a subset of $\mathbb{I}$, we see that $c_0(\Gamma)$ is a closed subspace of $B_1(\mathbb{I})$ because $c_0(\Gamma)$ is the uniform closure of the set of simple functions which are finitely supported on Γ. Therefore, by the above remark, $\ell_c^\infty(\Gamma)$ is a closed subspace of $(B_1(\mathbb{I}))_w$.

The space $\ell_c^\infty(\Gamma)$, for arbitrary Γ, was first examined by Day in 1955 [81], where it was designated by the notation $m_0(\Gamma)$. As mentioned above on page 86, this space is always separably injective, but it is not injective when Γ is uncountable (see [20, Example 2.4] and [247, p. 1722]). Indeed, in this case, $\ell_c^\infty(\Gamma)$ is not complemented in $\ell^\infty(\Gamma)$ (as was first shown in [199]).

The Banach space $\ell_c^\infty(\Gamma)$ does, however, share some properties with injective $C(K)$ spaces. For example, $\ell_c^\infty(\Gamma)$ is a Grothendieck space, as defined in Definition 2.1.1. This is a particular instance of the fact that $B_1(K)$ is always a Grothendieck space. Indeed, since $c_0(\Gamma) \sim C(\Gamma_\infty)$, we have

$$B_1(\Gamma_\infty) \sim C(\Gamma_\infty)_w \sim c_0(\Gamma)_w = \ell_c^\infty(\Gamma).$$

Hence $\ell_c^\infty(\Gamma)$ is isomorphic to the Grothendieck space $B_1(\Gamma_\infty)$; we recall that the Grothendieck property is an isomorphic invariant of Banach spaces.

The Banach space $\ell_c^\infty(\Gamma)$ is studied further in [146]; in this paper the infinite-dimensional, complemented subspaces of $\ell_c^\infty(\Gamma)$ are classified. □

The first use of E_w *as a Banach space*, in [75], isolates an isomorphic invariant of Banach spaces that distinguishes certain non-separable $C(K)$ spaces; the Banach space property involves the way in which E is embedded in E_w.

Definition 6.7.2. Let E be a Banach space. Then E is *Baire complemented* if E is a complemented subspace of E_w.

Thus a Banach space E is Baire complemented if and only if there is a bounded projection from E_w onto E. Clearly the property of being Baire complemented is an isomorphic invariant for the class of all Banach spaces. It was shown in [75] that all the Banach spaces $B_\alpha(\mathbb{I})$ for countable ordinals α fail to be Baire complemented (note that this is a stronger statement than clause (iii) in Theorem 3.3.7). However, in the case where $\alpha = \omega_1$, the space $B_{\omega_1}(\mathbb{I})$ is Baire complemented (a more general, and stronger, statement is given in Theorem 3.3.8). This distinguishes isomorphically the spaces $B_\alpha(\mathbb{I})$ for $1 \leq \alpha < \omega_1$ from the space $B_{\omega_1}(\mathbb{I}) = B^b(\mathbb{I})$.

We now relax the requirements of Definition 6.7.2 so that they are less restrictive on the assumed map from E_w to E: instead of a bounded projection which restricts to the identity on E, we require only a map which restricts to an embedding of E into itself. The following concept is new.

Definition 6.7.3. The Banach space E is *Baire large* if there is a bounded operator $T : E_w \to E$ such that $T \mid E$ is an embedding of E into itself.

Thus any Baire complemented space is clearly Baire large, and so $B_{\omega_1}(\mathbb{I})$ is Baire large.

Our motivation for introducing this concept is that the following holds.

Proposition 6.7.4. *Let K be a non-empty, compact space, and take ordinals α, β with $0 \leq \alpha < \beta \leq \omega_1$. Assume that $B_\alpha(K)$ is not Baire large. Then there is no bounded operator from $B_\beta(K)$ into $B_\alpha(K)$ which restricts to an embedding of $B_\alpha(K)$ into itself. In particular, $B_\alpha(K)$ and $B_\beta(K)$ are not isomorphic.*

Proof. The weakly Cauchy sequences which define $B_\alpha(K)_w$ clearly converge pointwise on K, and this defines a map $B_\alpha(K)_w \to B_{\alpha+1}(K) \subset B_\beta(K)$ which leaves the elements of $B_\alpha(K)$ fixed. Assume towards a contradiction that there exists a map from $B_\beta(K)$ to $B_\alpha(K)$ with the stated properties. Then the composition defines a map $B_\alpha(K)_w \to B_\alpha(K)$ to show that $B_\alpha(K)$ is Baire large, a contradiction. □

Of course this proposition focuses our attention on the question whether any of the spaces $B_\alpha(K)$ are Baire large. Clearly a Banach space E which is injective is Baire large, i.e., if E is not Baire large, then E is not injective. We intend to prove the following. We now write B_α for $B_\alpha(\mathbb{I})$.

Theorem 6.7.5. *The Baire class B_1 is not Baire large. For each ordinal β with $2 \le \beta < \omega_1$, there is no bounded operator from B_β into B_1 which is injective on B_1, and hence B_β and B_1 are not isomorphic and B_1 is not complemented in B_β.*

Question 1a: *Is it true that each space $B_\alpha(\mathbb{I})$ for $2 \le \alpha < \omega_1$ fails to be Baire large?*

In view of Example 6.7.1 and Proposition 6.7.4, it suffices for Theorem 6.7.5 to prove the following theorem, originally given in [77, Remark, p. 336].

Theorem 6.7.6. *Suppose that Γ is an uncountable set and that $T : \ell_c^\infty(\Gamma) \to B_1$ is a bounded operator. Then T is determined by a countable subset of Γ, in the sense that there exists a countable subset Γ_0 of Γ such that $Tz = 0$ whenever $z \in \ell_c^\infty(\Gamma)$ with $z \mid \Gamma_0 = 0$. In particular, $T\delta_s = 0$ for $s \in \Gamma \setminus \Gamma_0$, and T is not injective on $c_0(\Gamma)$.*

Before proving Theorem 6.7.6, we note the following immediate corollary. This is a new proof, at least for $\aleph_0 < |\Gamma| \le \mathfrak{c}$, of the known fact that the space $c_0(\Gamma)$ is not injective (see Corollary 2.4.13).

Corollary 6.7.7. *Suppose that Γ is an uncountable set with cardinality at most $\mathfrak{c}$. Then $c_0(\Gamma)$ is not Baire large.*

Proof. We note simply that $c_0(\Gamma) \subset B_1(\mathbb{I})$ and $c_0(\Gamma)_w = \ell_c^\infty(\Gamma)$. The result now follows immediately from Theorem 6.7.6 and the definition of Baire large. □

Take Γ to be the discrete set $\mathbb{I}_d$. Then Theorem 6.7.6 of course implies that there must be a countably supported function on $\mathbb{I}$ which is not in the class B_1. The typical example of such a function is the characteristic function $\chi_{\mathbb{Q}\cap\mathbb{I}}$ of the rationals in $\mathbb{I}$; the standard proof of this fact uses the requirement that any pointwise limit of a sequence of continuous functions must have at least one point of continuity, and $\chi_{\mathbb{Q}\cap\mathbb{I}}$ has no point of continuity. The proof of Theorem 6.7.6 which is given below turns on this crucial property.

Before proving Theorem 6.7.6, we require a lemma. Throughout, Γ is an uncountable set. For $x \in \ell_c^\infty(\Gamma)$ with $\|x\|_\infty = 1$, set

$$F_x = \{y \in \ell_c^\infty(\Gamma) : \|y\| = 1 \text{ and } y(t) = x(t) \text{ whenever } x(t) \ne 0\}.$$

Observe that $y \in F_x$ if and only if $F_y \subset F_x$.

Lemma 6.7.8. *Take $x \in \ell_c^\infty(\Gamma)$ with $\|x\| = 1$. Suppose that (ρ_n) is a sequence of bounded, convex functions on F_x. Then there exists a fixed $y \in F_x$ such that all the functions ρ_n are constant on F_y.*

Proof. We first prove the lemma for a single ρ. This part of the proof was suggested by the argument in [81, pp. 521–522].

Choose inductively a sequence (x_n) with $x_1 = x$ and $F_x = F_{x_1} \supset F_{x_2} \supset F_{x_3} \supset \cdots$ such that, for each $n \in \mathbb{N}$, we have $\|x_n\| = 1$ and

$$\rho(x_{n+1}) \ge \frac{3}{4}M_n + \frac{1}{4}m_n\,,$$

where $M_n = \sup \rho(F_{x_n})$ and $m_n = \inf \rho(F_{x_n})$ for $n \in \mathbb{N}$. Clearly the sequence (x_n) is pointwise convergent on Γ; set $y = \lim x_n$, so that $F_y = \bigcap_{n=1}^{\infty} F_{x_n}$.

We *claim* that ρ is constant on F_y. Indeed, for $n \in \mathbb{N}$ and $\varepsilon > 0$, choose $z \in F_{x_n}$ with $\rho(z) < m_n + 2\varepsilon$. Then $2x_n - z \in F_{x_n}$ and $x_n = (2x_n - z)/2 + z/2$. By the convexity of ρ, we have

$$\rho(x_n) \leq \frac{1}{2}\rho(2x_n - z) + \frac{1}{2}\rho(z) < \frac{1}{2}M_n + \frac{1}{2}m_n + \varepsilon,$$

so that $m_n \geq 2\rho(x_n) - M_n$. For $n > 1$, we have $\rho(x_n) \geq 3M_{n-1}/4 + m_{n-1}/4$, so that

$$M_n - m_n \leq M_n - 2\rho(x_n) + M_n \leq 2M_{n-1} - \frac{3}{2}M_{n-1} - \frac{1}{2}m_{n-1} = \frac{1}{2}(M_{n-1} - m_{n-1}).$$

Hence, by induction, $M_n - m_n \leq (M_1 - m_1)/2^{n-1}$ $(n \in \mathbb{N})$, and so $M_n - m_n \to 0$ as $n \to \infty$. Thus ρ is constant on $\bigcap_{n=1}^{\infty} F_{x_n} = F_y$, proving the claim.

For a sequence (ρ_n) of bounded, convex functions on F_x, choose inductively $y_0 = x$ and $y_n \in F_{y_{n-1}}$, so that ρ_n is constant on F_{y_n}; the latter choice is possible by the claim. Then all the ρ_n are constant on the set $F_y = \bigcap_{n=1}^{\infty} F_{y_n}$, where y is the pointwise limit of the sequence (y_n).

This proves the lemma. □

Proof of Theorem 6.7.6. Let Γ be an uncountable set, and take $T : \ell_c^\infty(\Gamma) \to B_1$ to be a bounded operator. The desired countable set Γ_0 will be the support set of an element x in the unit sphere of $\ell_c^\infty(\Gamma)$ such that T is constant on F_x. For, in this case, for each $z \in \ell_c^\infty(\Gamma)$ with $\|z\| = 1$ and $z \mid \Gamma_0 = 0$, we have $x + z \in F_x$, so that $T(x+z) = T(x)$, i.e., $Tz = 0$. We describe such an element x.

For $\varepsilon > 0$ and $y \in \ell_c^\infty(\Gamma)$ with $\|y\| = 1$, define

$$S_{\varepsilon,y} = \left\{ t \in \mathbb{I} : \sup_{z \in F_y} Tz(t) - \inf_{z \in F_y} Tz(t) < \varepsilon \right\}.$$

Our first *claim* is the following: for each closed subset K of $\mathbb{I}$, each $\varepsilon > 0$, and each $x \in \ell_c^\infty(\Gamma)$ with $\|x\| = 1$, there exists $y \in F_x$ such that $S_{\varepsilon,y}$ contains a non-empty, relatively open subset of K.

To prove the first claim, fix K, ε, and x as stated. Let $\{G_1, G_2, \dots\}$ be a base for the topology of $\mathbb{I}$, and define functions $\overline{\varphi}_m$ and $\underline{\varphi}_m$ on F_x (for m with $G_m \cap K \neq \emptyset$) by

$$\overline{\varphi}_m(z) = \sup\{(Tz)(t) : t \in K \cap G_m\}, \quad \underline{\varphi}_m(z) = \inf\{(Tz)(t) : t \in K \cap G_m\}.$$

Then $\overline{\varphi}_m$ and $-\underline{\varphi}_m$ are bounded, convex functions on F_x. By Lemma 6.7.8, there is an element $y \in F_x$ such that all these functions are constant on F_y. Since $Ty \in B_1$, it follows from Baire's theorem that $Ty \mid K$ has a point of continuity relative to K. Hence there is an $m_0 \in \mathbb{N}$ such that $K \cap G_{m_0} \neq \emptyset$ and $\overline{\varphi}_{m_0}(y) - \underline{\varphi}_{m_0}(y) < \varepsilon$.

We now show that we have $K \cap G_{m_0} \subset S_{\varepsilon,y}$. Since $\overline{\varphi}_{m_0}$ and $\underline{\varphi}_{m_0}$ are constant on F_y, we have, for each $b \in K \bigcap G_{m_0}$, the inequality

$$\sup_{z\in F_y}(Tz)(b) - \inf_{z\in F_y}(Tz)(b) \leq \sup_{z\in F_y}\overline{\varphi}_{m_0}(z) - \inf_{z\in F_y}\underline{\varphi}_{m_0}(z) = \overline{\varphi}_{m_0}(y) - \underline{\varphi}_{m_0}(y) < \varepsilon\,,$$

and hence $b \in S_{\varepsilon,y}$. This proves the first claim.

Our second *claim* is the following: for each $\varepsilon > 0$ and $x \in \ell_c^\infty(\Gamma)$ with $\|x\| = 1$, there exists $y \in F_x$ such that $S_{\varepsilon,y} = \mathbb{I}$.

Fix such ε and x. By the first claim applied to the case where $K = \mathbb{I}$, there exists $m_1 \in \mathbb{N}$ such that G_{m_1} is contained in each set $S_{\varepsilon,z}$ for $z \in F_x$, and we can suppose that m_1 is the minimum element of $\mathbb{N}$ with this property. Also choose $y_1 \in F_x$ such that $G_{m_1} \subset S_{\varepsilon,y_1}$. Next let $m_2 > m_1$ be the minimum element of $\mathbb{N}$ such that G_{m_2} is contained in some set $S_{\varepsilon,z}$, where $z \in F_{y_1}$. Choose then $y_2 \in F_{y_1}$ such that $G_{m_2} \subset S_{\varepsilon,y_2}$. Continuing, we obtain by induction a sequence (y_k) with $y_1 \in F_x$ and such that $y_{k+1} \in F_{y_k}$ $(k \in \mathbb{N})$ and also a sequence $(G_{m_k} : k \in \mathbb{N})$ of basic open subsets of $\mathbb{I}$ with $G_{m_k} \subset S_{\varepsilon,y_k}$ $(k \in \mathbb{N})$. Set

$$G = \bigcup\{G_{m_k} : k \in \mathbb{N}\}\,.$$

Now the sequence (y_k) converges pointwise; we set $y = \lim y_k$ and finally we shall see that $S_{\varepsilon,y} = \mathbb{I}$.

Since $S_{\varepsilon,y_0} \subset S_{\varepsilon,y}$ whenever $y \in F_{y_0}$, each $G_{m_k} \subset S_{\varepsilon,y_k} \subset S_{\varepsilon,y}$, we see that $G \subset S_{\varepsilon,y}$. It now suffices to show $G = \mathbb{I}$. Assume to the contrary that $G \subsetneq \mathbb{I}$. Then $K := \mathbb{I} \setminus G$ is non-empty. By the first claim, some set $S_{\varepsilon,z}$ with $z \in F_y$ contains a non-empty set of the form $K \cap G_m$. But

$$G_m = (K \cap G_m) \cup (G_m \setminus K) \subset S_{\varepsilon,z} \cup G \subset S_{\varepsilon,z} \cup S_{\varepsilon,y} = S_{\varepsilon,z}\,.$$

Since $z \in F_{y_k}$ for all $k \in \mathbb{N}$, the set G_m must have been included in the subsequence $(G_{m_k} : k \in \mathbb{N})$, and hence $G_m \subset G$, contradicting the fact that $K \cap G = \emptyset$. This proves the second claim.

The theorem now follows. Indeed, let $\varepsilon_n \searrow 0$. By the second claim, we can choose (y_n) with $y_{n+1} \in F_{y_n} \subset F_x$ and $S_{\varepsilon_n,y_n} = \mathbb{I}$ for each $n \in \mathbb{N}$. Take $x = \lim_{n\to\infty} y_n$, so that $Ty = Tx$ for $y \in F_x$. This shows that the operator T is constant on the set F_x, as required, and so concludes the proof of Theorem 6.7.6. □

6.8 Injectivity of $C(K)$-spaces

Let K be a non-empty, locally compact space. We recall once again that the space $C_0(K)$ is not injective whenever $C_0(K)$ contains c_0 as a closed, complemented subspace, and hence whenever K contains a convergent sequence of distinct points, and also whenever K is not pseudo-compact (Theorem 2.4.12). We also recall from Theorem 3.3.7(iii) that, for each ordinal α with $1 \leq \alpha < \omega_1$ and each uncountable Polish space X, the Baire class $B_\alpha(X)$ is not injective; this also holds for the space $B_{\omega_1}(\mathbb{I})$ (see Corollary 6.8.10, below).

First we make some remarks about some more spaces $C(K)$ that are not injective.

Definition 6.8.1. A Banach space E has the *Sobczyk property* if every isomorphic copy of c_0 in E is complemented.

Thus, by Proposition 2.4.6(i) and Theorem 2.4.15, $C(K)$ is not injective and not isomorphically a dual space whenever K is infinite and $C(K)$ has the Sobczyk property: examples of such spaces are given in [65, Corollary 2.5] and [102]. As we noted in Proposition 2.4.7, $C(K)$ contains no complemented subspace that is isomorphic to c_0 if and only if $C(K)$ is a Grothendieck space. Thus, by Theorem 2.4.15, $C(K)$ is a Grothendieck space whenever $C(K)$ is injective, a result already proved in Corollary 4.5.10. It is proved in [157, Theorem 2] that $C(K, C(L))$ is never a Grothendieck space when K and L are infinite, compact spaces, and hence we have the following result, noting that $C(K \times L) \cong C(K, C(L))$.

Proposition 6.8.2. *Let K and L be infinite, compact spaces. Then $C(K \times L)$ is not injective.* □

We shall now relate Stonean spaces to the injectivity and to the projection property of spaces of continuous functions. The following result was proved by Bade in [23, Theorem 4.20] and [24, Theorem 7.20]; see also [166, §11] and [247, Theorem 2.1], where some history of the theorem is given. For a short exposition of the main equivalences, see [126].

Theorem 6.8.3. *Let K be a non-empty, compact space. Then the following are equivalent:*

(a) *the Banach space $C(K)$ is 1-injective;*

(b) *K is projective;*

(c) *K is Stonean;*

(d) *K is a retract of βK_d;*

(e) *the Banach space $C(K)$ is a P_1-space;*

(f) *the Banach lattice $C(K)$ is Dedekind complete;*

(g) *$C(K)$ is 1-complemented in $C(K)''$.*

Proof. (b) ⇔ (c) ⇔ (d) This is the equivalence of (a), (b) and (d) in Theorem 1.6.3.

(c) ⇔ (f) This is Theorem 2.3.3.

(a) ⇔ (e) This is Proposition 2.5.9.

(a) ⇒ (b) Take compact spaces L and M and continuous surjections $\theta : L \to M$ and $\varphi : K \to M$. Then the maps $\varphi^\circ : C(M) \to C(K)$ and $\theta^\circ : C(M) \to C(L)$ are unital C^*-embeddings which are isometries. By hypothesis, there exists a linear map

$$\widetilde{T} : C(L) \to C(K)$$

with $\left\|\widetilde{T}\right\| = 1$ and $\widetilde{T} \circ \theta^\circ = \varphi^\circ$. Clearly $\widetilde{T}(1_L) = 1_K$ and $\widetilde{T}$ is an isometry, and so, by Theorem 6.1.2, there is a homeomorphism $\eta : K \to L$ such that $\widetilde{T} = \eta^\circ$. We have $\varphi^\circ = (\theta \circ \eta)^\circ$, and so $\varphi = \theta \circ \eta$. Thus K is a projective space.

(c) $\Rightarrow$ (a) This is Theorem 2.5.11.

(a) $\Rightarrow$ (g) This is trivial.

(g) $\Rightarrow$ (e) There is a bounded projection Q of $C(K)''$ onto $C(K)$ with $\|Q\| = 1$. Take a Banach space E such that $C(K)$ is a closed subspace of E. Then $C(K)''$ is a closed subspace of E''. As earlier, we identify $C(K)''$ with $C(\widetilde{K})$. Since $\widetilde{K}$ is hyper-Stonean, the above implication (c) $\Rightarrow$ (a) shows that there is a bounded projection P of E'' onto $C(\widetilde{K})$ with $\|P\| = 1$. The map $R := Q \circ (P \mid E)$ is a bounded projection of E onto $C(K)$ with $\|R\| = 1$, and so $C(K)$ is 1-complemented in E. □

There is another interesting result about complemented subspaces of $C(K)$ spaces that we discuss briefly. Consider the following property of a Banach space E: *Each closed subspace of E that is isomorphic to E is complemented in E.* Clearly each injective Banach space has this property, and so $C(K)$, for a compact space K, has the property whenever K is Stonean. On the other hand, it was first proved by Amir in [8] that $C(\mathbb{I})$ does not have this property, and it follows from results of Baker in [26] that $C([0,\alpha])$ does not have this property for each ordinal $\alpha \geq \omega^\omega$. Indeed, the following interesting theorem of Baker is a special case of [27, Theorem 4.6]. Recall that Δ denotes the Cantor set.

Theorem 6.8.4. *There is a continuous surjection $\eta : \Delta \to \Delta$ such that the image of the isometric homomorphism $\eta^\circ : C(\Delta) \to C(\Delta)$ is a closed subalgebra of $C(\Delta)$ that is not complemented as a Banach space.* □

Corollary 6.8.5. *Let K be an uncountable, compact, metrizable space. Then $C(K)$ contains a closed, uncomplemented subspace E such that E is isomorphic to $C(K)$.*

Proof. By Milutin's theorem, Theorem 6.1.7, $C(K)$ is isomorphic to $C(\Delta)$. By the theorem, $C(\Delta)$ has the stated property. But the property is an isomorphic invariant, and so $C(K)$ has the property. □

The next theorem, together with its history, is given in full in the real-valued case in [3, Theorem 4.3.7]; the proof is due to Kelley [154], extending results of Goodner [119] and Nachbin [188]. The proof in the complex-valued case seems to be technically rather different from that in the real-valued case; it was first given by Hasumi [134] and extended by Cohen [55]. This approach was clarified and expounded in Bade's notes [24], and we essentially follow that account here. There are somewhat later accounts by Semadeni in [225, §25.5] and in Lacey's book [166, §11]; see also [126].

Theorem 6.8.6. *Let E be a 1-injective Banach space. Then E is isometrically isomorphic to $C(K)$ for some Stonean space K.*

We require two lemmas. Recall that the circled hull $\mathrm{ci}\,U$ of a set U was defined on page 8. Throughout, E is a Banach space and the topology on $B = E'_{[1]}$ is the (relative) weak* topology.

Lemma 6.8.7. *There is a subset U of $\mathrm{ex}\,B$ such that $\overline{\mathrm{ci}\,U} = \overline{\mathrm{ex}\,B}$ and such that $\mathrm{ci}\{\lambda\} \cap \overline{U} = \{\lambda\}$ $(\lambda \in U)$.*

Proof. A subset A of a linear space is said to be *deleted* if $\zeta a \in A$ for each $a \in A$ and all but exactly one point $\zeta \in \mathbb{T}$.

Let $\mathscr{F}$ be the family of all non-empty, deleted, open subsets of $\overline{\mathrm{ex}\,B}$, and order $\mathscr{F}$ by inclusion. Clearly the union of an increasing chain of subsets of $\mathscr{F}$ belongs to $\mathscr{F}$, and so it follows from Zorn's lemma that $\mathscr{F}$ has a maximal element, say W.

We first *claim* that each non-empty, open, circled subset A of $\overline{\mathrm{ex}\,B}$ contains a member of $\mathscr{F}$. Indeed, set $D = \mathbb{D} \setminus [0,1)$, an open set in $\mathbb{C}$. Choose $\lambda_0 \in W$ and $x_0 \in E$ with $\langle x_0, \lambda_0 \rangle \in D$, and set $V = \{\lambda \in A : \langle x_0, \lambda \rangle \in D\}$. Then $V \in \mathscr{F}$.

It is now clear that $\mathrm{ci}\,W$ is dense in $\overline{\mathrm{ex}\,B}$. For otherwise $\overline{\mathrm{ex}\,B} \setminus \overline{\mathrm{ci}\,W}$ is a non-empty, open, circled set, and so contains a member of $\mathscr{F}$. Since the union of two disjoint sets in $\mathscr{F}$ belongs to $\mathscr{F}$, this contradicts the maximality of W in $\mathscr{F}$.

It follows that $\mathrm{ex}\,B \cap \mathrm{ci}\,W$ is dense in $\overline{\mathrm{ex}\,B}$.

Now define

$$U = \{\lambda \in \mathrm{ex}\,B \setminus W : \zeta\lambda \in W \ (\zeta \in \mathbb{T} \setminus \{1\})\}.$$

Then $\mathrm{ci}\,U = \mathrm{ex}\,B \cap \mathrm{ci}\,W$, and so $\overline{\mathrm{ci}\,U} = \overline{\mathrm{ex}\,B}$. Note that $\overline{U} \cap W = \emptyset$. For each $\lambda \in U$ and $\zeta \in \mathbb{T} \setminus \{1\}$, we have $\zeta\lambda \notin \overline{U}$, and so $\mathrm{ci}\{\lambda\} \cap \overline{U} = \{\lambda\}$.

Thus U has the required properties. □

Now take U as in Lemma 6.8.7, and set $L = \overline{U}$, a non-empty, compact space with $\mathrm{ci}\,L = \overline{\mathrm{ex}\,B}$. By Corollary 2.6.4, the map

$$J : x \mapsto \kappa_E(x) \mid L, \quad E \to C(L),$$

is an isometric embedding. Let (G_L, π_L) be the Gleason cover of L, as in Definition 1.6.6, and set $K = G_L$ and $\pi = \pi_L$, so that $\pi : K \to L$ is an irreducible surjection. Then K is a Stonean space, and the map $\pi^\circ : C(L) \to C(K)$ is an isometric embedding, and hence $\pi^\circ \circ J : E \to C(K)$ is an isometric embedding; we regard E as a closed subspace of $C(K)$.

Recall that essential and rigid extensions of a Banach space were defined in Definition 2.5.13.

Lemma 6.8.8. *The space $C(K)$ is an essential extension of E.*

Proof. Let G be a Banach space, and take $T \in \mathscr{B}(C(K), G)$ to be a contraction such that $T \mid E$ is an isometry. We must show that $\|Tf\| = |f|_K$ for each $f \in C(K)$.

Take $f \in C(K)$, say with $|f|_K = 1$. We may suppose that there exists $t_0 \in K$ with $f(t_0) = 1$. Take $\varepsilon > 0$, and set

$$V_\varepsilon = \{t \in K : |f(t) - 1| < \varepsilon\},$$

so that V_ε is a non-empty, open subset of K. Set $V = \{y \in L : F_y \subset V_\varepsilon\}$, where again $F_y = \pi^{-1}(\{y\})$. By Proposition 1.4.21(ii), V is non-empty and open in L, and $\pi^{-1}(V)$ is dense in V_ε.

We take $U \subset \mathrm{ex}\, B$ to be as specified above, with $\overline{U} = L$. Thus there exists an element $\lambda \in U \cap V$, and hence with $\pi^{-1}(\{\lambda\}) \subset V_\varepsilon$. Set $S = \pi^{-1}(\{\lambda\})$.

We now identify K with the set of point masses in $M(K)$, so that $\overline{\mathrm{co}}\, S$ is a compact subset of $M(K)_{[1]}$. We have

$$|\langle f, \mu\rangle| > 1 - \varepsilon \quad (\mu \in \overline{\mathrm{co}}\, S).$$

We denote by $R : M(K) \to E'$ the restriction map that is the dual of the embedding of E in $C(K)$, so that $S = R^{-1}(\{\lambda\}) \cap K$. However, it follows from Lemma 6.8.7 that, in fact, $S = R^{-1}(\{\lambda\}) \cap \mathrm{ci}\, K$. Indeed, suppose that $\zeta \in \mathbb{T}$ and $t \in K$ with $R(\zeta\delta_t) = \lambda$. Then $\zeta^{-1}\lambda = R(\delta_t) = \pi(t) \in K$, and so $\zeta^{-1} = 1$, whence $\zeta = 1$.

Since $\lambda \in \mathrm{ex}\, B$, the set $R^{-1}(\{\lambda\}) \cap M(K)_{[1]}$ is a closed face in $M(K)_{[1]}$, and so

$$\begin{aligned}\overline{\mathrm{co}} S &= \overline{\mathrm{co}}\left(R^{-1}(\{\lambda\}) \cap \mathrm{ci}\, K\right) = \overline{\mathrm{co}}\left(R^{-1}(\{\lambda\}) \cap \mathrm{ex}\, M(K)_{[1]}\right) \\ &= \overline{\mathrm{co}}\, \mathrm{ex}\left(R^{-1}(\{\lambda\}) \cap M(K)_{[1]}\right) = R^{-1}(\{\lambda\}) \cap M(K)_{[1]}.\end{aligned}$$

There exists $\nu \in G'$ with $\|\nu\| = 1$ and $(T \mid E)'(\nu) = \lambda$. Thus

$$(R \circ T')(\nu) = (T \mid E)'(\nu) = \lambda,$$

and so $\mu := T'\nu$ is a measure in $R^{-1}(\{\lambda\}) \cap M(K)_{[1]}$. Hence $T'\nu \in \overline{\mathrm{co}}\, S$, and so

$$\|Tf\| \geq |\langle Tf, \nu\rangle| = |\langle f, \mu\rangle| > 1 - \varepsilon.$$

This holds true for each $\varepsilon > 0$, and so $\|Tf\| = 1$, as required. □

Proof of Theorem 6.8.6. Take $K = G_L$ to be the Stonean space specified above, so that $C(K)$ is an essential extension of E. By Proposition 2.5.14, $C(K)$ is a rigid extension of E. Since E is 1-injective, there is a contractive projection $P \in \mathscr{B}(C(K))$ with $P(C(K)) = E$. By the definition of a rigid extension, $P = I_{C(K)}$, and so $E = C(K)$. □

The characterization of Banach spaces which are injective spaces, or, equivalently, P_λ-spaces for some $\lambda \geq 1$, seems to be an open problem; it is a conjecture that each such space is isomorphic to a 1-injective space and hence to a space $C(K)$ for some Stonean space K.

Question 3: *Let E be an injective Banach space. Is E isomorphic to a 1-injective space and hence to $C(K)$ for a Stonean space K?*

The following important result was first proved by Rosenthal in [212], and an exposition is given in [174, p. 192]. The latter proof depends on a combinatorial lemma of Rosenthal whose proof was omitted from [174]; soon afterwards a short proof of this lemma was provided by Kupka [163]. For a self-contained proof of the theorem, see Theorem 1.14 of [20].

Theorem 6.8.9. *Let E be an injective space that contains a closed subspace isomorphic to $c_0(\Gamma)$ for some non-empty set Γ. Then E also contains a closed subspace isomorphic to $\ell^\infty(\Gamma)$.* □

A related result is proved by Haydon in [135]: every injective Banach space that is isomorphically a bidual space is isomorphic to a space of the form $\ell^\infty(\Gamma)$ for some set Γ.

The following remark of Bade is contained in [25, p. 10]. The result was also later proved by Argyros in [13]; the proof of [13] was extended in [38] to determine when the space of bounded functions with the Baire property on a compact space is injective.

Corollary 6.8.10. *The space $B_{\omega_1}(\mathbb{I}) = B^b(\mathbb{I})$ is not injective.*

Proof. Assume towards a contradiction that the space B_{ω_1} is injective. Since B_{ω_1} contains a copy of $c_0(\Gamma)$, where $\Gamma = \mathbb{I}_d$, it follows from Theorem 6.8.9 that B_{ω_1} contains a closed subspace of cardinality $2^{\mathfrak{c}}$. However, $|B_{\omega_1}| = \mathfrak{c}$, and so this is not possible. □

Example 6.8.11. There is a basically disconnected compact space K such that $C(K)$ is not injective. Indeed, take $K = \Phi_b(\mathbb{I})$, the character space of $B^b(\mathbb{I})$. We have noted in Proposition 3.3.3(i) that K is basically disconnected. By the above corollary, $C(K)$ is not injective. □

We now begin our account that seeks to specify the topological properties of K when $C(K)$ is an injective Banach space. We know that K is Stonean whenever $C(K)$ is 1-injective; we now consider the case where $C(K)$ is just λ-injective for some $\lambda > 1$. Example 6.9.8, below, will show that, in this case, K is not necessarily Stonean. In this regard, we note that it is a well-known open question whether a compact F-space K for which $C(K)$ is injective must necessarily be Stonean (see [212, p. 20]).

Definition 6.8.12. Let K and L be two non-empty, compact spaces, and suppose that $\eta : K \to L$ is a continuous surjection. A point $y \in L$ is a *multiple point* for η if the fibre F_y has more than one point. A continuous map

$$\mu : L \to (M(K), \sigma(M(K), C(K)))$$

averages the map η if

$$\mu(y)(F_y) = 1 \quad (y \in L). \tag{6.10}$$

Suppose that $\mu : L \to M(K)$ averages the map η. Then we may suppose that the range of μ is contained in $M_{\mathbb{R}}(K)$. Recall that averaging operators were defined in Definition 3.2.5.

Let K and L be non-empty, compact spaces, and suppose that $T : C(K) \to C(L)$ is a continuous linear operator. Then $T' : M(L) \to M(K)$ is a linear operator which is continuous with respect to the respective weak* topologies, and so the function

$$\theta_T : y \mapsto T'(\delta_y), \quad L \to (M(K), \sigma(M(K), C(K))), \tag{6.11}$$

is a continuous function.

Theorem 6.8.13. *Let K and L be two non-empty, compact spaces, and suppose that $\eta : K \to L$ is a continuous surjection.*

(i) *A map $T \in \mathscr{B}(C(K), C(L))$ is an averaging operator for η if and only if*

$$\theta_T(y)(\eta^{-1}(B)) = \delta_y(B) \quad (y \in L, B \in \mathfrak{B}_L). \tag{6.12}$$

(ii) *Suppose that there is an averaging operator for η. Then there exists a map $\mu : L \to M(K)$ which averages η.*

Proof. (i) It is clear that T is an averaging operator for η if and only if

$$\int_K (f \circ \eta) \, \mathrm{d}(\theta_T(y)) = \langle \eta^\circ(f), \theta_T(y) \rangle = f(y) \quad (f \in C(L), y \in L). \tag{6.13}$$

Suppose that T is an averaging operator. By equation (4.7), this formula also holds for $f = \chi_B$, where $B \in \mathfrak{B}_L$, and so we have the desired formula (6.12).

Conversely suppose that (6.12) holds. Then $\langle \eta^\circ(\chi_B), \theta_T(y) \rangle = \chi_B(y)$ $(y \in L)$ for each $B \in \mathfrak{B}_K$, and so (6.13) follows.

(ii) Take $T \in \mathscr{B}(C(K), C(L))$ to be an averaging operator for η, and then define $\mu = \theta_T : L \to M(K)$. That μ averages η is an immediate consequence of (6.12). □

The critical feature of the averaging map, namely equation (6.10), was crystallized by Ditor [88], [90, p. 197], and Bade [24, Corollary 6.2]. We shall now use the averaging map to prove a theorem of Amir [9] on injective Banach spaces of type $C(K)$ (see Theorem 6.8.15, below). The proof is due to Cohen, Labbe, and Wolfe in [58].

In the following proof, we use the function θ defined by

$$\theta(\zeta) = |1 - \zeta| - |\zeta| \quad (\zeta \in \mathbb{C}).$$

Lemma 6.8.14. *Let K and L be two non-empty, compact spaces with an irreducible surjection $\eta : K \to L$, and suppose that $\mu : L \to M_{\mathbb{R}}(K)$ averages η. Take $x \in K$ and $y \in L$ with $\eta(x) = y$. Then*

$$\|\mu(y)\| + 1 \leq \sup_{v \in V} \|\mu(v)\| - \theta(\mu(y)(\{x\})) \quad (V \in \mathscr{N}_y). \tag{6.14}$$

Proof. We set $y = \eta(x)$, $\mu_0 = \mu(y) \in M_{\mathbb{R}}(K)$, and F_z for the fibre $\eta^{-1}(\{z\})$ when $z \in L$.

Take $V \in \mathcal{N}_y$, and let $\varepsilon > 0$ be given. Then there are open sets W and U in K with

$$x \in W \subset \overline{W} \subset U \subset \overline{U} \subset \eta^{-1}(V) \quad \text{and} \quad |\mu_0|\left(\overline{U} \setminus \{x\}\right) < \varepsilon .$$

Next choose $f \in C_{\mathbb{R}}(K)$ with $\chi_{\overline{W}} \leq f \leq \chi_U$; this implies that $\int_K f \,\mathrm{d}\mu_0 = \int_U f \,\mathrm{d}\mu_0$.

Since the map η is irreducible, every non-empty, open set in K contains a fibre F_z for some $z \in L$, and hence there is a net (y_α) in V converging to y such that $F_{y_\alpha} \subset W$ for each α; we write K_α for F_{y_α} and μ_α for $\mu(y_\alpha)$, so that $\mu_\alpha(K_\alpha) = 1$ by (6.10). Hence $|\mu_\alpha|(K_\alpha) \geq 1$ for each α.

We have

$$\begin{aligned}
\mu_0(\{x\}) + \int_{U\setminus\{x\}} f \,\mathrm{d}\mu_0 &= \int_U f \,\mathrm{d}\mu_0 = \lim_\alpha \int_U f \,\mathrm{d}\mu_\alpha \\
&= \lim_\alpha \left(\int_{U\setminus K_\alpha} f \,\mathrm{d}\mu_\alpha + \int_{K_\alpha} f \,\mathrm{d}\mu_\alpha \right) = \lim_\alpha \int_{U\setminus K_\alpha} f \,\mathrm{d}\mu_\alpha + 1 .
\end{aligned}$$

Consequently, the limit $\lim_\alpha \int_{U\setminus K_\alpha} f \,\mathrm{d}\mu_\alpha$ exists and

$$\lim_\alpha \int_{U\setminus K_\alpha} f \,\mathrm{d}\mu_\alpha = \mu_0(\{x\}) + \int_{U\setminus\{x\}} f \,\mathrm{d}\mu_0 - 1 . \tag{6.15}$$

Also $\liminf_\alpha |\mu_\alpha|(K \setminus \overline{U}) \geq |\mu_0|(K \setminus \overline{U})$ by Proposition 4.1.13. Therefore

$$\begin{aligned}
\sup_{v\in V} \|\mu(v)\| &\geq \liminf_\alpha \|\mu_\alpha\| = \liminf_\alpha \left(|\mu_\alpha|(K_\alpha) + |\mu_\alpha|(\overline{U} \setminus K_\alpha) + |\mu_\alpha|(K \setminus \overline{U}) \right) \\
&\geq 1 + \liminf_\alpha |\mu_\alpha|(U \setminus K_\alpha) + |\,\mu_0|(K \setminus \overline{U}) \\
&\geq 1 + \lim_\alpha \left| \int_{U\setminus K_\alpha} f \,\mathrm{d}\mu_\alpha \right| + |\mu_0|(K \setminus \overline{U}) \\
&= 1 + \left| 1 - \mu_0(\{x\}) - \int_{U\setminus\{x\}} f \,\mathrm{d}\mu_0 \right| + \|\,\mu_0\| - |\mu_0|(\overline{U} \setminus \{x\}) - |\,\mu_0(\{x\})|
\end{aligned}$$

using (6.15). Now

$$\begin{aligned}
\theta(\mu_0(\{x\})) &= |1 - \mu_0(\{x\})| - |\mu_0(\{x\})| \\
&\leq \left| 1 - \mu_0(\{x\}) - \int_{U\setminus\{x\}} f \,\mathrm{d}\mu_0 \right| + \left| \int_{U\setminus\{x\}} f \,\mathrm{d}\mu_0 \right| - |\mu_0(\{x\})| ,
\end{aligned}$$

and so

$$\begin{aligned}
\sup_{v\in V} \|\mu(v)\| &\geq 1 + \theta(\mu_0(\{x\})) + \|\mu_0\| - |\mu_0|(\overline{U} \setminus \{x\}) - \left| \int_{U\setminus\{x\}} f \,\mathrm{d}\mu_0 \right| \\
&\geq 1 + \theta(\mu_0(\{x\})) + \|\mu_0\| - 2\varepsilon .
\end{aligned}$$

This holds true for each $\varepsilon > 0$, and so equation (6.14) follows. □

When $F_y = \{x\}$, so that $\theta(\mu(\{x\})) = -1$, no new information is gained from the above lemma. However, take $n \in \mathbb{N}$ with $n \geq 2$ and suppose that $|F_y| \geq n$. Then we *claim* that there exists $x \in F_y$ with $\mu(\{x\}) \leq 1/n$. For $x \in F_y$, set $c_x = |\mu|(\{x\})$. Then $\Sigma\{c_x : x \in F_y\} \leq \|\mu\| < \infty$. In the case where F_y is infinite, there exists $x \in F_y$ such that $c_x \leq 1/n$, and then $\mu(\{x\}) \leq 1/n$. In the case where F_y is finite,

$$1 = \mu(F_y) \geq |F_y| \min\{\mu(\{x\}) : x \in F_y\} \geq n \min\{\mu(\{x\}) : x \in F_y\}.$$

Thus the claim holds.

Suppose that $y \in L$ is a multiple point for η, take $x \in F_y$ with $\mu(\{x\}) \leq 1/2$, and set $\zeta = \mu(\{x\})$. Then $\theta(\zeta) \geq 0$, and so, by equation (6.14), we have

$$\|\mu(y)\| + 1 \leq \sup_{v \in V} \|\mu(v)\| \quad (V \in \mathcal{N}_y). \tag{6.16}$$

Now let M denote the set of multiple points for η. Assume towards a contradiction that $\overline{M}$ contains a non-empty, open set, say V. For $v \in V$, there is a net (y_α) in $M \cap V$ with $v = \lim_\alpha y_\alpha$. By Proposition 4.1.13 and equation (6.16), we have

$$\|\mu(v)\| \leq \liminf_\alpha \|\mu(y_\alpha)\| \leq \sup_{y \in V} \|\mu(y)\| - 1,$$

and so $\sup_{v \in V} \|\mu(v)\| \leq \sup_{y \in V} \|\mu(y)\| - 1$, a contradiction. Thus M is nowhere dense in K.

We now come to *Amir's theorem*.

Theorem 6.8.15. *Let K be a non-empty, compact space, and suppose that $C(K)$ is an injective space. Then K contains a dense, open, extremely disconnected subset.*

Proof. We apply the above argument with L taken to be the space K, K taken to be the Gleason cover G_K of K, and η taken to be the irreducible map π_K of Definition 1.6.6.

Since the space $C(K)$ is injective, the space $\pi_K^\circ(C(K))$ is a complemented subspace of $C(G_K)$, and so, by Theorem 6.8.13(ii), there exists $\mu : K \to M(G_K)$ which averages π_K.

Let M denote the set of multiple points for π_K. Since M is nowhere dense, the set $V := K \setminus \overline{M}$ is a dense, open subset of K and $U := \pi_K^{-1}(V)$ is an extremely disconnected space which is carried bijectively by π_K onto V.

It suffices to show that, on U, the map π_K is an open mapping. Let W be an open subset of U, and assume to the contrary that $\pi_K(W)$ is not open. Take $y \in W$ and a net (x_α) in $K \setminus \pi_K(W)$ with $x_\alpha \to \pi_K(y)$ in K. For each α, there exists $y_\alpha \in G_K \setminus W$ with $x_\alpha = \pi_K(y_\alpha)$. Since $G_K \setminus W$ is compact, we may, by passing to a convergent subnet, suppose that (y_α) converges to some point $z \in G_K \setminus W$. Then

$$\pi_K(z) = \lim_\alpha x_\alpha = \pi_K(y).$$

Since $\pi_K(y) \notin M$ and $z \neq y$, this is a contradiction. Hence π_K is indeed an open map on U. □

It is easy to construct a compact space K such that D_K is countable, dense and open, and hence extremely disconnected, but such that K is not totally disconnected.

Corollary 6.8.16. (i) *Let K be an infinite, compact space, and suppose that $C(K)$ is an injective space. Then K contains a clopen subspace that is Stonean.*

(ii) *Let K be a compact space that is locally connected and without isolated points. Then $C(K)$ is not injective.*

(iii) *Let K be an infinite, compact space with only finitely many clopen subsets. Then $C(K)$ is not injective.*

(iv) *Let S be an infinite set. Then $C(S^*)$ is not injective.*

Proof. (i) By Theorem 6.8.15, there is a non-empty, open, extremely disconnected subspace U of K. Take $x \in U$. Then x has a clopen neighbourhood which is a Stonean space.

(ii), (iii) In both cases, it is clear that K does not contain any non-empty, open, extremely disconnected subspace.

(iv) By Corollary 1.5.6, the space S^* does not contain a non-empty, clopen, Stonean subspace, and so, by (i), $C(S^*)$ is not injective. □

Example 6.8.17. It follows from clause (iv) of Corollary 6.8.16 that, for each infinite set S, the space $C(S^*)$ is not isomorphic to $C(\beta S)$ because $C(\beta S)$ is 1-injective by Proposition 2.5.5 and injectivity is an isomorphic invariant; a stronger result was given in Corollary 2.2.25.

In Corollary 4.5.10, we noted that each injective space is a Grothendieck space. We now see that there are compact spaces K such that $C(K)$ is a Grothendieck space, but $C(K)$ is not injective. Indeed, take $K = S^*$ for an infinite set S, so that K is an infinite F-space, or take K to be an infinite, connected, compact F-space, as on page 31. By Corollary 4.5.9, $C(K)$ is a Grothendieck space, but $C(K)$ is not injective by clauses (iv) and (iii), respectively, of Corollary 6.8.16. □

Corollary 6.8.18. *Let K be an infinite, homogeneous, compact space. Then $C(K)$ is not injective.*

Proof. Assume to the contrary that $C(K)$ is injective. Then Theorem 6.8.15 shows that K contains a dense, open, extremely disconnected subspace, say U. Each point of U has a clopen, extremely disconnected neighbourhood. Since K is homogeneous, each point of K has such a neighbourhood, and so K is Stonean and hence an F-space. However, by Theorem 1.5.15, an infinite, compact F-space is not homogeneous, a contradiction. □

In particular, consider an infinite, compact group G. Then $C(G)$ is not injective, a result already proved in Theorem 4.4.2.

Let L be a compact space such that $C(L)$ is λ-injective. Set $K = G_L$ and $\eta = \pi_L$, and regard $C(L)$ as a closed subspace of $C(K)$. Then there is a bounded projection P of $C(K)$ onto $C(L)$ with $\|P\| = \lambda$. As in Theorem 6.8.13(ii), set

$$\mu = \theta_P : L \to M(K),$$

so that μ averages the map η; again, we may suppose that $\mu \in M_{\mathbb{R}}(K)$. We have $\mu(y) = P'(\delta_y)$ $(y \in L)$, and so $\sup_{v \in L} \|\mu(v)\| \leq \lambda$.

We *claim* that, in the case where $\lambda < 3$, there exists $n \in \mathbb{N}$ such that

$$|F_y| < n \quad (y \in L). \tag{6.17}$$

Indeed, take $n \in \mathbb{N}$ such that $3 - 2/n > \|P\|$, and assume to the contrary that there exists $y \in L$ with $|F_y| \geq n$. As above, there exists $x \in F_y$ with $\mu(y)(\{x\}) \leq 1/n$, and hence $\theta(\mu(y)(F_y)) \geq 1 - 2/n$. Since $\|\mu(y)\| \geq 1$, it follows from equation (6.14) that $3 - 2/n \leq \|P\|$, a contradiction.

In the case where $\lambda < 2$, the above argument (with $n = 2$) also gives a contradiction, and so η is injective and hence a homeomorphism. Thus, $C(L) = C(K)$ and $C(L)$ is 1-injective. This shows that $C(L)$ is 1-injective whenever it is λ-injective for some $\lambda < 2$.

The above remarks are contained in [143]; more general results are contained in [242]. For some partial positive results which apply in the case where K satisfies CCC and $C(K)$ is a P_λ-space for some $\lambda < 3$, see the article of Wolfe [243] from 1978, which is based on his thesis as a student of Bade. This article contains some examples when $C(K)$ is an injective space and concludes with a list of interesting questions; it seems that none has been resolved in the intervening 37 years.

The results of this section leave open the following question.

Question 4: *Let K be a compact space such that $C(K)$ is an injective Banach space. Is K totally disconnected? Is $C(K)$ isomorphic to $C(L)$ for some Stonean space L?*

We do have a partial result. Recall from Definition 2.2.26 that $c(X)$ denotes the Souslin number of a topological space X.

Theorem 6.8.19. *Let L be an infinite, compact space such that $C(L)$ is injective. Suppose that K is a Stonean space such that there is a continuous surjection $\eta : K \to L$, and assume that L is not totally disconnected. Then:*

(i) *the codimension of $\eta^\circ(C(L))$ in $C(K)$ is at least $\mathfrak{c}$;*

(ii) *the space L contains a family of $\mathfrak{c}$ non-empty, pairwise-disjoint, open subsets, and so $c(L) \geq \mathfrak{c}$.*

Proof. (i) We regard $C(L)$ as a C^*-subalgebra of $C(K)$; since $C(L)$ is injective, there is a bounded projection, say P, of $C(K)$ onto $C(L)$, and P is an averaging operator for η. The map $P' : M(L) \to M(K)$ is an embedding, and $y \mapsto P'(\delta_y)(V)$, $L \to \mathbb{C}$, is continuous for each $V \in \mathfrak{C}_K$ because P' is weak*-weak*-continuous.

For each $y \in L$, choose $x_y \in K$ with $\eta(x_y) = y$, and set $v_y = P'(\delta_y) - \delta_{x_y} \in M(K)$, so that $\|v_y\| \leq \|P\| + 1$. By equation (6.12),

$$v_y(\eta^{-1}(B)) = 0 \quad (y \in L, B \in \mathfrak{B}_L). \tag{6.18}$$

Since L is not totally disconnected, there is a connected subset C of L containing two distinct points, say a and b; choose $f \in C(L,\mathbb{I}) \subset C(K,\mathbb{I})$ with $f(a) = 0$ and $f(b) = 1$, so that f is constant on the fibres in K, and define

$$W_t = \{x \in K : f(x) \in [0,t)\} \quad (t \in (0,1]),$$

so that each W_t is open in K and $\overline{W_t} \in \mathfrak{C}_K$ because K is Stonean. The sets

$$H_t := \{x \in K : f(x) = t\}$$

for $t \in (0,1]$ form a pairwise-disjoint family of compact subsets of K, and so, for each $y \in L$, we have $|\nu_y|(H_t) = 0$ save for countably many values of t and

$$\sum\{|\nu_y|(H_t) : t \in (0,1]\} \leq \|P\| + 1.$$

Take $y \in L$ and $t \in (0,1]$. By (6.18), $\nu_y(W_t) = 0$, and so $\nu_y(\overline{W_t}) = \nu_y(\overline{W_t} \setminus W_t)$. But $\overline{W_t} \setminus W_t \subset H_t$, and so $|\nu_y|(\overline{W_t}) = 0$ save for countably many values of t and

$$\sum\{|\nu_y|(\overline{W_t}) : t \in (0,1]\} \leq \|P\| + 1. \tag{6.19}$$

Set

$$T = \{t \in (0,1) : \nu_a(\overline{W_t}) = \nu_b(\overline{W_t}) = 0\}.$$

Then $(0,1) \setminus T$ is countable, and so $|T| = \mathfrak{c}$. We note that, for each $t \in T$, we have $P'(\delta_a)(\overline{W_t}) = 1$ and $P'(\delta_b)(\overline{W_t}) = 0$.

Take $t \in T$, and assume towards a contradiction that $\overline{W_t}$ is a union of fibres. Then, by (6.18), $\nu_y(\overline{W_t}) = 0$ $(y \in L)$, and so $\{P'(\delta_y)(\overline{W_t}) : y \in L\} = \{0,1\}$, a contradiction because $\{P'(\delta_y)(\overline{W_t}) : y \in C\} = [0,1]$. Thus, there exist $u_t \in \overline{W_t}$ and $v_t \in K \setminus \overline{W_t}$ with $\eta(u_t) = \eta(v_t)$. It follows that $u_t, v_t \in \overline{W_s}$ for each $s \in T$ with $s > t$ and that $u_t, v_t \notin \overline{W_s}$ for each $s \in T$ with $s < t$. Write χ_t for the characteristic function of $\overline{W_t}$ when $t \in T$. We have shown that $\chi_t(u_t) = 1$ and $\chi_t(v_t) = 0$; $\chi_s(u_t) = \chi_s(v_t) = 1$ for $s \in T$ with $s > t$; and $\chi_s(u_t) = \chi_s(v_t) = 0$ for $s \in T$ with $s < t$.

Let $Q : C(K) \to C(K)/C(L)$ be the quotient map. Suppose that $n \in \mathbb{N}$, that $t_1, \dots, t_n \in T$ with $t_1 < \cdots < t_n$, that $\alpha_1, \dots, \alpha_n \in \mathbb{C}$, and that $g \in C(L)$ with

$$\alpha_1 \chi_{t_1} + \cdots + \alpha_n \chi_{t_n} = g.$$

Take $i \in \mathbb{N}_n$. Then, evaluating the functions in this equation at the points u_{t_i} and v_{t_i}, and recalling that $g(u_{t_i}) = g(v_{t_i})$, we see that $\alpha_i = 0$. Thus the set $\{Q(\chi_t) : t \in T\}$ is linearly independent in $C(K)/C(L)$, and so the codimension of $C(L)$ in $C(K)$ is at least $|T| = \mathfrak{c}$.

(ii) Choose α with $0 < \alpha < 1/2$ and $y \in L$, write $\psi_t(y) = P'(\delta_y)(\overline{W_t})$ $(t \in T)$, so that $\psi_t \in C(L)$, and set

$$A_y = \{t \in T : |\psi_t(y)| > \alpha\} \cap \{t \in T : |\psi_t(y) - 1| > \alpha\}.$$

Take $t \in A_y$. Since $\nu_y(\overline{W_t})$ is either $\psi_t(y)$ or $\psi_t(y) - 1$, it follows that $|\nu_y(\overline{W_t})| > \alpha$. By (6.19), A_y is finite and $\alpha\,|A_y| \leq \|P\| + 1$.

For each subset H of T, set $U_H = \{y \in L : H \subset A_y\}$, so that U_H is an open subset of L. In the case that $U_H \neq \emptyset$, the set H is finite with $\alpha |H| \leq \|P\| + 1$. Let $\mathscr{F}$ denote the family of finite subsets H of T that are maximal with respect to the condition that $U_H \neq \emptyset$. We see that, for H_1 and H_2 in $\mathscr{F}$ with $H_1 \neq H_2$, we have $U_{H_1 \cup H_2} = U_{H_1} \cap U_{H_2} = \emptyset$.

Again take $t \in A_y$. Since $\psi_t(a) = 1$ and $\psi_t(b) = 0$, there is a point c in the connected set C with $\psi_t(c) = 1/2$, and so $t \in A_c$ and $U_{\{t\}} \neq \emptyset$.

We *claim* that $\bigcup\{H : H \in \mathscr{F}\} = T$. Assume to the contrary that this is not the case, and take $t \in T$ with $t \notin H$ $(H \in \mathscr{F})$. Then $U_{\{t\}} = \emptyset$ by the maximality of $\mathscr{F}$, a contradiction. Thus the claim holds.

It follows that $|\mathscr{F}| = |T| = \mathfrak{c}$ and that $\{U_H : H \in \mathscr{F}\}$ is a family of non-empty, pairwise-disjoint, open subsets of L. Hence $c(L) \geq \mathfrak{c}$. □

The above theorem extends the following corollary of Wolfe [243, Corollary 1.4].

Corollary 6.8.20. *Let L be an infinite, compact space such that $C(L)$ is injective and L satisfies* CCC. *Then L is totally disconnected.*

Proof. We apply Theorem 6.8.19 with K taken to be the Gleason space G_L. □

6.9 The isomorphic theory

Our aim (which is not achieved) is to characterize topologically the compact spaces K such that $C(K)$ is isomorphically a dual space. Of course, in the case where K is hyper-Stonean, we know from Theorem 6.4.1 that $C(K)$ is even isometrically isomorphic to the dual of a Banach space, and so our desired topological condition will be weaker than 'K is hyper-Stonean'.

We begin with an easy example which shows that a space of the form $C_0(K)$, for K locally compact, can be isomorphically, but not isometrically, a dual space. Examples with K compact will be given in Examples 6.9.8 and 6.9.10.

Example 6.9.1. As usual we identify ℓ^∞ and $C(\beta\mathbb{N})$. For $x \in \beta\mathbb{N}$, consider the maximal ideal
$$M_x = \{f \in C(\beta\mathbb{N}) : f(x) = 0\} = C_0(\beta\mathbb{N} \setminus \{x\}).$$
Now take $p \in \mathbb{N}^*$. Then we *claim* that $M_p \sim \ell^\infty$. Indeed, define
$$Tf = (2f(p), f(2) - f(p), f(3) - f(p), \dots) \quad (f \in M_1),$$
so that $Tf \in M_p$ $(f \in M_1)$ and $T \in \mathscr{B}(M_1, M_p)$ with $\|T\| = 2$, and define
$$Sg = (0, g(2) + g(1)/2, g(3) + g(1)/2, \dots) \quad (g \in M_p),$$

so that $Sg \in M_1$ $(g \in M_p)$ and $S \in \mathscr{B}(M_p, M_1)$ with $\|S\| = 3/2$. Clearly S is the inverse of T, and so $M_1 \sim M_p$ with $d(M_1, M_p) \leq 3$. Since $M_1 \cong \ell^\infty$, the claim follows, with $d(\ell^\infty, M_p) \leq 3$.

The above claim that $M_p \sim \ell^\infty$ also follows from the more general (unproved) Theorem 2.4.19 (ii). □

For the general theory, we first give an observation, apparently originally made by Pełczyński (see [143, Note 14, p. 45]).

Proposition 6.9.2. *Let K be a non-empty, locally compact space such that $C_0(K)$ is isomorphically a dual Banach space. Then $C_0(K)$ is an injective space.*

Proof. Suppose that $C_0(K) \sim E'$ for a Banach space E. By Corollary 2.4.5, $C_0(K)$ is complemented in $C_0(K)'' = C(\widetilde{K})$; $\widetilde{K}$ is Stonean, and so, by Theorem 2.5.11, $C(\widetilde{K})$ is 1-injective and hence injective. Thus $C_0(K)$ is injective by Proposition 2.5.3. □

Corollary 6.9.3. *Let K be a non-empty, compact space such that $C(K)$ is isomorphically a dual space, and suppose that L is a non-empty, clopen subspace of K. Then $C(L)$ is injective.*

Proof. By Proposition 6.9.2, $C(K)$ is an injective space. But it is clear that $C(L)$ is a complemented subspace of $C(K)$, and so $C(L)$ is injective by Proposition 2.5.3. □

We shall see in Example 6.9.12, below, that, in the above situation, $C(L)$ is not necessarily a dual space.

The next result is a converse to Proposition 6.9.2 in a special case.

Proposition 6.9.4. *Let K be an infinite, separable, locally compact space such that $C_0(K)$ is an injective space. Then $C_0(K)$ is isomorphic to ℓ^∞, and hence is isomorphically a bidual Banach space.*

Proof. The space $C_0(K)$ is infinite dimensional, and it is isomorphic to a closed subspace of ℓ^∞ because K is separable. Since $C_0(K)$ is an injective space, it is complemented in ℓ^∞. By Theorem 2.4.19(ii), ℓ^∞ is a prime Banach space, and so $C_0(K)$ is isomorphic to the bidual space ℓ^∞. □

Corollary 6.9.5. *Let K be an infinite, separable Stonean space. Then $C(K) \sim \ell^\infty$.* □

Theorem 6.9.6. *Let K be a non-empty, compact space such that $C(K)$ is isomorphically a dual Banach space. Then K contains a dense, open, extremely disconnected subset.*

Proof. This is immediate from Theorem 6.8.15 and Proposition 6.9.2. □

Corollary 6.9.7. *Let K be a compact space that is either locally connected and without isolated points or of the form S^* for S an infinite set or an infinite, compact space with only finitely many clopen subsets. Then $C(K)$ is not isomorphically a dual space.*

Proof. This follows as before. □

The following question is related to Question 4.

Question 5: *Let K be a compact space such that $C(K)$ is isomorphically a dual space. Is K totally disconnected? Does there exist a Stonean space L such that $C(K)$ is isomorphic to $C(L)$? Does there exist a hyper-Stonean space L such that $C(K)$ is isomorphic to $C(L)$?*

At this stage, we shall give only some illustrative examples.

We shall first show that $C(K)$ can be isomorphically a dual space (and hence injective) for a compact space K without K being Stonean.

Example 6.9.8. We describe a totally disconnected, compact space K such that $C(K) \sim \ell^\infty$, so that $C(K)$ is isomorphically a bidual space, but such that K is not an F-space.

Let p, q be two points of $\beta\mathbb{N}$, with $p \neq q$, and consider the closed subalgebra

$$W = \{f \in C(\beta\mathbb{N}) : f(p) = f(q)\}$$

of $C(\beta\mathbb{N})$. We *claim* that $W \sim \ell^\infty$. Indeed, let $\beta\mathbb{N} = A \cup B \cup C$ where A, B, and C are pairwise-disjoint, clopen subsets of $\beta\mathbb{N}$ such that $p \in A$, $q \in B$, and $C = \{r\}$ for some $r \in \mathbb{N}$; set $V = C(\beta\mathbb{N} \setminus C)$. For $f \in V$, define $Tf \in \ell^\infty$ by

$$Tf = f\chi_A + (f + f(p) - f(q))\chi_B + (f(p) - f(q))\chi_C.$$

Clearly $(Tf)(p) = (Tf)(q)$, so that $T \in \mathscr{B}(V, W)$ (with $\|T\| = 3$). Suppose that $f \in V$ and $Tf = 0$. Then $f\chi_A = 0$ and $f(p) = f(q)$, and so $f\chi_B = 0$, whence $f = 0$. Thus T is injective. To verify that T is surjective, take $g \in W$, and define

$$f = g\chi_A + (g - g(r))\chi_B,$$

so that $f \in V$. We see that $f(p) = g(p)$ and $f(q) = g(q) - g(r) = g(p) - g(r)$, so that $f(p) - f(q) = g(r)$. It follows that $Tf \mid A = g \mid A$, that $Tf \mid B = g \mid B$, and that $(Tf)(r) = g(r)$, and hence $Tf = g$. Thus T is an isomorphism from V onto W. Since $V \cong \ell^\infty$, the claim follows. (The claim also follows from Theorem 2.4.19 (ii).)

Now take K to be the quotient space of $\beta\mathbb{N}$ obtained by identifying the points p and q. Then W is isometrically isomorphic to $C(K)$. Finally, in the case where $p, q \in \mathbb{N}^*$, the space K fails to be an F-space because the common point $p = q$ is in the closure of the two disjoint cozero sets $A \cap \mathbb{N}$ and $B \cap \mathbb{N}$. In particular, K is not Stonean, and so $C(K)$ is not isometrically a dual space. □

Example 6.9.9. There is a totally disconnected, compact space K such that $C(K)$ is not isomorphically a dual space. Indeed, take K to be Δ, the Cantor set, so that K is indeed an infinite and totally disconnected, compact metric space. By Corollary 2.4.17, $C(\Delta)$ is not isomorphically a dual space. □

Our next example shows that there is a compact space K such that $C(K)$ is isomorphic to a dual space and K is Stonean, but K is not hyper-Stonean because $N(K) = \{0\}$.

Example 6.9.10. Indeed, our space K is $G_{\mathbb{I}}$, the Gleason cover of $\mathbb{I}$, as described in §1.5 and in Examples 1.7.16 and 1.7.17: as remarked in Example 1.7.16, $G_{\mathbb{I}}$ is an infinite, separable Stonean space without isolated points. By Corollary 6.9.5, $C(G_{\mathbb{I}})$ is isomorphically a bidual space.

As in Example 5.1.4(ii), $N(G_{\mathbb{I}}) = \{0\}$, and so $G_{\mathbb{I}}$ is not hyper-Stonean, and hence $C(G_{\mathbb{I}})$ is not isometrically a dual space. □

Now we exhibit a compact space L such that $C(L)$ is injective, but not isomorphically a dual space.

Example 6.9.11. Let K be a non-empty, compact space satisfying CCC, and suppose that $C(K)$ is isomorphically a dual space. In [211], Rosenthal showed that there is a strictly positive measure on K. He moreover observed (*ibid.*, page 228) that Gaifman had described a Stonean space L satisfying CCC and carrying no strictly positive measure (for a thorough treatment, see [63, Theorem 6.23]). Nowadays one would refer instead to the remarkable example, described by Talagrand in [235, Theorem 1.2], of a Stonean space L carrying a strictly positive *submeasure*, and so satisfying CCC, and carrying no strictly positive measure. For an exposition of this example, see [107, Remark 394N(c)]. Thus, in this case, $C(L)$ is not isomorphically a dual space, but $C(L)$ is even 1-injective by Theorem 6.8.3, (c) $\Rightarrow$ (a). □

The following example is based on a discussion with Tomasz Kania.

Example 6.9.12. Let L be a non-empty, Stonean space such that $C(L)$ is injective, but not isomorphically a dual space (the space L in the above example has these properties). By Proposition 2.2.14(i), there is a non-empty (necessarily infinite) set S such that $C(L) \cong G$, where G is a closed subspace of $\ell^\infty(S) = C(\beta S)$. Since $C(L)$ is injective, G is a complemented subspace of $C(\beta S)$. Let K be the compact space that is the disjoint union of L and βS, so that K is Stonean. Then L is a clopen subspace of K, and

$$C(K) = C(L) \oplus_\infty C(\beta S).$$

We apply Theorem 2.4.9; for this, we take $E = C(\beta S)$ and $F = C(K)$, so that $F = C(L) \oplus_\infty E$. Certainly E is a complemented subspace of F. Also $F \cong G \oplus_\infty E$, and $G \oplus_\infty E$ is a complemented subspace of $E \oplus_\infty E \sim E$, so that F is isomorphic to a complemented subspace of E. Finally, we *claim* that $E \cong \ell^\infty(E)$. Indeed, we can partition S into infinitely many subsets S_n for $n \in \mathbb{N}$ such that $|S_n| = |S|$ $(n \in \mathbb{N})$, and then the map

$$f \mapsto (f \mid S_n), \quad \ell^\infty(S) \to \ell^\infty(\ell^\infty(S)),$$

is an isometric isomorphism.

Thus it follows from Theorem 2.4.9 that $C(K) \sim C(\beta S)$. Since $C(\beta S) \cong c_0(S)''$, the space $C(\beta S)$ is a isometrically a bidual space, and so $C(K)$ is isomorphically a bidual space.

Now we see that the Banach space $C(L)$ is not necessarily isomorphically a dual space when L is a clopen subspace of a compact space K and $C(K)$ is isomorphically a bidual space. □

Proposition 6.9.13. *There exist compact spaces K and L such that $C(L)$ is isometrically isomorphic to a 1-complemented subspace of $C(K)$ and $C(K)$ is isometrically a dual space, but $C(L)$ is not even isomorphically a dual space.*

Proof. Let L be a Stonean space. By Theorem 2.5.11, $C(L) \cong F$, where F is a closed subspace of $C(\beta L_d)$; the latter is isometrically a dual space. Example 6.9.11 shows that L is not necessarily isomorphically a dual space. □

Proposition 6.9.14. *There is a compact space X such that $C(X)$ is isometrically a bidual space and such that a certain 1-complemented subspace of $C(X)$ is not isomorphically a dual space.*

Proof. Again, let $E = C(L)$ be such that E is 1-injective, but not isomorphically a dual space, as in Example 6.9.11. Set $X = \widetilde{L}$, so that $E'' = C(X)$. Then $C(X)$ is isometrically a bidual space and E is 1-complemented in $C(X)$. □

6.10 $C(X)$ as a bidual space

In the Preface we raised a question concerning the topology of compact spaces X for which $C(X)$ is a bidual space. We hope to show that X is necessarily of the form $\widetilde{K}$ for a locally compact space K. More precisely, we ask the following question.

Question 6: *Let X be a non-empty, compact space. Suppose that $C(X)$ is isometrically isomorphic to the bidual E'' of some Banach space E. Does there exist a locally compact space K such that $C(X)$ is isometrically isomorphic to the bidual space $C_0(K)'' = M(K)' = C(\widetilde{K})$ and hence such that X is homeomorphic to $\widetilde{K}$?*

Suppose that there is a locally compact space K such that $C_0(K)'' \cong C(X)$. Then it follows from equation (6.7) that $C(L)'' \cong C(X)$ for a compact space L.

We attempt to resolve this question in this final section. We shall show that, in the general case, X is at least homeomorphic to a clopen subspace of a space of the form $\widetilde{K}$; in the special case in which X is infinite and we hypothesize that $C(X)$ is isometrically isomorphic to the bidual E'' of a *separable* Banach space E, we can show that X is homeomorphic to either $\beta\mathbb{N} = \widetilde{\mathbb{N}}$ or to $\widetilde{\mathbb{I}}$.

The hypothesis that $E'' \cong C(X)$ is of course equivalent to assuming that E' is an $L^1(\mu)$ space for some (possibly infinite) positive measure μ, which is to say that E is what is called an 'L^1-predual space' or a *Lindenstrauss space* (in the case of real-valued spaces) in the Banach space literature.

We now suppose that X is a non-empty, compact space such that $C(X)$ is isometrically isomorphic to the bidual E'' of some Banach space E. We shall in fact throughout regard E as a closed linear subspace of $C(X)$. Certainly this implies that $C(X)$ is isometrically the dual of E', and so X is hyper-Stonean. Further, by Theorem 6.4.2, the von Neumann algebra $C(X)$ has a strongly unique predual, namely,

$C(X)_* = N(X)$, and so we identify E' with $N(X)$. Thus we are considering when $N(X)$ is isometrically a dual space. In the case where $x \in D_X$, the corresponding element δ_x belongs to $N(X)_{[1]}$; we recall that, by Proposition 4.7.12, each extreme point of $E'_{[1]} = N(X)_{[1]}$ has the form $\zeta\delta_x$ for some $\zeta \in \mathbb{T}$ and $x \in D_X$.

We first remark that $N(X)$ may have many isometric preduals. For example, take $X = \beta\mathbb{N}$. Then $N(X) = \ell^1$, and we have remarked in §6.3 that ℓ^1 has a multitude of isometric preduals no two of which are pairwise isomorphic. We also recall that it does not follow from the fact that F is a Banach space such that $F'' \cong C(X)$ for a compact space X that F has the form $C_0(K)$ for some locally compact space K. Indeed, we have remarked in §6.3 that there are Banach spaces F with $F' \cong \ell^1$, but such that F is not isomorphic to any space of the form $C_0(K)$. However, this does not give a counter-example to our question, above: $C(\beta\mathbb{N})$ has an obvious isometric pre-bidual, namely, c_0.

The weak* topology $\sigma = \sigma(N(X), E)$ is, of course, such that $(N(X)_{[1]}, \sigma)$ is compact and $(N(X), \sigma)$ is a locally convex space that is a complex Riesz space. However, we have seen on page 196 that the positive cone $N(X)^+$ may not be closed in $(N(X), \sigma)$, and so we cannot say that $(N(X)^+_{[1]}, \sigma)$ is compact; this precludes some tempting proofs.

We also note that Godefroy [115, p. 175] stated that it is 'unknown whether a dual space which is a Banach lattice has at least one predual which is a Banach lattice'. More relevant to our investigation is a weaker version of this question: Let X be a non-empty, compact space such that $N(X)$ is isometrically isomorphic to the dual E' of some Banach space E. Does there exist a Banach lattice F such that the dual Banach lattice F' is Banach-lattice isometric to the Banach lattice $N(X)$? Assume that the answer to this question is 'yes', so that the Banach lattice F'' is Banach-lattice isometric to the Banach lattice $C(X)$. Does it then follow that there exists a locally compact space K such that $C(X)$ is isometrically isomorphic to the bidual space $C_0(K)''$? Unfortunately we cannot answer this question either in full generality.

In [171, Chapter 6], Lindenstrauss discusses (non-zero) Banach spaces E such that E'' is a P_1-space, equivalently, such that E'' is 1-injective. Suppose that E is such a space. Then, by Theorem 6.8.6, there is a non-empty, Stonean space X such that $E'' \cong C(X)$, and then X is hyper-Stonean. Conversely, suppose that $E'' \cong C(X)$ for a non-empty, compact space X. Then X is hyper-Stonean, and so, by Theorem 6.8.3, E'' is 1-injective. Thus our hypothesis that X is a non-empty, compact space such that $C(X) \cong E''$ for a Banach space E is the same as Lindenstrauss's hypothesis that E'' is a P_1-space. Theorem 6.1 of [171] gives 12 properties of a Banach space E that are all equivalent to the fact that E'' is a P_1-space. For example, clause (5) of [171, Theorem 6.1] states that, for every Banach space G, every closed subspace F of G, and every $T \in \mathscr{B}(F, E)$, there is an extension $\widetilde{T} \in \mathscr{B}(G, E'')$ of $\kappa_E \circ T$ with $\left\|\widetilde{T}\right\| = \|T\|$. Theorem 6.6 of [171] shows that our hypothesis in Question 6 'almost characterizes $C(K)$–spaces'; we shall obtain a version of this theorem below.

Our first result is the following. Recall from Theorem 6.5.3 that, in the case where K is a compact space and $X = \widetilde{K}$, we can identify K with D_X as sets.

Theorem 6.10.1. *Let X be an infinite, compact space such that $C(X)$ is isometrically a bidual space. Then D_X is infinite.*

Proof. Take E to be a Banach space with $C(X) \cong E''$, and then set $\sigma = \sigma(N(X), E)$ and $B = N(X)_{[1]}$, so that B is a compact, convex subspace of the locally convex space $(N(X), \sigma)$. It follows from the Krein–Milman theorem, Theorem 2.6.1, that $B = \overline{\text{co}}(\text{ex}\, B)$, and so $B = \overline{\text{co}}\{\zeta\delta_x : \zeta \in \mathbb{T}, x \in D_X\}$. In particular, $D_X \neq \emptyset$.

Assume towards a contradiction that D_X is finite. Then $\text{aco}\{\delta_x : x \in D_x\}$ is $\|\cdot\|$-compact, and so it is equal to $N(X)_{[1]}$. Thus $N(X)$ and $C(X)$ are finite-dimensional spaces, a contradiction of the fact that X is infinite. Thus D_X is infinite. □

Example 6.10.2. The above theorem shows that the compact space $X := \widetilde{\mathbb{I}} \setminus \beta\mathbb{I}_d$ is a hyper-Stonean space such that $C(X)$ is not isometrically a bidual space. Indeed, by Proposition 5.1.2(i), X is a hyper-Stonean space. But clearly $D_X = \emptyset$. □

In the case where $X = \beta\mathbb{N}$, so that $C(X) \cong c_0''$, we see that $D_X = \mathbb{N}$, and so D_X is countable. In the case where $X = \widetilde{\mathbb{I}}$, so that $C(X) \cong C(\mathbb{I})''$, we see that $|D_X| = \mathfrak{c}$. We shall show that one of these two alternatives for $|D_X|$ always obtains whenever $C(X) \cong E''$ for a separable Banach space E.

Before giving the proof, we remark that the result is obvious in the presence of (CH). It is also easy in the case of real Banach spaces. Indeed, suppose that X is an infinite, compact space such that $C_{\mathbb{R}}(X)$ is isometrically the bidual of a real, separable Banach space E. Then $E'_{[1]} = N_{\mathbb{R}}(X)_{[1]}$ is compact and metrizable in its weak* topology. By Proposition 2.1.9, $\text{ex}\, N_{\mathbb{R}}(X)_{[1]}$ is a G_δ-set, so by Proposition 1.4.14 it is either countable or of cardinality $\mathfrak{c}$. But, just as in Proposition 4.7.12, $\text{ex}\, N_{\mathbb{R}}(X)_{[1]} = \{\pm\delta_x : x \in D_X\}$, hence either D_X is countable or $|D_X| = \mathfrak{c}$. Thus the result holds in this 'real' case. The difficulty in the complex case is that the set $N(X)^+_{[1]}$ is not necessarily closed in $(N(X), \sigma(N(X), E))$, and so it is not clear that $\text{ex}\, N(X)^+_{[1]}$, which is identified with D_X, is a Borel set.

Theorem 6.10.3. *Let X be an infinite, compact space such that $C(X) \cong E''$ for a separable Banach space E. Then D_X is either countable or of cardinality $\mathfrak{c}$.*

Proof. We define an equivalence relation $\sim$ on the space $Y := \text{ex}\, N(X)_{[1]}$ by setting $\mu \sim \nu$ whenever $\mu = \zeta\nu$ for some $\zeta \in \mathbb{T}$. The space Y, as a G_δ-set in the compact metric space $N(X)_{[1]}$ (Proposition 2.1.9), is itself a Polish space with respect to the relative topology from $N(X)_{[1]}$ (as proved in Proposition 1.4.12). Moreover, it follows from the continuity of multiplication by scalars from the compact space $\mathbb{T}$ that each equivalence class is compact and that $\sim$, as a subset of $Y \times Y$, is closed in the product topology. We know from Proposition 1.4.13 that the equivalence relation $\sim$ has either countably many or exactly $\mathfrak{c}$ equivalence classes. Applying this to the space $(Y, \sim)$, we note that $D_X \subset Y$ and D_X has exactly one point in each equivalence class of Y, and so $|D_X| = \mathfrak{c}$ whenever D_X is uncountable. □

We now obtain a partial result towards the solution of Question 6.

Theorem 6.10.4. *Let X be a non-empty, compact space for which $C(X) \cong E''$ for a Banach space E. Then there is a compact space K such that X is homeomorphic to a clopen subspace of $\widetilde{K}$.*

Proof. Again, we regard E as being a closed subspace of $C(X)$. We take F to be the closed, unital C^*-subalgebra of $C(X)$ generated by $E \cup \{1_X\}$. As on page 101, we define $x \sim y$ for $x,y \in X$ by setting $x \sim y$ if $f(x) = f(y)$ $(f \in F)$, so that $\sim$ is an equivalence relation on X, and we define $K = X/\sim$, so that K is a compact space. Thus there is a unital C^*-isomorphism $\iota : F \to C(K)$.

Now, for $\mu \in N(X)$, define

$$(T\mu)(\iota(f)) = \langle f, \mu \rangle \quad (f \in F).$$

Then $T\mu \in (\iota(F))'$ with $\|T\mu\| = \|\mu\|$, and so $T\mu \in C(K)' = N(\widetilde{K})$. Further, the map $T \in \mathscr{B}(N(X), N(\widetilde{K}))$ is a linear isometry and $T(N_{\mathbb{R}}(X)) \subset N_{\mathbb{R}}(\widetilde{K})$.

Next suppose that $\mu \in N(X) \cap P(X)$, and set $\nu = T\mu \in N(\widetilde{K}) \cap P(\widetilde{K})$. Then we *claim* that

$$T \mid L^1(\mu) : L^1(\mu) \to L^1(\nu)$$

is an isometric isomorphism. Indeed, take $\rho \in L^1(\mu)$. Then

$$\|\rho\| = \sup\{|\langle f, \rho \rangle| : f \in F_{[1]}\} = \sup\{|\langle \iota(f), T\rho \rangle| : f \in F_{[1]}\} = \|T\rho\|,$$

so that $T \mid L^1(\mu)$ is an isometry. Now take a function $f \in F$, so that $f\mu \in L^1(\mu)$ and $\iota(f)\nu \in L^1(\nu)$. Then $T(f\mu) = \iota(f)\nu \in L^1(\nu)$. Since measures of the form $\iota(f)\nu$ form a dense subset of $L^1(\nu)$, the range of $T \mid L^1(\mu)$ is dense in $L^1(\nu)$, and hence $T \mid L^1(\mu)$ is a surjection onto $L^1(\nu)$. This gives the claim.

We have proved that $L^1(\mu) \cong L^1(T\mu)$ $(\mu \in N(X) \cap P(X))$, and this implies that $L^\infty(\mu) \cong L^\infty(T\mu)$ $(\mu \in N(X) \cap P(X))$, and hence the two spaces Φ_μ and Φ_ν are homeomorphic to each other.

Now let $\mathscr{F} = \{\mu_\alpha : \alpha \in A\}$ be a maximal singular family in $N(X) \cap P(X)$ such that $\mathscr{F}$ contains $\{\delta_x : x \in D_X\}$, as in §4.6. As in §5.1, we can identify each Φ_{μ_α} with $S_\alpha = \operatorname{supp} \mu_\alpha$; the family $\{S_\alpha : \alpha \in A\}$ is a pairwise-disjoint family of clopen subsets of X, and $U_{\mathscr{F}} = \bigcup\{S_\alpha : \alpha \in A\}$ is a dense, open subset of X with $\beta U_{\mathscr{F}} = X$. Since $\mu_\alpha \perp \mu_\beta$ whenever $\alpha, \beta \in A$ with $\alpha \neq \beta$, it follows from Corollary 4.2.6 (with E taken to be $N(X)$) that $T\mu_\alpha \perp T\mu_\beta$ whenever $\alpha, \beta \in A$ with $\alpha \neq \beta$. In a similar way, $\Phi_{T\mu_\alpha}$ is homeomorphic to $\operatorname{supp} T\mu_\alpha$ for each $\alpha \in A$ and $\{\operatorname{supp} T\mu_\alpha : \alpha \in A\}$ is a family of pairwise-disjoint, clopen subsets of $\widetilde{K}$; we set

$$L = \overline{\bigcup\{\operatorname{supp} T\mu_\alpha : \alpha \in A\}} = \beta\left(\bigcup\{\operatorname{supp} T\mu_\alpha : \alpha \in A\}\right),$$

so that L is a clopen subspace of $\widetilde{K}$.

For each $\alpha \in A$, denote by h_α a homeomorphism from Φ_{μ_α} onto $\Phi_{T\mu_\alpha}$, and then define

$$h : \bigcup\{\Phi_{\mu_\alpha} : \alpha \in A\} \to \bigcup\{\Phi_{T\mu_\alpha} : \alpha \in A\} \subset \widetilde{K}$$

by requiring that $h \mid \Phi_{\mu_\alpha} = h_\alpha$ $(\alpha \in A)$. Then h is well defined and h is a continuous embedding of $U_{\mathscr{F}}$ onto $\bigcup\{\operatorname{supp} T\mu_\alpha : \alpha \in A\}$ in $\widetilde{K}$. The maps h and h^{-1} extend to continuous maps from X onto L and from L onto X, respectively, and so there is a homeomorphism $h : X \to L$.

This completes the proof. □

Consider the special case in which $X = \beta\mathbb{N}$ and $E = c_0$, so that indeed we have $C(X) \cong E''$. Now F is the space c, regarded as a subalgebra of $C(\beta\mathbb{N})$, and the equivalence relation $\sim$ identifies points of $\mathbb{N}^*$ (and leaves the points of $\mathbb{N}$ untouched). Thus the compact space K of the theorem is $\mathbb{N}_\infty$, the one-point compactification of $\mathbb{N}$. The space of continuous functions on K is c. The first dual of this space is $\ell^1(\mathbb{N} \cup \{\infty\})$, and the bidual is $\ell^\infty(\mathbb{N} \cup \{\infty\})$, whose character space is $\widetilde{K} = \beta((\mathbb{N}_\infty)_d)$, a larger space than $V = X = \beta\mathbb{N}$. Thus it is not always the case that $V = \widetilde{K}$ in the above theorem. Of course, in this case, $\widetilde{K}$ is homeomorphic to X. We do not know, in the general case, whether one can always find a locally compact space K such that X is homeomorphic to $\widetilde{K}$, but we can resolve this question in the special case in which E is separable: see Theorem 6.10.8, given below.

However, before doing this, we shall prove a variant of the above theorem in the non-separable case; the variant is obtained by assuming stronger hypotheses on the space E of the theorem. The key extra assumption that we shall make is the apparently innocent one that, in the above notation, $1_X \in E$. We first make some remarks on the consequences of this assumption.

Suppose that $1_X \in E$. Then

$$N(X)^+ = \{\mu \in N(X) : \langle 1_X, \mu \rangle = \|\mu\|\},$$

and so the positive cone $N(X)^+$ is closed in $(N(X), \sigma)$. Further, set

$$S = N(X) \cap P(X) = \{\mu \in N(X)^+ : \langle 1_X, \mu \rangle = \|\mu\| = 1\}, \tag{6.20}$$

so that S is a non-empty, convex subset of $N(X)^+_{[1]}$ that is also closed in $(N(X)_{[1]}, \sigma)$, and hence S too is compact. In the terminology of Example 1.7.15, the set S is a Choquet simplex in the locally convex space $(N(X), \sigma)$.

In the course of the following proof, we shall need to identify the space $A(S)$ of continuous, affine functions on (S, σ) (as defined on page 8) with E, and we indicate how to do this.

Indeed, take an element $\theta \in A(S)$. We first extend θ to be a function on $N(X)^+$ by setting $\theta(\nu) = r\theta(\nu/r)$ for $\nu \in N(X)^+ \setminus \{0\}$, where $r = \|\nu\| > 0$, and $\theta(0) = 0$. It is easily checked that θ is additive on $N(X)^+$ and that $\theta(\alpha\nu) = \alpha\theta(\nu)$ for $\alpha > 0$ and $\nu \in N(X)^+$. Further, this extension is continuous on the space $(N(X)^+, \sigma)$. To see this, suppose that (ν_i) is a net in $N(X)^+$ such that $\lim_i \nu_i = \nu$ in $(N(X)^+, \sigma)$. In the case where $\nu \neq 0$, it is immediate that $\lim_i \theta(\nu_i) = \theta(\nu)$. In the case where $\nu = 0$, the values of the net $|\theta(\nu_i / \|\nu_i\|)|$ are bounded by $|\theta|_S$ and

$$\lim_i \|\nu_i\| = \lim_i \langle 1_X, \nu_i \rangle = 0,$$

and so $\lim_i \theta(\nu_i) = 0$.

We next extend θ to be a function on $N_{\mathbb{R}}(X)$ by setting $\theta(\nu) = \theta(\nu_1) - \theta(\nu_2)$, whenever $\nu \in N(X)$ has the form $\nu = \nu_1 - \nu_2$, where $\nu_1, \nu_2 \in N(X)^+$. Clearly θ is well defined on $N_{\mathbb{R}}(X)$ and θ is a (real) linear functional on $N_{\mathbb{R}}(X)$. Now suppose that (ν_i) is a net in $N_{\mathbb{R}}(X)_{[1]}$ with $\lim_i \nu_i = \nu$ in $(N_{\mathbb{R}}(X)_{[1]}, \sigma)$. Then (ν_i^+) and (ν_i^-) are nets in $N_{\mathbb{R}}(X)_{[1]}$, and we may suppose by passing to subnets that they converge. Thus $\lim_i \theta(\nu_i) = \theta(\nu)$. By Theorem 2.1.4(iv), the functional θ is continuous on $N_{\mathbb{R}}(X)$.

Finally we extend θ to $N(X)$ by setting

$$\theta(\nu_1 + \mathrm{i}\nu_2) = \theta(\nu_1) + \mathrm{i}\theta(\nu_2) \quad (\nu_1, \nu_2 \in N_{\mathbb{R}}(X));$$

again θ is a continuous (complex) linear functional on $N(X)$.

Thus each element of $A(S)$ is the restriction to S of an element of the space $(E', \sigma(E', E))' = E$.

Theorem 6.10.5. *Let X be a non-empty, compact space for which $C(X) \cong E''$ for a Banach space E that is a closed subspace of $C(X)$. Suppose further that $1_X \in E$ and that $K := \operatorname{ex} N(X)_{[1]}^+$ is compact in $(N(X), \sigma(N(X), E))$. Then $E \cong C(K)$ and X is homeomorphic to the space $\widetilde{K}$.*

Proof. Again we set $\sigma = \sigma(N(X), E)$ and $S = N(X) \cap P(X)$. We have remarked that (S, σ) is compact because $1_X \in E$, and so S is a Choquet simplex in the locally convex space $(N(X), \sigma)$. By hypothesis, (K, σ) is compact, and, by Proposition 4.7.12, $K = \{\delta_x : x \in D_X\}$, and so $D_X \neq \emptyset$.

Each $f \in E$ is a continuous function on (K, σ), and we *claim* that the map

$$f \mapsto f \mid K, \quad E \to C(K),$$

is a linear isometry. Indeed, suppose that $f \in E$ and $|f(x)| \leq 1$ $(x \in K)$. Then $|\langle f, \lambda \rangle| \leq 1$ for $\lambda \in \overline{\operatorname{aco}} K = E'_{[1]}$, and so $|f|_X \leq 1$, giving the claim. Thus we can regard E as a closed linear subspace of $C(K)$.

We now *claim* that $E = C(K)$. Indeed, assume to the contrary that $E \subsetneq C(K)$. Then there exists $\nu \in M_{\mathbb{R}}(K)$ such that $\nu \mid E = 0$ and $\nu \neq 0$. We write $\nu = \nu^+ - \nu^-$, where $\nu^+, \nu^- \in M(L)^+$ and $\|\nu\| = \|\nu^+\| + \|\nu^-\|$. Then

$$0 = \langle 1_X, \nu \rangle = \langle 1_X, \nu^+ \rangle - \langle 1_X, \nu^- \rangle = \|\nu^+\| - \|\nu^-\|,$$

so $\|\nu^+\| = \|\nu^-\|$; we may suppose that $\|\nu^+\| = \|\nu^-\| = 1$, so that $\nu^+, \nu^- \in P(K)$.

Define

$$\mu(f) = \langle f, \nu^+ \rangle = \langle f, \nu^- \rangle \quad (f \in E).$$

Then $\mu \in N(X) \cap P(X) = S$.

Now $\nu^+, \nu^- \in M(S)^+$. Further, $\nu^+ \approx \varepsilon_\mu$ (in the notation of equation (4.5)). To see this, we need to prove that $\langle \theta, \nu^+ \rangle = \langle \theta, \varepsilon_\mu \rangle$ for each continuous, affine function θ on S. By the preliminary remark, $\theta = f \mid S$ for an element $f \in E$, and in this case the equality holds by our definition. Similarly, $\nu^- \approx \varepsilon_\mu$.

By Theorem 4.1.12, ν^+ and ν^- are maximal measures on the Choquet simplex S, and so, by the uniqueness statement in Theorem 4.1.12, $\nu^+ = \nu^-$. Thus $\nu = 0$, a contradiction, and so $E = C(K)$, as claimed.

It follows that $E'' = C(\widetilde{K}) = C(X)$, and hence X is homeomorphic to $\widetilde{K}$. □

Let X be a hyper-Stonean space. Then X is homeomorphic to $\widetilde{K}$ for some compact space K if and only if there is a locally convex topology on $N(X)$ for which both $P(X) \cap N(X)$ and D_X are compact. For real scalars, this is easily proved using Theorem 2.2.28(ii) and Theorem 6.10.5; the complex case then follows immediately.

Let X be a non-empty, compact space for which $C_{\mathbb{R}}(X) \cong E''$ for a real Banach space E. In [171, Theorem 6.6], Lindenstrauss proved that, in this case, $E \cong C_{\mathbb{R}}(K)$, where $K = \mathrm{ex}\,N_{\mathbb{R}}(X)^+_{[1]}$, under the extra hypotheses that $E_{[1]}$ has an extreme point (which is implied by our hypothesis in Theorem 6.10.5 that $1_X \in E$) and that K is $\sigma(N_{\mathbb{R}}(X), E)$-compact (as in our Theorem 6.10.5). Thus, the real case of our theorem is implied by that of Lindenstrauss; our proof is different and covers the complex case as well.

A complex Banach space whose dual space is isometrically isomorphic to a complex L^1-space is called a *complex Lindenstrauss space*. Using what the Mathematical Review calls 'some technically intricate lemmas whose proofs exhibit considerable virtuosity', Hirsberg and Lazar in [140] proved that a complex Lindenstrauss space whose unit ball has at least one extreme point is isometric to the space of complex-valued, continuous, affine functions on a Choquet simplex. The work of Hirsberg and Lazar is described in [18, Chapter 4, §9] and [166, pp. 245–247]. See also [170].

In the setting of Theorem 6.10.5, it is also tempting to 'add a 1 to E'. This is possible in a special case.

Proposition 6.10.6. *Let X be a non-empty, compact space, and let E be a closed subspace of $C(X)$ such that $E'' \cong C(X)$. Suppose that $N(X)^+$ is closed in the topology $\sigma(N(X), E)$. Then there is a compact space $Y = X \cup \{x_0\}$ and a closed subspace F of $C(Y)$ such that $F'' \cong C(Y)$ and $1_Y \in F$.*

Proof. The result is trivial if $1_X \in E$, and so we suppose that this is not the case. Again we set $\sigma = \sigma(N(X), E)$.

Choose a point $x_0 \notin X$, and take Y to be the disjoint union of X and $\{x_0\}$. We regard E as a subset of $C(Y)$ by setting $f(x_0) = 0$ $(f \in E)$, and we set $F = E \oplus \mathbb{C}1_Y$ as a closed subspace of $(C(Y), |\cdot|_Y)$, so that $1_Y \in F$.

The dual space of F is isomorphic to $N(Y)$; we denote by $|||\cdot|||$ the corresponding norm on $N(Y)$, so that

$$|||\nu||| = \sup\{|\langle g, \nu\rangle| : g \in F_{[1]}\} \quad (\nu \in N(Y)).$$

Clearly $|||\nu||| \le \|\nu\|$ $(\nu \in N(Y))$. For $\nu \in N(Y)^+$, we have

$$\|\nu\| = \langle 1_Y, \nu\rangle \le |||\nu|||\,,$$

and so $|||\nu||| = \|\nu\|$. Set $\tau = \sigma(N(Y), F)$.

We *claim* that $\{\nu \in N(Y) : \|\nu\| \le 1\}$ is compact in $(N(Y), \tau)$. First, let (ν_i) be a net in $\{\nu \in N(Y)^+ : \|\nu\| \le 1\}$. Then there exist (μ_i) in $N(X)^+$ and (α_i) in $\mathbb{R}^+$ such that $\nu_i = \mu_i + \alpha_i\delta_{x_0}$ and $\|\nu_i\| = \|\mu_i\| + \alpha_i \le 1$ for each i. Since the set $\{\nu \in N(Y) : |||\nu||| \le 1)$ is compact in $(N(Y), \tau)$, there exist $\mu \in N(X)$ and $\beta \in \mathbb{R}$ such that $\nu_i \to \mu + \beta\delta_{x_0}$ in $(N(Y), \tau)$. By evaluating at elements of E, we see that $\mu_i \to \mu$ in $(N(X), \sigma)$, and so, by our hypothesis that $N(X)^+$ is closed in $(N(X), \sigma)$, it follows that $\mu \in N(X)^+$. By passing to a subnet, we may suppose that $(\|\mu_i\|)$ and (α_i) converge in $\mathbb{R}$. Since $\|\mu\| \le \lim_i \|\mu_i\|$, we see that $0 \le \lim_i \alpha_i \le \beta$, and so $\mu + \beta\delta_{x_0} \in N(Y)^+$. Thus

$$\|\mu + \beta\delta_{x_0}\| = |||\mu + \beta\delta_{x_0}||| \le 1\,.$$

This establishes that $\{\nu \in N(Y)^+ : \|\nu\| \le 1\}$ is compact in $(N(Y), \tau)$. Our claim follows easily.

As in Theorem 2.2.28(i), $N(Y) \cong F'$, and so $F'' \cong C(Y)$. □

The following result is not obviously weaker than that of [171, Theorem 6.6].

Theorem 6.10.7. *Let X be a non-empty, compact space for which $C(X) \cong E''$ for a Banach space E that is a closed subspace of $C(X)$. Suppose further that $N(X)^+$ is closed and that $K := \operatorname{ex} N(X)^+_{[1]}$ is compact in $(N(X), \sigma(N(X), E))$. Then $E \cong C(K)$ and X is homeomorphic to the space $\widetilde{K}$.*

Proof. In the notation of the previous proposition, $D_Y = D_X \cup \{x_0\}$ is compact in $(N(Y), \sigma(N(Y), F))$, and so, by Theorem 6.10.5, there is a compact space

$$L = \operatorname{ex} N(Y) \cap P(Y) = D_X \cup \{x_0\}$$

such that $F \cong C(L)$ and $\widetilde{L}$ is homeomorphic to Y. It now follows that K has the required properties. □

We now turn to the solution to Question 6 in the special case in which the Banach space E is separable.

Theorem 6.10.8. *Let X be an infinite, compact space such that $C(X)$ is isometrically isomorphic to the bidual space of a separable Banach space. Then exactly one of the following two cases holds:*

(1) *D_X is countable and infinite, X is homeomorphic to $\widetilde{\mathbb{N}_\infty}$ and to $\beta\mathbb{N}$, $C(X)$ is isometrically isomorphic to*

$$C(\mathbb{N}_\infty)'' \cong c_0'' \cong \ell^\infty,$$

and $N(X)$ is isometrically isomorphic to ℓ^1;

(2) *D_X has cardinality $\mathfrak{c}$, X is homeomorphic to $\widetilde{\mathbb{I}}$, $C(X)$ is isometrically isomorphic to $C(\mathbb{I})''$ and $N(X)$ is isometrically isomorphic to $M((\mathbb{I})$.*

The analogue of this result for the space $C_{\mathbb{R}}(X)$ of real-valued, continuous functions on X is already contained in an old paper of H. Elton Lacey [165] in a slightly more general form; the result is also given in the text [166, §22, Theorem 5], and H.-U. Hess gives a stronger result in [136]. However it seems that our proof is quite direct and elementary and avoids the appeal to some deep results in Banach-space theory that Lacey and Hess make. Lacey et al. make no comment on the complex case.

The analogous question in the isomorphic (not isometric) theory of Banach spaces was resolved in a similar way by Stegall [229].

The proof of Theorem 6.10.8 will proceed through several preliminary results. Some steps do not need the full hypotheses of the theorem. In the first results, we write $[f]$ for the element of $L^1(X,\nu)$ that corresponds to $f \in C(X)$, and regard E as a subspace of $C(X)$.

Lemma 6.10.9. *Let X be a non-empty, compact space such that $C(X) \cong E''$ for a Banach space E. Take $\nu \in N(X) \cap P(X)$. Then, for each $g \in C(X)_{[1]}$ and $\varepsilon > 0$, there exists $f \in E_{[1]}$ with $\|[g]-[f]\|_1 < \varepsilon$.*

Proof. For each $h \in C(X)$, the measure $h\nu$ is in $N(X)$. Given $n \in \mathbb{N}$ and functions $h_1,\dots,h_n \in C(X)$, there exists $f \in E_{[1]}$ such that $|\langle [g]-[f], h_i\nu\rangle| < \varepsilon$ $(i \in \mathbb{N}_n)$, and so $\{[g] : g \in E_{[1]}\}$ is weakly dense in $C(X)_{[1]}$. By Mazur's theorem, Theorem 2.1.4(iv), our claim follows. □

Corollary 6.10.10. *Let X be a non-empty, compact space such that $C(X)$ is isometrically isomorphic to the bidual space of a separable Banach space. Take $\nu \in N(X) \cap P(X)$. Then the Banach space $(L^1(X,\nu), \|\cdot\|_1)$ is separable.*

Proof. Since $\{[f] : f \in C(X)\}$ is a $\|\cdot\|_1$-dense linear subspace of $L^1(X,\nu)$, it suffices to show that $(C(X), \|\cdot\|_1)$ is separable.

Take a separable Banach space E with $C(X) \cong E''$. Then $\|[f]\|_1 \leq \|f\|$ $(f \in E)$, and so the result follows from the lemma. □

As in Definition 5.3.1, the hyper-Stonean space of the unit interval is denoted by $\mathbb{H}$, so that $\mathbb{H}$ is the character space of the commutative C^*-algebra $L^\infty(\mathbb{I})$.

Corollary 6.10.11. *Let X be an infinite, compact space such that $C(X)$ is isometrically isomorphic to the bidual of a separable Banach space. Then* supp ν *is homeomorphic to $\mathbb{H}$, and so $C(\text{supp}\,\nu) \cong L^\infty(\mathbb{I})$, for each $\nu \in N_c(X) \cap P(X)$.*

Proof. By Corollary 6.10.10, the Banach space $(L^1(X,\nu), \|\cdot\|_1)$ is separable; by Theorem 4.4.14, $L^1(X,\nu)$ is Banach-lattice isometric to $L^1(\mathbb{I},m)$. By Theorem 5.3.2, Φ_ν and $\mathbb{H}$ are homeomorphic, and so the result follows from Corollary 4.7.6. □

For the remainder of this section, we shall suppose that X is an infinite, compact space such that $C(X) \cong E''$ for some separable Banach space E. We identify E'' with $C(X)$ and regard E as a subspace of $C(X)$ via its canonical embedding into E''. Recall that the space X is hyper-Stonean, from Theorem 6.10.1 that D_X is infinite, and that $\overline{D_X}$ is a clopen subset of X (being the closure of the open set D_X). We shall again identify E' with $N(X)$, the normal measures on X, and set

$$\sigma = \sigma(N(X), E),$$

the weak* topology on $N(X)$.

We consider now the space $N_c(X)$ of continuous, normal measures on X. First, there are the following three possibilities:

(A) $N_c(X) = \{0\}$, i.e., there are no non-zero, normal, continuous measures on X;

(B) there exists some $\mu \in N_c(X)^+$ with $\operatorname{supp} \mu = X \setminus \overline{D_X}$;

(C) there exists an uncountable family of mutually singular measures in $N_c(X)^+$.

Notice that, if (A) and (C) do not hold and $\mathscr{F}$ is a maximal family of mutually singular members of $N_c(X)^+$, then $\mathscr{F}$ is countable, say $\mathscr{F} = \{\mu_1, \mu_2, \dots\}$. Since X is hyper-Stonean, the union of the supports of the measures in $N_c(X)$ is dense in $X \setminus \overline{D_X}$. Hence the measure

$$\mu = \sum_{n=1}^{\infty} \mu_n / 2^n \|\mu_n\|$$

has $\operatorname{supp} \mu = X$ by the maximality of $\mathscr{F}$, and so (B) holds. Thus, (A), (B), and (C) exhaust all possibilities.

In the case (C), we fix $\mathscr{F}$ to be an uncountable family of mutually singular measures in $N_c(X)^+$. We may suppose that $\|\mu\| = 1$ for each $\mu \in \mathscr{F}$; this implies that $\|\mu - \nu\| = 2$ whenever $\mu, \nu \in \mathscr{F}$ with $\mu \neq \nu$, and so $N_c(X)$ is not separable.

Turning to the isolated points of X, there are two possibilities:

(I) D_X is countable (and infinite);

(II) D_X is uncountable.

Thus in combination there are apparently six possible cases of $\{A, B, C\}$ and $\{I, II\}$, and we examine these in turn.

First suppose that D_X is countable. Then, by Corollary 4.7.15, $N(X) \cong \ell^1$: take $E = c_0$, a separable Banach space, so that $E'' = C(\beta\mathbb{N})$, giving $X = \beta\mathbb{N}$ as one of the two cases stated in the theorem. This remark also shows that cases (I-B) and (I-C) do not occur.

For the remainder of the proof, we consider the case where D_X is uncountable.

The Case (II-C) does occur: take $E = C(\mathbb{I})$, a separable Banach space, so that $E'' = C(\widetilde{\mathbb{I}})$, and this gives $X = \widetilde{\mathbb{I}}$ as one the two cases stated in the theorem. Indeed, D_X is now equipotent to $\mathbb{I}$, and so $|D_X| = \mathfrak{c}$. We shall show in Theorem 6.10.13, below, that, whenever Case (II-C) holds, D_X has cardinality $\mathfrak{c}$, that X is homeomorphic to $\widetilde{\mathbb{I}}$, and that $C(X)$ is isometrically isomorphic to $C(\mathbb{I})''$.

We now seek to eliminate Cases (II-A) and (II-B). This will be achieved in Theorem 6.10.13, below, which will follow from the next key theorem. We give a direct and self-contained proof of this result suggested by, and somewhat simpler than, Rosenthal's proof of Proposition 3 in [213, p. 371].

Theorem 6.10.12. *Let X be a compact space such that $C(X) \cong E''$ for some separable Banach space E and such that D_X is uncountable. Then there is an isometric embedding of $M(\Delta)$ into $N(X)$.*

Proof. The space $(N(X)_{[1]}, \sigma)$ is compact and metrizable; we suppose that its topology is specified by a metric d. We regard D_X as a subset of $N(X)_1$. By Corollary 1.4.16, there is an infinite, dense-in-itself subset D of D_X, and we set $\Omega = \overline{D}$, so that $\Omega \subset N(X)_{[1]}$ and Ω is an infinite, perfect set. Each non-degenerate, closed ball (for the metric d) in Ω contains infinitely many points in D.

It follows from Theorem 2.1.7(i) that $(C(\Omega), |\cdot|_\Omega)$ is separable; fix $\{h_n : n \in \mathbb{Z}^+\}$ to be a dense subset of this space.

We describe (1) an inductive process which will produce a subspace K of Ω such that K is homeomorphic to the Cantor set Δ, and (2) a bounded linear surjection $R : E \twoheadrightarrow C(K)$ such that $T = R'$ is an isometric embedding of $M(K)$ into $N(X)$. The space K will be the intersection of a decreasing sequence $(K_n : n \in \mathbb{Z}^+)$ of compact subsets of Ω, where each K_n is the union of 2^n pairwise-disjoint, closed balls in (Ω, d). The induction makes use of the following observation.

Given $m \in \mathbb{N}$, distinct points $d_1, \dots, d_m$ in D, $\alpha_1, \dots, \alpha_m \in \mathbb{C}$, and $\varepsilon > 0$, there exists $f \in E$ such that

$$|f(d_j) - \alpha_j| < \varepsilon \quad (j \in \mathbb{N}_m)$$

and $|f|_X = \max\{|\alpha_j| : j \in \mathbb{N}_m\}$.

To see that this holds, choose $h \in C(X)$ such that $h(d_j) = \alpha_j \ (j \in \mathbb{N}_m)$ and also $h(x) = 0 \ (x \in X \setminus \{d_1, \dots, d_m\})$, so that

$$\|h\| = |h|_X = \max\{|\alpha_j| : j \in \mathbb{N}_m\}.$$

Using the fact that $E_{[1]}$ is weak*-dense in $C(X)_{[1]}$, approximate h at the points d_j by an appropriate $f \in E$.

Starting with the function h_0, choose a point $t_0 \in \Omega \cap D$. By the observation, there exists $f_0 \in E$ with $|f_0(t_0) - h_0(t_0)| < 1$ and $|f_0|_X \leq |h_0(t_0)|$. Then choose a non-degenerate, closed ball K_0 in Ω, with centre t_0, with diameter less than 1, and such that $|f_0(t) - h_0(t)| < 1 \ (t \in K_0)$. Thus $|f_0 - h_0|_{K_0} < 1$ and $|f_0|_X \leq |h_0|_{K_0}$.

For each $\varepsilon \in \mathbb{Z}_2^{<\omega}$, we shall define inductively non-degenerate, closed balls K_ε contained in K_0 to satisfy the conditions (i)–(iii), below. Here we are writing K_n for $\bigcup\{K_\varepsilon : \varepsilon \in \mathbb{Z}_2^n\}$. The conditions are:

(i) the two sets $K_{\varepsilon-}$ and $K_{\varepsilon+}$ are disjoint subsets of K_ε;

(ii) for each $n \in \mathbb{N}$ and $\varepsilon \in \mathbb{Z}_2^n$, we have $\operatorname{diam} K_\varepsilon < 1/2^n$;

(iii) for each $n \in \mathbb{N}$, there exist $f_{n,1}, \dots, f_{n,n} \in E$ such that $|f_{n,i} - h_i|_{K_n} < 1/2^n$ and $|f_{n,i}|_X \leq |h_i|_{K_n}$ for each $i \in \mathbb{N}_n$.

The choice of the initial balls $K_{(0)}$ and $K_{(1)}$ in K_0 and the initial function $f_{1,1}$ is as in the inductive step, given below.

Now take $n \in \mathbb{N}$, and assume inductively that the non-degenerate, closed balls K_ε and the functions $f_{m,1}, \dots, f_{m,m} \in E$ have been constructed for all $\varepsilon \in \mathbb{Z}_2^m$ for $m \le n$ and hence that we have the set K_n.

For each $\varepsilon \in \mathbb{Z}_2^n$, choose two distinct point s_ε and t_ε in $D \cap \operatorname{int} K_\varepsilon$. By the observation, there exist functions $f_{n+1,1}, \dots, f_{n+1,n+1} \in E$ such that

$$\max\{|f_{n+1,i}(s_\varepsilon) - h_i(s_\varepsilon)|, |f_{n+1,i}(t_\varepsilon) - h_i(t_\varepsilon)|\} < \frac{1}{2^{n+1}} \quad (i \in \mathbb{N}_{n+1})$$

and $|f_{n+1,i}|_X \le |h_i|_{K_n}$ $(i \in \mathbb{N}_{n+1})$. For each $\varepsilon \in \mathbb{Z}_2^n$, choose non-degenerate, closed balls $K_{\varepsilon-}$ and $K_{\varepsilon+}$ in Ω, with centres s_ε and t_ε, respectively, such that $K_{\varepsilon-}$ and $K_{\varepsilon+}$ are disjoint subsets of K_ε, such that

$$\max\{\operatorname{diam} K_{\varepsilon-}, \operatorname{diam} K_{\varepsilon+}\} < \frac{1}{2^{n+1}},$$

and such that

$$|f_{n+1,i}(t) - h_i(t)| < \frac{1}{2^{n+1}} \quad (t \in K_{\varepsilon-} \cup K_{\varepsilon+}, i \in \mathbb{N}_{n+1}).$$

The set K_{n+1} is the union of the 2^{n+1} sets $K_{\varepsilon-}$ and $K_{\varepsilon+}$ for $\varepsilon \in \mathbb{Z}_2^n$. We see that $|f_{n+1,i} - h_i|_{K_{n+1}} < 1/2^{n+1}$ and $|f_{n+1,i}|_X \le |h_i|_{K_{n+1}}$ for each $i \in \mathbb{N}_{n+1}$. This continues the inductive construction.

It follows from (i) and (ii), as in Corollary 1.4.16, that there is a homeomorphic embedding from the Cantor set Δ into K_0; we identify Δ with its image, and so

$$\Delta = \bigcap \left\{ \bigcup \{K_\varepsilon : \varepsilon \in \mathbb{Z}_2^n\} : n \in \mathbb{N} \right\}.$$

Take $i \in \mathbb{N}$ and $\eta > 0$, and set $U = \{t \in K_0 : |h_i(t)| < |h_i|_\Delta + \eta\}$, so that U is an open neighbourhood of Δ. There exists $n \in \mathbb{N}$ such that $K_n \subset U$ and $1/2^n < \eta$. By (iii), there exists $f \in E$ such that $|f - h_i|_{K_n} < 1/2^n$ and $|f|_X \le |h_i|_{K_n}$. Hence $|f - h_i|_\Delta < \eta$ and $|f|_X \le |h_i|_\Delta + \eta$.

Define the restriction map $R : f \mapsto f \mid \Delta$, $E \to C(\Delta)$. Then T is a continuous linear contraction, and we have the dual map

$$T = R' : C(\Delta)' = M(\Delta) \to E' = N(X).$$

Clearly $\|T\mu\| \le \|\mu\|$ $(\mu \in M(\Delta))$. We *claim* that $\|T\mu\| \ge \|\mu\|$ $(\mu \in M(\Delta))$. To see this, take $\mu \in M(\Delta)$. We observe that, for each $\eta > 0$, there is some $i \in \mathbb{N}$ such that $|h_i|_\Delta < 1 + \eta$ and $|\langle h_i, \mu \rangle| \ge \|\mu\|$. Then choose $f \in E$ with $|f - h_i|_\Delta < \eta$ and $|f|_X \le |h_i|_\Delta + \eta < 1 + 2\eta$. We have

$$|\langle f, T\mu \rangle| = |\langle Rf, \mu \rangle| = |\langle f, \mu \rangle| \ge |\langle h_i, \mu \rangle| - |\langle f - h_i, \mu \rangle| \ge (1 - \eta)\|\mu\|,$$

and so $(1+2\eta)\|T\mu\| \geq (1-\eta)\|\mu\|$. This holds true for each $\eta > 0$, and so the claim follows. We conclude that $T : M(\Delta) \to N(X)$ is an isometric embedding. □

Theorem 6.10.13. *Let X be an infinite, compact space such that $C(X)$ is isometrically isomorphic to the bidual space of a separable Banach space. Suppose that D_X is uncountable. Then $|D_X| = \mathfrak{c}$, X is homeomorphic to $\widetilde{\mathbb{I}}$, and $C(X)$ is isometrically isomorphic to $C(\mathbb{I})''$.*

Proof. Let E be a separable Banach space such that $C(X) \cong E''$.

Since D_X is uncountable, it follows from Theorem 6.10.3 that $|D_X| = \mathfrak{c}$.

By Theorem 6.10.12, there is an isometric embedding $T : M(\Delta) \to N(X)$. By Corollary 4.2.6, $T\mu \perp T\nu$ whenever $\mu \perp \nu$ and, by Corollary 4.2.8, $T\mu \in N_c(X)$ whenever $\mu \in M_c(\Delta)$. By Proposition 4.6.2(ii), $M_c(\Delta)_{[1]} \cap P(\Delta)$ contains a singular family of cardinality $\mathfrak{c}$, and then, using T, we obtain a singular family of cardinality $\mathfrak{c}$ consisting of measures in the unit sphere of $N_c(X)$; in fact, we may suppose that each measure belongs to $N_c(X) \cap P(X)$. We enlarge this family to a maximal singular family $\mathscr{F} \subset N_c(X)^+_{[1]}$, so that $|\mathscr{F}| \geq \mathfrak{c}$. However $|\mathscr{F}| \leq |N(X)_{[1]}| = \mathfrak{c}$ since $N(X)_{[1]}$ is a compact, metrizable space in the weak* topology, hence $|\mathscr{F}| = \mathfrak{c}$.

Recalling that the support of a normal measure on a Stonean space is clopen (Corollary 4.7.10(i)), we see that $X \setminus \overline{D_X}$ contains a pairwise-disjoint family of $\mathfrak{c}$ clopen subspaces. Since X is hyper-Stonean, the maximality of $\mathscr{F}$ makes the union of these support sets dense in $X \setminus \overline{D_X}$. By Corollary 6.10.11, each of these support sets is homeomorphic to $\mathbb{H}$. By Theorem 6.5.4, X is homeomorphic to $\widetilde{\mathbb{I}}$, and so $C(X)$ is isometrically isomorphic to $C(\mathbb{I})''$. □

With Theorem 6.10.13, we conclude the proof of Theorem 6.10.8.

Theorem 6.10.13 states only what is required for the purpose of achieving a direct analysis of the situation where $C(X)$ is isometrically the bidual of a separable space (Theorem 6.10.8). Our objective was to obtain the result as easily as possible; with more machinery more is possible.

For example, it follows easily from Theorem 6.10.8 that the Banach space

$$F_D := L^1(\mathbb{I}) \oplus_1 \ell^1(D)$$

is not isometrically the dual of a separable space when D is uncountable. In fact, F_D is not even isomorphically a dual of any Banach space. The impressive thesis of Hagler [127], significantly extending earlier work of Pełczyński, contains the following. Suppose that E is a Banach space. Then E' contains an isomorphic copy (respectively, a complemented isomorphic copy) of $L^1(\mathbb{I})$ if and only if E' contains an isomorphic copy (respectively, a complemented isomorphic copy) of $M(\mathbb{I})$. The same statements are true if 'isomorphic' be replaced by 'isometric'. This work also shows that E' contains an isomorphic copy of $L^1(\mathbb{I})$ if and only if E contains an isomorphic copy of ℓ^1. (Proofs were later published in [128, Theorem 5], [130, Theorem 1], [86, Theorem 2], and [129, Theorem 1]). However, by Corollary 4.6.9, $M(\mathbb{I})$ is not isomorphic to any closed subspace of the space F_D, so justifying our statement above. The theorem of Pełczyński–Hagler (the uncomplemented isomorphic case) is expounded in [93, Chapter IV], along with additional equivalences, and as Théorème II.5 of Chapitre 7 of [169], which treats the case of complex scalars (but does not mention the contribution of Hagler).

Finally, we consider the following question.

Question 7 *Let A be a unital C^*-algebra, and suppose that, as a Banach space, A is isometrically isomorphic to the bidual of a Banach space E. Does it follow that there is a C^*-algebra B such that A is C^*-isomorphic to $(B'', \Box)$? Is this true in the special case in which E is separable?*

We have partially resolved the above question in the special case that A is commutative; the general case, when the C^*-algebra A may be non-commutative, is a task for the future.

6.11 Summary

We now summarize some results that we have described concerning the question when a space $C_0(K)$ is injective, when it is (isomorphically or isometrically) the dual of a Banach space, and when it is the bidual of a Banach space.

1. The non-empty, compact spaces K such that $C(K)$ is 1-injective are characterized in Theorem 6.8.3. This is the case if and only if K is Stonean, equivalently, projective, if and only if $C(K)$ is Dedekind complete. For example, the spaces $\ell^\infty(S) = C(\beta S)$ are 1-injective for each non-empty set S (Proposition 2.5.5).
2. Suppose that K is a non-empty, compact space such that $C(K)$ is injective. Then K contains a dense, open, extremely disconnected subset (Theorem 6.8.15).
3. For the following locally compact spaces K, the space $C_0(K)$ is not injective:

   K contains a convergent sequence of distinct points (Corollary 2.4.17);

   K is an infinite, compact, metric space;

   K is an infinite, discrete space (Corollary 2.4.13);

   K is locally compact, but not pseudo-compact (Theorem 2.4.12);

   K is an infinite, locally compact group (Theorem 4.4.3);

   K is locally connected and has no isolated points (Corollary 6.8.16(ii));

   $K = S^*$ for an infinite set S (Corollary 6.8.16(iv));

   K is an infinite, compact space with only finitely many clopen subsets (Corollary 6.8.16(iii));

   K is infinite, compact, and homogeneous (Corollary 6.8.18);

   K is such that $C_0(K)$ is not a Grothendieck space (Corollary 4.5.10);

   K is a product of two infinite, compact spaces (Proposition 6.8.2).

4. There are compact spaces K such that $C(K)$ is a Grothendieck space, but is not injective (Example 6.8.17).
5. For each ordinal α with $1 \leq \alpha < \omega_1$ and each uncountable Polish space X, the Baire class $B_\alpha(X) = C(\Phi_\alpha(X))$ is not injective (Theorem 3.3.7(iii)).
6. The space $B_{\omega_1}(\mathbb{I}) = B^b(\mathbb{I}) = C(\Phi_{\omega_1}(\mathbb{I}))$ is not injective (Corollary 6.8.10).
7. Let K be compact. The question when $C(K)$ is isometrically a dual space was resolved in Theorem 6.4.1. This occurs if and only if K is hyper-Stonean, and then $C(K)$ has a strongly unique isometric predual, namely $N(K)$ (Theorem 6.4.2).
8. Let K be a hyper-Stonean space, and suppose that E is a Banach space such that $E' \sim C(K)$. Then it does not follow that $E \sim N(K)$; a counter-example, with $K = \beta\mathbb{N}$, is given in Example 6.4.3.
9. A space $C_0(K)$ is injective whenever it is isomorphically a dual space (Theorem 6.9.2). Thus $C_0(K)$ is not isomorphically a dual space whenever $C_0(K)$ is not injective and hence for all the examples listed above. However, there are Stonean spaces K such that $C(K)$ is a 1-injective space, but not isomorphically a dual space (Example 6.9.11), and there are pseudo-compact (and hence locally compact) spaces K that are not compact such that $C_0(K)$ is isomorphically a dual space, and hence $C_0(K)$ is injective (Example 6.9.1).
10. There is a compact space K such that $C(K)$ is isomorphically a bidual space, but K has a clopen subspace L such that $C(L)$ is not isomorphically a dual space. (Example 6.9.12).
11. There are compact spaces K such that $C(K)$ is isomorphically, but not isometrically, a dual space. For such an example, K can be totally disconnected, but not an F-space, and hence not Stonean (Example 6.9.8), and K can be a separable Stonean space without isolated points (Example 6.9.10); $C(K)$ can be isomorphically a dual space, and hence injective, but not 1-injective (Example 6.9.8).
12. There is hyper-Stonean space X such that $C(X)$ is not isometrically a bidual space (Example 6.10.2).
13. Let X be an infinite, compact space such that $C(X) \cong E''$ for a Banach space E. Then there are a compact space K and a clopen subspace V of $\widetilde{K}$ such that X is homeomorphic to V (Theorem 6.10.4). In the case where the Banach space E is separable, either X is homeomorphic to $\beta\mathbb{N}$, and $C(X) \cong C(\beta\mathbb{N}) = c_0''$, or X is homeomorphic to $\widetilde{\mathbb{I}}$, and $C(X) \cong C(\widetilde{\mathbb{I}}) \cong C(\mathbb{I})''$ (Theorem 6.10.8).

6.12 Open questions

We collect here all the open questions that we have raised in the text.

Questions 1 and 1a: *Are any or all of the Banach spaces $B_\alpha(\mathbb{I})$ and $B_\beta(\mathbb{I})$ pairwise isomorphic in the cases where $2 \leq \alpha < \beta < \omega_1$? Is it true that each space $B_\alpha(\mathbb{I})$ for $2 \leq \alpha < \omega_1$ fails to be Baire large?* See pages 105 and 214.

Question 2: *Let E be a C^*-algebra . Is E_w necessarily a Grothendieck space?* See page 212.

Question 3: *Let E be an injective Banach space. Is E isomorphic to a 1-injective space, and hence to $C(K)$ for a Stonean space K?* See page 220.

Question 4: *Let K be a compact space such that $C(K)$ is an injective Banach space. Is K totally disconnected? Is $C(K)$ isomorphic to $C(L)$ for some Stonean space L?* See page 226.

Question 5: *Let K be a compact space such that $C(K)$ is isomorphically a dual space. Is K totally disconnected? Does there exist a Stonean space L such that $C(K)$ is isomorphic to $C(L)$? Does there exist a hyper-Stonean space L such that $C(K)$ is isomorphic to $C(L)$?* See page 230.

Question 6: *Let X be a non-empty, compact space. Suppose that $C(X)$ is isometrically isomorphic to the bidual E'' of some Banach space E. Does there exist a locally compact space K such that $C(X)$ is isometrically isomorphic to the bidual space $C_0(K)'' = M(K)' = C(\widetilde{K})$, and hence such that X is homeomorphic to $\widetilde{K}$?* See page 232.

Question 7: *Let A be a unital C^*-algebra, and suppose that, as a Banach space, A is isometrically isomorphic to the bidual of a Banach space E. Does it follow that there is a C^*-algebra B such that A is C^*-isomorphic to $(B'', \Box)$? Is this true in the special case in which E is separable?* See page 245.

References

1. Y. A. Abramovich, C. D. Aliprantis, in *An Invitation to Operator Theory*. Graduate Studies in Mathematics, vol. 50 (American Mathematical Society, Providence, 2002)
2. C. A. Akemann, The dual space of an operator algebra. Trans. American Math. Soc., **126**, 286–302 (1967)
3. F. Albiac, N. J. Kalton, *Topics in Banach Space Theory*. Graduate Texts in Mathematics, vol. 233 (Springer, New York, 2006)
4. E. M. Alfsen, *Compact Convex Sets and Boundary Integtrals*. Ergebnisse der Mathematik und ihrer Granzgebiete, Band 57 (Springer, New York, 1971)
5. C. D. Aliprantis, O. Burkinshaw, *Positive Operators*, 2nd printing (with revised section numbers, page numbers, and references) (Springer, Dordrecht, 2006)
6. G. R. Allan, *Introduction to Banach Spaces and Algebras*. Oxford Graduate Texts in Mathematics, vol. 20 (Oxford University Press, Oxford, 2011)
7. D. E. Alspach, A quotient of $C(\omega^\omega)$ which is not isomorphic to a subspace of $C(\alpha)$, $\alpha < \omega_1$. Israel J. Math. **35**, 49–60 (1980)
8. D. Amir, Continuous function spaces with the bounded extension property. Bull. Res. Counc. Israel Sect. F **101**, 133–138 (1962)
9. D. Amir, Projections onto continuous function spaces. Proc. American Math. Soc. **15**, 396–402 (1964)
10. D. Amir, On isomorphisms of continuous function spaces. Israel J. Math. **3**, 205–210 (1965)
11. R. Arens, Operations induced in function classes. Monatsh. Math. **55**, 1–19 (1951)

H.G. Dales et al., *Banach Spaces of Continuous Functions as Dual Spaces*,
CMS Books in Mathematics, DOI 10.1007/978-3-319-32349-7

12. R. Arens, The adjoint of a bilinear operation. Proc. American Math. Soc. **2**, 839–848 (1951)

13. S. A. Argyros, On the space of bounded measurable functions. Quarterly Journal Math. **34**, 129–132 (1983)

14. S. A. Argyros, J. F. Castillo, A. S. Granero, M. Jiménez, J. Moreno, Complementation and embeddings of $c_0(I)$ in Banach spaces. Proc. London Math. Soc. **85** (3), 742–768 (2002)

15. S. A. Argyros, G. Godefroy, H. P. Rosenthal, Descriptive set theory and Banach spaces, in *Handbook of the Geometry of Banach Spaces*, vol. 1, ed. by W. B. Johnson, J. Lindenstrauss (North-Holland/Elsevier, Amsterdam, 2003), pp. 1007–1069

16. S. A. Argyros, R. G. Haydon, A hereditarily indecomposable $\mathscr{L}_\infty$-space that solves the scalar-plus-compact problem. Acta Math. **206**, 1–54 (2011)

17. L. Asimov, A. J. Ellis, On Hermitian functionals on unital Banach algebras. Bull. London Math. Soc. **4**, 333–336 (1972)

18. L. Asimov, A. J. Ellis, *Convexity Theory and Its Applications in Functional Analysis*. London Mathematical Society Monographs, 1st Series, vol. 16 (Academic, London, 1980)

19. A. Avilés, F. Cabello Sánchez, J. M. F. Castillo, M. González, Y. Moreno, On separably injective Banach spaces. Advances Math. **234**, 192–216 (2013)

20. A. Avilés, F. Cabello Sánchez, J. M. F. Castillo, M. Gonzáles, Y. Moreno, *Separably Injective Banach Spaces*. Lecture Notes in Mathematics, vol. 2132 (Springer, Berlin, 2016)

21. A. Avilés, P. Koszmider, A continuous image of a Radon–Nikodým compact space which is not Radon–Nikodým. Duke Math. J. **162**, 2285–2299 (2013)

22. W. G. Bade, On Boolean algebras of projections and algebras of operators. Trans. American Math. Soc. **80**, 345–360 (1955)

23. W. G. Bade, *The Space of all Continuous Functions on a Compact Hausdorff Space* (University of California, Berkeley, 1957). Library call no. QA689.B16

24. W. G. Bade, *The Banach Space* $C(S)$. Lecture Note Series, vol. 26 (Matematisk Institut, Aarhus Universitët, Aarhus, 1971)

25. W. G. Bade, Complementation problems for the Baire classes. Pacific J. Math. **45**, 1–11 (1973)

26. J. W. Baker, Some uncomplemented subspaces of $C(X)$ of the type $C(Y)$. Studia Math. **36**, 85–103 (1970)

27. J. W. Baker, Uncomplemented $C(X)$-subalgebras of $C(X)$. Trans. American Math. Soc. **186**, 1–15 (1973)

28. R. Balbes, P. Dwinger, *Distributive Lattices* (University of Missouri Press, Columbia, 1974)

29. B. Balcar, P. Simon, Disjoint refinements, in *Handbook of Boolean Algebras*, vol. 2, ed. by J. D. Monk (North-Holland/Elsevier, Amsterdam, 1989), pp. 334–386

30. S. Banach, *Théorie des Opérations Linéaires*. Monografie Matematyczne, vol. 1 (Instytut Matematyczny Polskiej Akademii Nauk, Warsaw, 1932)

31. E. Behrends, R. Danckwerts et al., *L^p-Structure in Real Banach Spaces*. Lecture Notes in Mathematics, vol. 613 (Springer, New York, 1977)

32. Y. Benyamini, J. Lindenstrauss, A predual of ℓ^1 which is not isomorphic to a $C(K)$ space. Israel J. Math. **13**, 246–254 (1972)

33. Y. Benyamini, J. Lindenstrauss, *Geometric Nonlinear Functional Analysis*, vol. 48 (American Mathematical Society Colloquium Publications, Providence, 2000)

34. J. F. Berglund, H. D. Junghenn, P. Milnes, *Analysis on Semigroups; Function Spaces, Compactifications, Representations*. Canadian Mathematical Society Series of Monographs and Advanced Texts (Wiley, New York, 1989)

35. C. Bessaga, A. Pełczyński, On extreme points in separable conjugate spaces. Israel J. Math. **4**, 262–264 (1966)

36. G. Birkhoff, *Lattice Theory*, 3rd edn. (American Mathematical Society, Providence, 1967)

37. E. Bishop, R. R. Phelps, A proof that every Banach space is subreflexive. Bull. American Math. Soc. **67**, 97–98 (1961)

38. J. L. Blasco, C. Ivorra, Injective spaces of real-valued functions with the Baire property. Israel J. Math. **91**, 341–348 (1995)

39. V. I. Bogachev, *Measure Theory*, vols. I, II (Springer, Berlin, 2007)

40. B. Bollobás, An extension to the theorem of Bishop and Phelps. Bull. London Math. Soc. **2**, 181–182 (1970)

41. F. F. Bonsall, J. Duncan, *Numerical Ranges of Operators on Normed Spaces and of Elements of Normed Algebras*. London Mathematical Society Lecture Note Series, vol. 2 (Cambridge University Press, Cambridge, 1971)

42. F. F. Bonsall, J. Duncan, *Complete Normed Algebras*. Ergebnisse der Mathematik und ihrer Grenzgebiete, Band 80 (Springer, New York, 1973)

43. J. Bourgain, Real isomorphic complex Banach spaces need not be complex isomorphic. Proc. American Math. Soc. **96**, 221–226 (1986)

44. J. Bourgain, F. Delbaen, A class of special $\mathscr{L}_\infty$ spaces. Acta Math. **145**, 155–176 (1980)

45. C. Brech, P. Koszmider, On universal Banach spaces of density continuum. Israel J. Math. **190**, 93–110 (2012)

46. C. Brech, P. Koszmider, ℓ^∞-sums and the Banach space ℓ^∞/c_0. Fund. Math. **224**, 175–185 (2014)

47. T. Budak, N. Işik, J. Pym, Minimal determinants of topological centres for some algebras associated with locally compact groups. Bull. London Math. Soc. **43**, 495–506 (2011)

48. F. C. Sánchez, J. M. F. Castillo, D. Yost, Sobczyk's theorems from A to B. Extracta Math. **15**, 391–420 (2000)

49. M. Cambern, On mappings of sequence spaces. Studia Math. **30**, 73–77 (1968)

50. M. Cambern, P. Greim, Uniqueness of preduals for spaces of continuous vector functions. Canadian Math. Bull. **31**, 98–104 (1988)

51. L. Candido, E. Galego, How far is $C(\omega)$ from the other $C(K)$ spaces? Studia Math. **217**, 123–138 (2013)

52. C. Caratheodory, Die homomorphieen von somen und die multiplikation von inhaltsfunktionen. Annalen Scuola Norm. Sup. Pisa Cl. Sci. **8** (2), 105–130 (1939)

53. B. Cengiz, On topological isomorphisms of $C_0(X)$ and the cardinal number of X. Proc. American Math. Soc. **72**, 105–108 (1978)

54. G. Choquet, Existence et unicite des representations integrales au moyen des points extremaux dans les cones convexes. Seminaire Bourbaki **139**, 1–15 (1956)

55. H. B. Cohen, Injective envelopes of Banach spaces. Bull. American Math. Soc. **70**, 723–726 (1964)

56. H. B. Cohen, A bound-two isomorphism between $C(X)$ Banach spaces. Proc. American Math. Soc. **50**, 215–217 (1975)

57. H. B. Cohen, C.-H. Chu, Topological conditions for bound-2 isomorphisms of $C(X)$. Studia Math. **113**, 1–24 (1995)

58. H. B. Cohen, M. A. Labbe, J. Wolfe, Norm reduction of averaging operators. Proc. American Math. Soc. **35**, 519–523 (1972)

59. D. L. Cohn, *Measure Theory*, 2nd edn. (Birkhäuser/Springer, New York, 2013)

60. W. W. Comfort, Topological groups, in *Handbook of Set Theoretic Topology*, ed. by K. Kunen, J.E. Vaughan (North Holland, New York, 1984), pp. 1143–1263

61. W. W. Comfort, S. Negrepontis, *The Theory of Ultrafilters* (Springer, Berlin, 1974)

62. W. W. Comfort, S. Negrepontis, *Continuous Pseudometrics*. Lecture Notes in Pure and Applied Mathematics, vol. 14 (Marcel Dekker, New York, 1975)

63. W. W. Comfort, S. Negrepontis, *Chain Conditions in Topology* (Cambridge University Press, Cambridge, 1982). Reprinted (with corrections) (2008)

64. J. B. Conway, Projections and retractions. Proc. American Math. Soc. **17**, 843–847 (1966)

65. C. Correa, D. V. Tausk, Compact lines and the Sobczyk property. J. Functional Anal. **266**, 5765–5778 (2014)

66. F. Cunningham Jr., L^1-structures in Banach spaces. Thesis, Harvard University (1953)

67. F. Cunningham Jr., L-structure in L-spaces. Trans. American Math. Soc. **95**, 274–299 (1960)

68. H. G. Dales, *Banach Algebras and Automatic Continuity*. London Mathematical Society Monographs, vol. 24 (Clarendon Press, Oxford, 2000)

69. H. G. Dales, William George Bade, 1924–2012. Bull. London Math. Soc. **45**, 875–888 (2013)

70. H. G. Dales, A. T.-M. Lau, The biduals of Beurling algebras. Mem. American Math. Soc. **177**, 191 (2005)

71. H. G. Dales, A. T.-M. Lau, D. Strauss, Banach algebras on semigroups and on their compactifications. Mem. American Math. Soc. **205**, 165 (2010)

72. H. G. Dales, A. T.-M. Lau, D. Strauss, Second duals of measure algebras. Diss. Math. (Rozprawy Matematyczne) **481**, 121 (2012)

73. H. G. Dales, M. E. Polyakov, Multi-normed spaces. Diss. Math. (Rozprawy Matematyczne) **488**, 165 (2012)

74. H. G. Dales, W. H. Woodin, *An Introduction to Independence for Analysts*. London Mathematical Society Lecture Note Series, vol. 115 (Cambridge University Press, Cambridge, 1987)

75. F. K. Dashiell Jr., Isomorphism problems for the Baire classes. Pacific J. Math. **52**, 29–43 (1974)

76. F. K. Dashiell Jr., Nonweakly compact operators from order-Cauchy complete $C(S)$ lattices, with application to Baire classes. Trans. American Math. Soc. **266**, 397–413 (1981)

77. F. K. Dashiell Jr., J. Lindenstrauss, Some examples concerning strictly convex norms on $C(K)$ spaces. Israel J. Math. **16**, 329–342 (1973)

78. K. R. Davidson, *C^*-Algebras by Example*. Fields Institute Monographs, vol. 6 (American Mathematical Society, Providence, 1996)

79. W. J. Davis, W. B. Johnson, A renorming of nonreflexive Banach spaces. Proc. American Math. Soc. **37**, 486–488 (1973)

80. M. Daws, R. Haydon, T. Schlumprecht, S. White, Shift invariant preduals of $\ell_1(\mathbb{Z})$. Israel J. Math. **192**, 541–585 (2012)

81. M. M. Day, Strict convexity and smoothness of normed spaces. Trans. American Math. Soc. **78**, 516–528 (1955)

82. M. M. Day, *Normed Linear Spaces* (Springer, New York, 1973)

83. J. Diestel, *Sequences and Series in Banach Spaces*. Graduate Texts in Mathematics, vol. 92 (Springer, New York, 1984)

84. J. Diestel, J. H. Fourie, J. Swart, *The Metric Theory of Tensor Products. Grothendieck's Résumé Revisited* (American Mathematical Society, Providence, 2008)

85. J. Diestel, J. J. Uhl Jr., *Vector Measures*. Mathematical Surveys, vol. 15 (American Mathematical Society, Providence, 1977)

86. S. J. Dilworth, M. Girardi, J. Hagler, Dual Banach spaces which contain an isometric copy of L_1. Bull. Pol. Acad. Sci. Math. **48**, 1–12 (2000)

87. R. P. Dilworth, The normal completion of the lattice of continuous functions. Trans. American Math. Soc. **68**, 427–438 (1950)

88. S. Z. Ditor, Linear operators of averaging and extension. Thesis, University of California at Berkeley (1968)

89. S. Z. Ditor, On a lemma of Milutin concerning operators in continuous function spaces. Trans. American Math. Soc. **149**, 443–452 (1970)

90. S. Z. Ditor, Averaging operators in $C(S)$ and lower semicontinuous sections of continuous maps. Trans. American Math. Soc. **175**, 195–208 (1973)

91. J. Dixmier, Sur certains espaces considérés par M. H. Stone. Summa Brasiliensis Math. **2**, 151–182 (1951)

92. L. Drewnowski, J. W. Roberts, On the primariness of the Banach space ℓ_∞/c_0. Proc. American Math. Soc. **112**, 949–957 (1991)

93. D. van Dulst, *Characterizations of Banach Spaces not Containing ℓ^1*. CWI Tract, vol. 59 (Stichting Mathematisch Centrum/Centrum voor Wiskunde en Informatica, Amsterdam, 1989), pp. iv+163

94. N. Dunford, J. T. Schwartz, *Linear Operators, Part I: General Theory* (Interscience Publishers, New York, 1957)

95. N. Dunford, J. T. Schwartz, *Linear Operators Part III: Spectral Operators* (Wiley, New York, 1971)

96. P. Dwinger, *Introduction to Boolean Algebras*, 2nd edn. (Physica-Verlag, Würtzburg, 1971)

97. R. E. Edwards, *Functional Analysis* (Holt/Rinehart/Winston, New York, 1965). Corrected republication (Dover, New York, 1995)

98. A. J. Ellis, Some applications of convexity theory to Banach algebras. Math. Scand. **33**, 23–30 (1973)

99. R. Engelking, *General Topology*. Monografie Matematyczne, vol. 60 (Polish Scientific Publishers, Warsaw, 1977). Revised and completed edition (Heldermann Verlag, Berlin, 1989)

100. M. Fabian, P. Habala, P. Hájek, V. Montesinos, V. Zizler, *Banach Space Theory*. Canadian Mathematical Society Books in Mathematics (Springer, New York, 2011)

101. F. J. Fernández-Polo, A. M. Peralta, A short proof of a theorem of Pfitzner. Quarterly Journal Math. **61**, 329–336 (2010)

102. J. Ferrer, P. Koszmider, W. Kubis, Almost disjoint families of countable sets and separable complementation properties. J. Math. Anal. Appl. **401**, 939–949 (2013)

103. B. Fishel, D. Papert, A note on hyperdiffuse measures. J. London Math. Soc. **39**, 245–254 (1964)

104. V. P. Fonf, J. Lindenstrauss, R. R. Phelps, Infinite dimensional convexity, in *Handbook of the Geometry of Banach Spaces*, vol. 1, ed. by W. B. Johnson, J. Lindenstrauss (North Holland/Elsevier, Amsterdam, 2001), pp. 599–670

105. D. Freeman, E. Odell, T. Schlumprecht, The universality of ℓ_1 as a dual space. Math. Annalen **351**, 149–186 (2011)

106. D. H. Fremlin, *Consequences of Martin's Axiom*. Cambridge Tracts in Mathematics, vol. 84 (Cambridge University Press, Cambridge, 1984)

107. D. H. Fremlin, *Measure Theory*, vol. 3, 2012 edn. This text is available at www.essex.ac.uk/maths/people/fremlin/mt.htm

108. D. H. Fremlin, G. Plebanek, Large families of mutually singular Radon measures. Bull. Pol. Acad. Sci. Math. **51**, 169–174 (2003)

109. I. Gasparis, A new isomorphic ℓ_1 predual not isomorphic to a complemented subspace of a $C(K)$ space. Bull. London Math. Soc. **45**, 789–799 (2013)

110. I. M. Gelfand, Abstrakte Functionen und Lineare Operatoren. Recueil Math. **4**, 235–286 (1938)

111. L. Gillman, M. Henriksen, Rings of continuous fumctions in which every finitely generated ideal is principal. Trans. American Math. Soc. **82**, 366–391 (1956)

112. L. Gillman, M. Jerison, *Rings of Continuous Functions* (van Nostrand Reinhold, New York, 1960). Reprinted as Graduate Texts in Mathematics, vol. 43 (Springer, New York, 1976)

113. S. Givant, P. R. Halmos, *Introduction to Boolean Algebras* (Springer, New York, 2009)

114. A. M. Gleason, Projective topological spaces. Illinois J. Math. **2**, 482–489 (1958)

115. G. Godefroy, Existence and uniqueness of isometric preduals: a survey, in *Banach Space Theory*, ed. by B.-L. Lin. Contemporary Mathematics, vol. 85 (American Mathematical Society, Providence, 1987), pp. 131–194

116. C. Goffman, *Real Functions* (Prindle/Weber/Schmidt, Boston, 1953)

117. K. R. Goodearl, Choquet simplexes and σ-convex faces. Pacific J. Math. **66**, 119–124 (1976)

118. K. R. Goodearl, *Partially Ordered Abelian Groups with Interpolation*. Mathematical Surveys and Monographs, vol. 20 (American Mathematical Society, Providence, 1986)

119. D. B. Goodner, Projections in normed linear spaces. Trans. American Math. Soc. **69**, 89–108 (1950)

120. H. Gordon, The maximal ideal space of a ring of measurable functions. American J. Math. **88**, 827–843 (1966)

121. Y. Gordon, On the distance coefficient between isomorphic function spaces. Israel J. Math. **8**, 391–397 (1970)

122. W. T. Gowers, A solution to Banach's hyperplane problem. Bull. London Math. Soc. **26**, 523–530 (1994)

123. F. P. Greenleaf, *Invariant Means on Topological Groups*. van Nostrand Mathematical Studies, vol. 16 (van Nostrand, New York, 1969)

124. A. Grothendieck, Sur les appplications lineaires faiblement compactes d'espaces du type $C(K)$. Canadian J. Math. **5**, 129–173 (1953)

125. A. Grothendieck, Une caractérisation vectorielle métrique des espaces L^1. Canadian Math. Bull. **7**, 552–561 (1955)

126. D. Hadwin, V. I. Paulsen, Injectivity and projectivity in analysis and topology. Sci. China Math. **54**, 2347–2359 (2011)

127. J. N. Hagler, Embeddings of L^1 spaces into conjugate Banach spaces. Thesis, University of California at Berkeley (1972)

128. J. N. Hagler, Some more Banach spaces which contain ℓ^1. Studia Math. **46**, 35–42 (1973)

129. J. N. Hagler, Complemented isometric copies of L_1 in dual Banach spaces. Proc. American Math. Soc. **130**, 3313–3324 (2002)

130. J. N. Hagler, C. Stegall, Banach spaces whose duals contain complemented subspaces isomorphic to $C[0,1]^*$. J. Functional Anal. **13**, 233–251 (1973)

131. P. Hájek, V. M. Santalucía, J. Vanderwerff, V. Zizler, *Biorthogonal Systems in Banach Spaces*. Canadian Mathematical Society Books in Mathematics (Springer, New York, 2008)

132. P. R. Halmos, *Measure Theory* (D. van Nostrand, New York, 1950)

133. P. Harmand, D. Werner, W. Werner, *M-Ideals in Banach Spaces and Banach Algebras*. Lecture Notes in Mathematics, vol. 1547 (Springer, New York, 1993)

134. M. Hasumi, The extension property of complex Banach spaces. Tôhoku Math. J. **10**, 135–142 (1958)

135. R. Haydon, On dual L^1-spaces and injective bidual spaces. Israel J. Math. **31**, 142–152 (1978)

136. H.-U. Hess, Some remarks on linear transformations between certain Banach spaces. Arch. Math. **36**, 342–347 (1981)

137. E. Hewitt, K. A. Ross, *Abstract Harmonic Analysis I*, 2nd edn. (Springer, Berlin, 1979)

138. E. Hewitt, K. Stromberg, *Real and Abstract Analysis* (Springer, New York, 1975)

139. N. Hindman, D. Strauss, *Algebra in the Stone–Čech Compactification, Theory and Applications* (Walter de Gruyter, Berlin, 1998). Second revised and extended edition (2012)

140. B. Hirsberg, A. J. Lazar, Complex Lindenstrauss spaces with extreme points. Trans. American Math. Soc. **186**, 141–150 (1973)

141. J. G. Hocking, G. S. Young, *Topology* (Addison-Wesley, Reading, 1961). Currently available in unabridged, corrected re-publication, Dover Books on Mathematics (Dover Publications, Mineola/New York, 1988)

142. A. Horn, A. Tarski, Measures in Boolean algebras. Trans. American Math. Soc. **64**, 467–497 (1948)

143. J. Isbell, Z. Semadeni, Projection constants and spaces of continuous functions. Trans. American Math. Soc. **107**, 38–48 (1963)

144. H. Jarchow, *Locally Convex Spaces* (B. G. Teubner, Stuttgart, 1981)

145. T. Jech, *Set Theory. The Third Millenium Edition, Revised and Expanded*. Springer Monographs in Mathematics (Springer, Berlin, 2002). Corrected 4th printing (2006)

146. W. B. Johnson, T. Kania, G. Schechtman, Closed ideals of operators on and complemented subspaces of Banach spaces of functions with countable support. Proc. American Math. Soc. **144**, 4471–4485 (2016)

147. W. B. Johnson, J. Lindenstrauss (ed.), *Handbook of the Geometry of Banach Spaces*, vols. 1 and 2 (North Holland/Elsevier, Amsterdam, 2001)

148. W. B. Johnson, J. Lindenstrauss, Basic concepts in the geometry of Banach spaces, in *Handbook of the Geometry of Banach Spaces*, vol. 1, ed. by W. B. Johnson, J. Lindenstrauss (North Holland/Elsevier, Amsterdam, 2001), pp. 1–84

149. R. V. Kadison, J. R. Ringrose, *Fundamentals of the Theory of Operator Algebras, vol. 1, Elementary Theory* (Academic, New York, 1983). Second printing: Graduate Studies in Mathematics, vol. 15 (American Mathematical Society, 1997)

150. R. V. Kadison, J. R. Ringrose, *Fundamentals of the Theory of Operator Algebras, vol. 2, Advanced Theory* (Academic, New York, 1986). Second printing: Graduate Studies in Mathematics, vol. 16 (American Mathematical Society, 1997)

151. N. J. Kalton, An elementary example of a Banach space not isomorphic to its complex conjugate. Canadian Math. Bull. **38**, 218–222 (1995)

152. N. J. Kalton, Hermitian operators on complex Banach lattices and a problem of Garth Dales. J. London Math. Soc. **86** (2), 641–656 (2012)

153. S. Kaplan, *The Bidual of C(X), I*. North-Holland Mathematical Studies, vol. 101 (North Holland, Amsterdam, 1985)

154. J. L. Kelley, Banach spaces with the extension property. Trans. American Math. Soc. **72**, 323–326 (1952)

155. J. L. Kelley, *General Topology* (Van Nostrand, Toronto, 1955)

156. J. L. Kelley, Measures on Boolean Algebras. Pacific J. Math. **9**, 1165–1177 (1959)

157. S. S. Khurana, Grothendieck spaces. Illinois J. Math. **22**, 79–80 (1978)

158. A. Kolmogorov, General measure theory and the calculus of probabilities. Trudy Kommunist Akad. Razd. mat. 8–21 (1929, in Russian). English transl.: Selected works of A. N. Kolmogorov, vol. II, ed. by A. N. Shiryayev (Kluwer Academic Publishers, Dordrecht, 1992), pp. 48–59

159. S. Koppelberg, General theory of Boolean algebras, in *Handbook of Boolean Algebras*, vol. I, ed. by J. D. Monk, R. Bonnet (North Holland/Elsevier, Amsterdam, 1989)

160. P. Koszmider, Banach spaces of continuous functions with few operators. Math. Annalen **330**, 151–183 (2004)

161. P. Koszmider, A survey on Banach spaces $C(K)$ with few operators. Rev. R. Acad. Cienc. Exact. Fís. Nat. Ser. A Math. RACSAM **104**, 309–326 (2010)

162. G. Köthe, Hebbare lokalkonvexe Räume. Math. Annalen **165**, 181–195 (1966)

163. J. Kupka, A short proof and a generalization of a measure theoretic disjointization lemma. Proc. American Math. Soc. **45**, 70–72 (1974)

164. K. Kuratowski, *Topology*, vol. 1 (Academic, New York, 1966)

165. H. E. Lacey, A note concerning $A^* = L_1(\mu)$. Proc. American Math. Soc. **29**, 525–528 (1971)

166. H. E. Lacey, *Isometric Theory of Classical Banach Spaces* (Springer, Berlin, 1974)

167. A. T.-M. Lau, V. Losert, Complementation of certain subspaces of $L_\infty(G)$ of a locally compact group. Pacific J. Math. **141**, 295–310 (1990)

168. D. R. Lewis, C. Stegall, Banach spaces whose duals are isomorphic to $\ell_1(\Gamma)$. J. Functional Anal. **12**, 177–187 (1973)

169. D. Li, H. Queffélec, *Introduction à l'étude des Espaces de Banach* (Societé Mathématique de France, Paris, 2004)

170. A. Lima, Complex Banach spaces whose duals are L_1 spaces. Israel J. Math. **24**, 59–72 (1976)

171. J. Lindenstrauss, Extensions of compact operators. Mem. American Math. Soc. **48**, 112 (1964)

172. J. Lindenstrauss, H. P. Rosenthal, The $\mathscr{L}_p$ spaces. Israel J. Math. **7**, 325–349 (1969)

173. J. Lindenstrauss, L. Tzafriri, On the complemented subspaces problem. Israel J. Math. **9**, 263–269 (1971)

174. J. Lindenstrauss, L. Tzafriri, *Classical Banach Spaces*. Lecture Notes in Mathematics, vol. 338 (Springer, Berlin, 1973).

175. J. Lindenstrauss, L. Tzafriri, *Classical Banach Spaces*, vol. I (Springer, Berlin, 1977)

176. J. Lindenstrauss, L. Tzafriri, *Classical Banach Spaces*, vol. II (Springer, Berlin, 1979)

177. V. Losert, M. Neufang, J. Pachl, J. Steprāns, Proof of the Ghahramani–Lau conjecture. Advances Math. **290**, 709–738 (2016)

178. J. Lukeš, J. Malý, I. Netuka, J. Spurný, *Integral Representation Theory* (Walter de Gruyter, Berlin, 2010)

179. W. A. J. Luxemburg, Is every integral normal? Bull. American Math. Soc. **73**, 685–688 (1967)

180. W. A. J. Luxemburg, A. C. Zaanan, *Riesz Spaces*, vol. I (North Holland, Amsterdam, 1971)

181. R. D. McWilliams, A note on weak sequential convergence. Pacific J. Math. **12**, 333–335 (1962)

182. D. Maharam, Decompositions of measure algebras. Proc. Natl. Acad. Sci. USA **28**, 142–160 (1942)

183. R. E. Megginson, *An Introduction to Banach Space Theory*. Graduate Texts in Mathematics, vol. 183 (Springer, New York, 1998)

184. P. Meyer-Nieberg, *Banach Lattices* (Springer, Berlin, 1991)

185. J. van Mill, An introduction to $\beta\omega$, in *Handbook of Set Theoretic Topology*, ed. by K. Kunen, J. E. Vaughan (North Holland/Elsevier, Amsterdam, 1984), pp. 503–567

186. A. W. Miller, Half of an inseparable pair. Real Anal. Exch. **32**, 179–194 (2006–2007)

187. G. A. Muñoz, Y. Sarantopoulos, A. Tonge, Complexifications of real Banach spaces, polynomials and multilinear maps. Studia Math. **134**, 1–33 (1999)

188. L. Nachbin, On the Hahn–Banach theorem. An. Acad. Bras. Cienc. **21**, 151–154 (1949)

189. I. Namioka, Neighborhoods of extreme points. Israel J. Math. **5**, 145–152 (1967)

190. I. Namioka, E. Asplund, A geometric proof of Ryll–Nardzewski's fixed point theorem. Bull. American Math. Soc. **73**, 443–445 (1967)

191. S. Negrepontis, Absolute Baire sets. Proc. American Math. Soc. **18**, 691–694 (1967)

192. E. Odell, H. P. Rosenthal, A double-dual characterization of separable Banach spaces containing ℓ^1. Israel J. Math. **20**, 375–384 (1975)

193. T. W. Palmer, Characterizations of C^*-algebras. Bull. American Math. Soc. **74**, 538–540 (1968)

194. T. W. Palmer, *Banach Algebras and the General Theory of $*$-Algebras, vol. I, Algebras and Banach Algebras* (Cambridge University Press, Cambridge, 1994)

195. T. W.Palmer, *Banach Algebras and the General Theory of $*$-Algebras, vol. II, $*$-Algebras* (Cambridge University Press, Cambridge, 2001)

196. A. Pełczyński, On the isomorphism of the spaces m and M. Bull. Acad. Pol. Sci. **6**, 695–696 (1958)

197. A. Pełczyński, On Banach spaces containing $L_1(\mu)$. Studia Math. **19**, 231–246 (1968)

198. A. Pełczyński, Linear extensions, linear averagings, and their applications to linear topological classification of spaces of continuous functions. Diss. Math. (Rozprawy Matematyczne) **63**, 92 (1968)

199. A. Pełczyński, V.N. Sudakov, Remarks on non-complemented subspaces of the space $m(S)$ Colloq. Math. **19**, 85–88 (1962)

200. H. Pfitzner, Weak compactness in the dual of a C^*-algebra is determined commutatively. Math. Annalen **298**, 349–371 (1994)

201. R. Phelps, *Lectures on Choquet's Theorem*. van Nostrand Mathematical Studies, vol. 7 (van Nostrand, Princeton, 1966). Second edition, Lecture Notes in Mathematics, vol. 1757 (Springer, Berlin, 2001)

202. R. S. Phillips, On linear transformations. Trans. American Math. Soc. **48**, 516–541 (1940)

203. R. S. Pierce, Countable Boolean algebras, in *Handbook of Boolean Algebras*, vol. III, ed. by J. D. Monk, R. Bonnet (North-Holland/Elsevier, Amsterdam, 1989), pp. 775–876

204. G. Plebanek, A construction of a Banach space $C(K)$ with few operators. Top. Appl. **143**, 217–239 (2004)

205. G. Plebanek, On isomorphisms of Banach spaces of continuous functions. Israel J. Math. **209**, 1–13 (2015)

206. G. Plebanek, A normal measure on a compact connected space. Unpublished notes (2014)

207. W. Ricker, *Operator Algebras Generated by Commuting Projections: A Vector Measure Approach*. Lecture Notes in Mathematics, vol. 1711 (Springer, Berlin, 1999)

208. M. A. Rieffel, Dentable subsets of Banach spaces, with applications to a Radon–Nikodým theorem, in *Functional Analysis Proceedings of a Conference held at the University of California*, Irvine, 1965, ed. by B. R. Gelbaum (Academic, London, 1967), pp. 71–77

209. C. A. Rogers, J. E. Jayne et al., *Analytic Sets* (Academic, London, 1980)

210. I. Rosenholtz, Another proof that any compact metric space is the continuous image of the Cantor set. American Math. Monthly **83**, 646–647 (1976)

211. H. P. Rosenthal, On injective Banach spaces and the spaces $L^\infty(\mu)$ for finite measures μ. Acta Math. **124**, 205–248 (1970)

212. H. P. Rosenthal, On relatively disjoint families of measures, with some applications to Banach space theory. Studia Math. **37**, 13–36 (1970)

213. H. P. Rosenthal, On factors of $C([0,1])$ with non-separable dual. Israel J. Math. **13**, 361–378 (1972); Correction: *ibid.* 21 (1975), 93–94

214. H. P. Rosenthal, The complete separable extension property. J. Operator Theory **43**, 329–374 (2000)

215. H. P. Rosenthal, The Banach spaces $C(K)$, in *Handbook of the Geometry of Banach Spaces*, vol. 2, ed. by W. B. Johnson, J. Lindenstrauss (North Holland/Elsevier, Amsterdam, 2003), pp. 1547–1602

216. H. L. Royden, *Real Analysis*, 3rd edn. (Prentice-Hall, Engelwood Cliffs, 1988)

217. W. Rudin, *Real and Complex Analysis*, 3rd edn. (McGraw-Hill, New York, 1986)

218. W. Rudin, *Functional Analysis*, 2nd edn. (McGraw-Hill, New York, 1991)

219. K. Saitô, J. D. M. Wright, C^*-algebras which are Grothendieck spaces. Rend. Circ. Mat. Palermo, Serie II **52**, 141–144 (2003)

220. K. Saitô, J. D. M. Wright, *Monotone Complete C^*-Algebras and Generic Dynamics*. Springer Monographs in Mathematics (Springer, London, 2015)

221. S. Sakai, A characterization of W^*-algebras. Pacific J. Math. **6**, 763–773 (1956)

222. S. Sakai, *C^*-Algebras and W^*-Algebras*. Ergebnisse der Mathematik und ihrer Grenzgebiete, vol. 60 (Springer, New York, 1971)

223. H. H. Schaefer, *Banach Lattices and Positive Operators* (Springer, Berlin, 1974)

224. G. L. Seever, Measures on F-spaces. Trans. American Math. Soc. **133**, 267–280 (1968)

225. Z. Semadeni, *Banach Spaces of Continuous Functions*. Monografie Matematyczne, vol. 55 (Instytut Matematyczny Polskiej Akademii Nauk, Warsaw, 1971)

226. S. Shelah, A. Usvyatsov, Banach spaces and groups—order properties and universal models. Israel J. Math. **152**, 245–270 (2006)

227. Y. A. Šreĭder, The structure of maximal ideals in rings of measures with convolution. Math. Sbornik (New Series) **27**, 297–318 (1950, in Russian). English translation in American Math. Soc. Trans. **81**, 365–391 (1953)

228. S. M. Srivastava, *A Course on Borel Sets*. Graduate Texts in Mathematics, vol. 180 (Springer, New York, 1998)

229. C. Stegall, Banach spaces whose duals contain $\ell_1(\Gamma)$ with applications to the study of dual $L_1(\mu)$ spaces. Trans. American Math. Soc. **176**, 463–477 (1973)

230. C. Stegall, The Radon–Nikodým property in conjugate Banach spaces. Trans. American Math. Soc. **206**, 213–223 (1975)

231. M. H. Stone, The theory of representations of Boolean algebras. Trans. American Math. Soc. **40**, 37–111 (1936)

232. D. P. Strauss, Extremally disconnected spaces. Proc. American Math. Soc. **18**, 305–309 (1967)

233. E. Szpilrajn, On the space of measurable sets. Annalen Soc. Pol. Math. **17**, 120–121 (1938)

234. M. Takesaki, *Theory of Operator Algebras I* (Springer, New York, 1979)

235. M. Talagrand, Maharam's problem. Annalen Math. **168**(2), 981–1009 (2008)

236. J. Terasawa, Spaces $N \cup R$ need not be strongly 0-dimensional. Bull. Acad. Pol. Sci. Sér. Sci. Math. Astonom. Phys. **25**, 279–281 (1977)

237. S. Todorcevic, *Topics in Topology*. Lecture Notes in Mathematics, vol. 1652 (Springer, Berlin, 1997)

238. W. A. Veech, Short proof of Sobczyk's theorem. Proc. American Math. Soc. **28**, 627–628 (1971)

239. R. C. Walker, *The Stone–Čech Compactification* (Springer, Berlin, 1974)

240. R. Whitley, Projecting m onto c_0. American Math. Mon. **73**, 285–286 (1966)

241. E. Wimmers, The Shelah P-point independence theorem. Israel J. Math. **43**, 28–48 (1982)

242. J. Wolfe, Injective Banach spaces of type $C(T)$. Israel J. Math. **18**, 133–140 (1974)

243. J. Wolfe, Injective Banach spaces of continuous functions. Trans. American Math. Soc. **235**, 115–139 (1978)

244. J. C. S. Wong, Abstract harmonic analysis of generalised functions on locally compact semigroups with applications to invariant means. J. Australian Math. Soc. A **23**, 84–94 (1977)

245. A. C. Zaanen, *Introduction to Operator Theory in Riesz Spaces* (Springer, Berlin, 1997)

246. M. Zippin, The separable extension problem. Israel J. Math. **26**, 372–387 (1977)

247. M. Zippin, Extension of bounded linear operators, in *Handbook of the Geometry of Banach Spaces*, vol. 2, ed. by W. B. Johnson, J. Lindenstrauss (North-Holland, Amsterdam, 2003), pp. 1703–1741

248. V. Zizler, Nonseparable Banach spaces, in *Handbook of the Geometry of Banach Spaces*, vol. 2, ed. by W. B. Johnson, J. Lindenstrauss (North-Holland, Amsterdam, 2003), pp. 1743–1816

Index of Terms

H.G. Dales et al., *Banach Spaces of Continuous Functions as Dual Spaces*,
CMS Books in Mathematics, DOI 10.1007/978-3-319-32349-7

Index of Symbols

H.G. Dales et al., *Banach Spaces of Continuous Functions as Dual Spaces*,
CMS Books in Mathematics, DOI 10.1007/978-3-319-32349-7

Printed in the United States
By Bookmasters